Next Generation Internet of Things

Distributed Intelligence at the Edge and Human Machine-to-Machine Cooperation

RIVER PUBLISHERS SERIES IN COMMUNICATIONS

Indexing: All books published in this series are submitted to the Web of Science Book Citation Index (BkCI), to CrossRef and to Google Scholar.

The "River Publishers Series in Communications" is a series of comprehensive academic and professional books which focus on communication and network systems. Topics range from the theory and use of systems involving all terminals, computers, and information processors to wired and wireless networks and network layouts, protocols, architectures, and implementations. Also covered are developments stemming from new market demands in systems, products, and technologies such as personal communications services, multimedia systems, enterprise networks, and optical communications.

The series includes research monographs, edited volumes, handbooks and textbooks, providing professionals, researchers, educators, and advanced students in the field with an invaluable insight into the latest research and developments.

For a list of other books in this series, visit www.riverpublishers.com

Next Generation Internet of Things

Distributed Intelligence at the Edge and Human Machine-to-Machine Cooperation

Editors

Ovidiu Vermesan

SINTEF, Norway

Joël Bacquet

EU, Belgium

Published, sold and distributed by:
River Publishers
Alsbjergvej 10
9260 Gistrup
Denmark

River Publishers
Lange Geer 44
2611 PW Delft
The Netherlands

Tel.: +4536953197
www.riverpublishers.com

ISBN: 978-87-7022-008-8 (Hardback)
978-87-7022-007-1 (Ebook)

Dedication

"We keep moving forward, opening new doors, and trying new things, because we are curious, and curiosity keeps leading us down new paths."

– Walt Disney

"Creativity takes courage."

– Henri Matisse

Acknowledgement

The editors would like to thank the European Commission for their support in the planning and preparation of this book. The recommendations and opinions expressed in the book are those of the editors and contributors, and do not necessarily represent those of the European Commission.

Ovidiu Vermesan

Joël Bacquet

Contents

Preface

Next Generation Internet of Things

Distributed Intelligence at the Edge and Human Machine-to-Machine Cooperation

The Internet of Things (IoT) and Industrial Internet of Things (IIoT) present opportunities for enterprises to improve efficiencies and enhance customer value.

IoT/IIoT continues to evolve with new technologies and applications, embedding ubiquitous hyperconnectivity (5G and beyond), edge/fog computing, distributed ledger technologies (DLTs) and artificial intelligence (AI). IoT/IIoT operates in a continuum that connects the physical, digital, virtual and cyber worlds through intelligent 'Digital Twins' by generating constant streams of data that can be harnessed into actionable distributed intelligence at the edge, with human, machine-to-machine cooperation to improve life, work and interaction with the everyday world.

Next-generation Tactile IoT/IIoT builds a real-time interactive system between the human and the machine and introduces a new evolution in human-machine (H2M) communication. Tactile IoT/IIoT enables the transfer of physical 'senses' (e.g. sense/touch, actuation, hepatic actions, etc.) in real-time form remotely and introduces a new paradigm shift to the skill-based/knowledge-based networks instead of content-based networks.

New solutions are needed for addressing IoT distributed end-to-end security because there are many more IoT/IIoT devices to secure, compared with traditional IT infrastructure devices. Many IoT/IIoT devices are embedded into systems that can affect physical health and safety, in addition to traditional communications or computing systems, and they introduce a complex management environment, with diverse technology profiles, processing capabilities, use-cases, and physical locations.

The chapters in the book present and discuss the next-generation IoT/IIoT research and innovation trends by addressing the enabling technologies, such as network technologies, edge/fog computing, distributed ledger technologies, AI, and the challenges for security, network management and integration for future IoT applications across industrial sectors.

The IoT/IIoT development ahead implies a human-centric approach that involves a judicious and dynamic balancing of collaboration and competition in IoT ecosystems in a common quest for advancing the digital transformation of industry and economy for the benefit of society and citizens.

Editors Biography

Dr. Ovidiu Vermesan holds a PhD degree in microelectronics and a Master of International Business (MIB) degree. He is Chief Scientist at SINTEF Digital, Oslo, Norway. His research interests are in the area of mixed-signal embedded electronics and cognitive communication systems. Dr. Vermesan received SINTEF's 2003 award for research excellence for his work on the implementation of a biometric sensor system. He is currently working on projects addressing nanoelectronics, integrated sensor/actuator systems, communication, cyber–physical systems and the IoT, with applications in green mobility, energy, autonomous systems and smart cities. He has authored or co-authored over 85 technical articles and conference papers. He is actively involved in the activities of the Electronic Components and Systems for European Leadership (ECSEL) Joint Technology Initiative (JTI). He has coordinated and managed various national, EU and other international projects related to integrated electronics. Dr. Vermesan actively participates in national, H2020 EU and other international initiatives by coordinating and managing various projects. He is the coordinator of the IoT European Research Cluster (IERC) and a member of the Steering Board of the Alliance for Internet of Things Innovation (AIOTI).

Joël Bacquet is a senior official of DG CONNECT of the European Commission, taking care of the research and innovation policy for the Internet of Things. Before working in this field, he was programme officer in "Future Internet Experimental Platforms", head of the sector "Virtual Physiological Human" in the ICT for health domain. From 1999 to 2003, he was head of the sector "networked organisations" in the eBusiness unit. He started working with the European Commission in 1993, in the Software Engineering Unit of the ESPRIT Programme. He started his carrier as visiting scientist for Quantel a LASER company in San José, California in 1981. From 1983 to 1987, he was with Thomson CSF (Thales) as software development engineer for a Radar System. From 1987 to 1991, he worked with the

European Space Agency as software engineer on the European Space shuttle and international Space platform programmes (ISS). From 1991 to 1993 he was with Eurocontrol where he was Quality manager of an Air Traffic Control system. He is an engineer in computer science from Institut Supérieur d'Electronique du Nord (ISEN) and he has a MBA from Webster University, Missouri.

List of Figures

List of Tables

1

IoT EU Strategy, State of Play and Future Perspectives

Mechthild Rohen

European Commission, Belgium

1.1 Introduction

Two years have passed since the publication of our Digitising European Industry (DEI) strategy, whose overall objective is to ensure that any industry in Europe, big or small, wherever situated and in any sector can fully benefit from digital innovations to upgrade its products, improve its processes and adapt its business models to the digital transformation. The underlying scenario is represented by the European platform of national initiatives on DEI, including digital innovation hubs, regulatory framework, skills and jobs, partnership and platforms (Figure 1.1).

IoT is at the heart of the digitisation process of the economy and society and it is an essential building block of the DEI strategy and the Digital Single Market strategy. Therefore, the overall goal is for Europe to be at the forefront of supplying innovative IoT solutions and to become the world's leading market for IoT products and services. As part of the DEI strategy, the goal for developing IoT leadership encompasses several building blocks funded under Horizon 2020:

- The IoT-European Platforms Initiative (IoT-EPI), addressing interoperability of IoT platforms, creating the ecosystem, using architectures, and integrating systems and networks for a multiplicity of novel applications;
- The Focus Area on IoT under Crosscutting Activities in the Horizon 2020 Work Programme 2016–2017, on experimentation with real-life solutions, tested at large scale with users; and

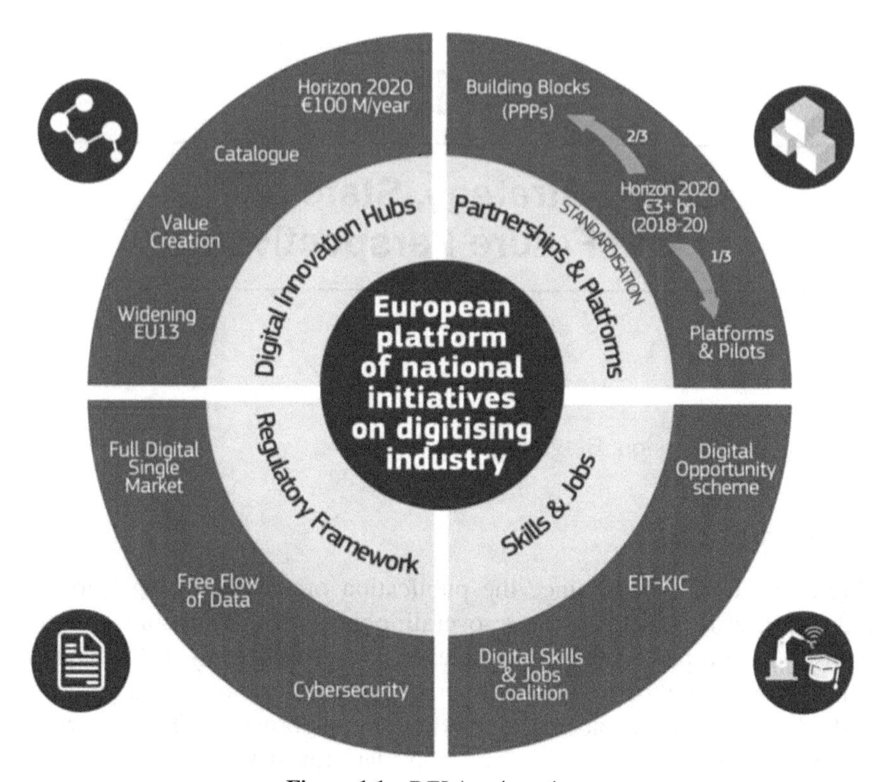

Figure 1.1 DEI 4 main actions.

- The Focus Area on Digitising and transforming European industry and services under the Horizon 2020 Work Programme 2018–2020, which supports the DEI strategy on digitization of industrial sectors, integrating digital technologies and innovation across societal challenges (Figure 1.2).

These building blocks are further elaborated below to provide with an overview of the state of play of the EU initiatives and activities.

1.2 Research and Innovation under Horizon 2020

The IoT-European Platforms Initiative (IoT-EPI) was formed to build a vibrant and sustainable IoT-ecosystem in Europe, maximising the opportunities for open platform development, interoperability and information sharing. At the core of the programme there are seven research and innovation

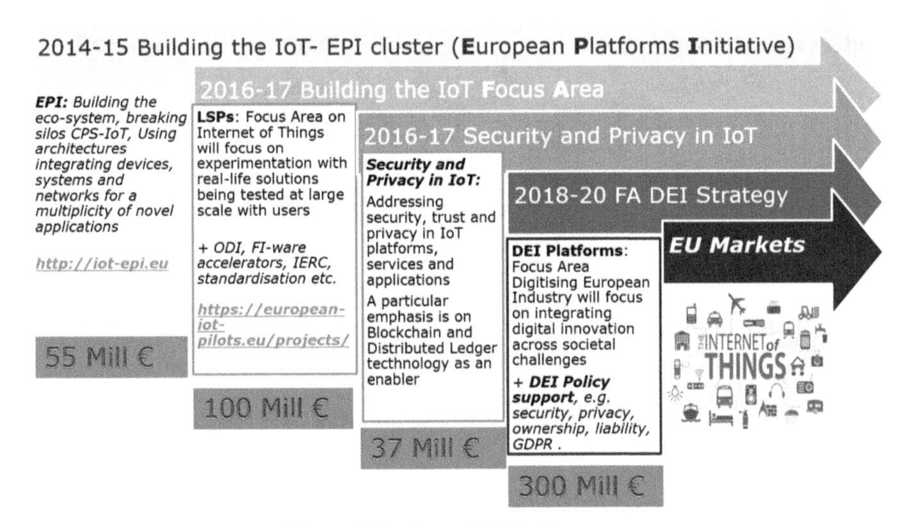

Figure 1.2 Overall EU IoT strategy.

projects and two coordination and support actions: Inter-IoT, BIG IoT, AGILE, SymbIoTe, TagItSmart, Vicinity, bIoTope, Be-IoT and UNIFY-IoT. With a total funding of EUR 50 million and a partner network of 120 organizations, these projects develop innovative solutions focusing on IoT architectures and semantic interoperability. Furthermore, they also foster technology adoption through the development of use cases in several industrial sectors, and community and business building activities. All projects ran within the time-frame of 2016–2018 – with one (Vicinity) extending until 2019.

The IoT-EPI projects are cooperating to define the research and innovation mechanisms and to identify opportunities for collaboration in IoT ecosystems to maximise the opportunities for common approaches to platform development, interoperability and information sharing. The common activities are organised under six task forces that are conceived and developed under IoT-EPI (Figure 1.3).

Each of the six Task Forces have produced major results in terms of research, but also in terms of policy, which has created a real impact on the European IoT market [1]. Some of the key results include:

- The analysis of IoT platforms showing a market growing rapidly, but still fragmented, with hundreds of different and incompatible platforms.

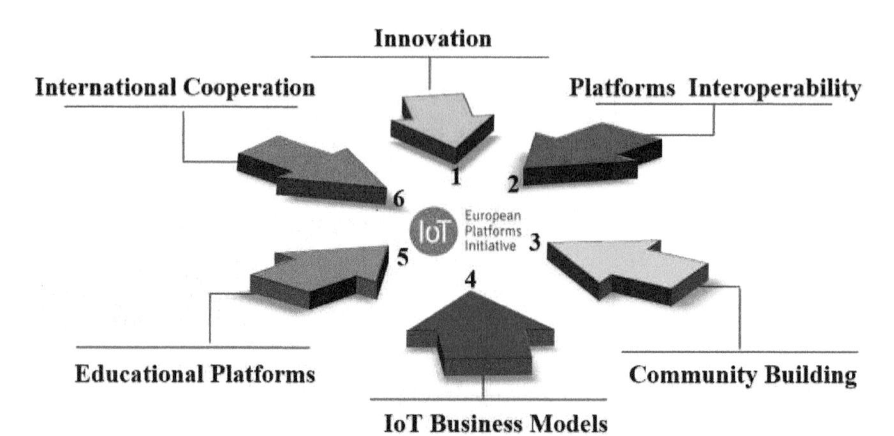

Figure 1.3 IoT-EPI Task Forces.

This report on IoT landscape has been further developed and published by the Alliance for Internet of Things Innovation (AIOTI) [2];

- The publication of a white paper on IoT platform interoperability, compiling the lessons learnt and results from the seven projects [4];
- The development of an open architecture and open IoT business model framework that has set the foundation of cooperation with the developers and entrepreneurs community, and that has mobilised SMEs and start-ups to join the ecosystem. In eleven open calls, with more than 100 external IoT-teams, the IoT-EPI has planned an investment of more than EUR 5.5 million until December 2018 to nurture an IoT ecosystem around the seven core projects;
- The development of policy recommendations for the uptake of IoT in Europe; and
- The set-up of an education platform using the results of the IoT-EPI projects.

Besides the IoT-EPI Task Forces, adequate security, trust and privacy are key issues to be tackled in connection with IoT, and therefore a specific cluster of project addressing these issues has been launched under Horizon 2020 in 2017. Seven projects have been selected with a total EU contribution of EUR 37 million in order to develop and test solutions providing IoT security, trust and privacy (ENACT, IoTCrawler, SecureIoT, BRAIN-IoT, SOFIE, CHARIOT, SEMIoTICS, SerIoT). The projects address the key issues of end-to-end security and trust in open IoT Platforms, as well as advanced concepts for IoT security and prevention of cyber-attacks, including blockchains and distributed ledger technology, which are tested in a set of ambitious use cases. In addition, the projects deploy open IoT platforms and include a strong contribution to upcoming open standards in IoT security.

1.3 Deployment – IoT Focus Area and Focus Area on Digitization

In order to foster the uptake of IoT in Europe and to enable the emergence of IoT ecosystems supported by open technologies, the European Commission launched an IoT Focus Area that supports the IoT European Large-Scale Pilots Programme (IoT-LSPs) on deployment of IoT at large in Europe. These IoT-LSPs started on 1 January 2017 and are funded with a budget of EUR 100 million. The IoT-LSPs cover the following domains:

- Smart living environments for ageing well (ACTIVAGE);
- Smart Farming and Food Security (IoF2020);
- Wearables for smart ecosystems (MONICA);
- Reference zones in EU cities (SYNCHRONICITY); and
- Autonomous vehicles in a connected environment (AUTOPILOT).

With these pilots, the European Commission is supporting the testing and experimentation of new IoT related technologies with the involvement of and result validation by end users. These pilots are expected to accelerate the standards setting across different business sectors, boosting further the IoT technology and provide input to policy developments, such as data protection, privacy and security.

Since January 2017, several successful results have been achieved. Each funded project is applying IoT approaches to specific real-life challenges across use cases, based on European relevance, technology readiness and socio-economic interest in Europe. More than 50 use cases have taken shape and are now fully running. This has also allowed the LSPs to work together in order to define common high-level architecture models. Another example is the well-defined and good cooperation among LSPs, which develop common mechanisms for the publication of open calls to enlarge their consortia with new partners, in particular SMEs. These open calls provide so-called cascading funds as financial support targeted to involve especially SMEs and start-ups to get access to pilot testing in an open and lean way.

In the Horizon 2020 Work Programme 2018–2020, the European Commission aims to use the strong concept of a Focus Area on Digitisation (*Digitising and transforming European industry and services*), accounting for EUR 250 million funding and forming a significant part of ICT calls in the Horizon 2020 Work Programme 2018–2020. Success in implementation of the Focus Area will depend to a large extent on the capacity to work across the digital, societal and industrial topics that are grouped under this Focus Area. The Focus Area requires close cooperation of different services across different DGs, namely CONNECT, GROW, RTD, AGRI and ENER, to ensure coherent policy setting across areas which so far were siloed economic and policy areas, e.g. to support Digitisation under the Energy Union or the Common Agriculture Policy.

Calls will close in November 2018 and resulting from this Focus Area, a further set of pilots will be launched in 2019 across different areas, such as health and care, energy efficiency, agriculture and industry 4.0. These pilots will accelerate standards setting across different business sectors, boosting further the investments and scalable market creation for IoT technology. This focus area will be funded by several parts of the Horizon 2020 programme, mainly by the Leadership in Enabling and Industrial Technologies and Societal Challenges pillars. Pilot activities will be supported in the areas of Smart Farming, Digitisation of Energy, Digital Health and Rural Platforms, as depicted in Figure 1.4.

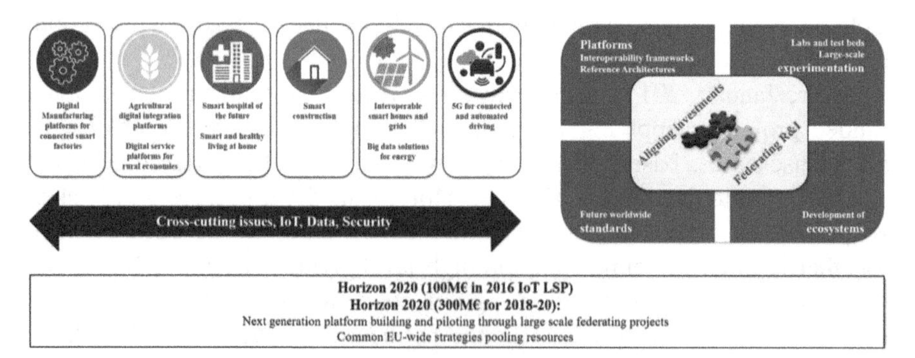

Figure 1.4 Focus Area on Digitization.

1.4 IoT within the Next Generation Internet – Preparing the Next Framework Programme for Research and Innovation

IoT continues to evolve rapidly, in particular in response to the major trends, such as the ever-increasing volumes of data generated and extraction of knowledge through smart data analytics, as well as increasing levels of automation and decision making, made possible by smart sensors, devices and actuators combined with machine learning and artificial intelligence.

In addition, there are new real-time requirements emerging, such as in industrial production and autonomous cars, which must be addressed. The capacity of communication networks is ever increasing with 5G deployment starting and processing power is still increasing exponentially, allowing for new distributed architectures (e.g. cognitive cloud, fog or edge computing) and solutions.

Also, new approaches to reduce or eliminate intermediaries, by building on blockchains and distributed ledgers, and the power to control access to and sharing of data, lead to new value chains and business models, which will open new opportunities for European companies and user communities.

This is all encapsulated in the vision of a Next Generation Internet, which is more human-centric in terms of identity, data protection and privacy, control and opportunity, addressing wider needs (from browsing to interconnection of billions of smart devices with new real-time requirements) and which is more secure and trusted by design as a critical infrastructure for society and business.

1.5 Conclusion

The digital revolution has only just started – and it is speeding up. Technology is entering into a critical phase, where connectivity and intelligence will permeate all areas of the physical world, with profound economic and social effects. Europe must maintain a leading position in the digital world and ensure that everyone whether businesses, public sector and citizens, can benefit from it. The European Commission has just published its proposal for the next EU budget for 2021–2027 [5], where there is a substantial focus on an ambitious investment in digital to make this a reality. This proposal includes a 60% increase in budget for Horizon Europe, the next EU Framework Programme for Research and innovation, as well as a proposed new programme, i.e. the Digital Europe Programme with a EUR 9 billion funding to support the large scale uptake of digital technologies, including digital skills. IoT and its future evolution is central to many of these efforts and Europe can take the lead in realising its potential.

References

[1] Digital Single Market – The Internet of Things, online at: https://ec.europa.eu/digital-single-market/en/internet-of-things

[2] Alliance for Internet of Things Innovation (AIOTI), online at: https://aioti.eu/

[3] IoT European Large-Scale Pilots Programme (IoT-LSPs), online at: https://european-iot-pilots.eu/

[4] IoT European Platforms Initiative (IoT-EPI), online at: https://iot-epi.eu/

[5] EU budget for the future, online at: https://ec.europa.eu/commission/sites/beta-political/files/budget-proposals-research-innovation-may-2018_en.pdf

2

Future Trends in IoT

Joël Bacquet, Rolf Riemenschneider and Peter Wintlev-Jensen

European Commission, Belgium

2.1 Introduction

The Next Generation Internet (NGI) initiative [1] aims at maintaining the European lead in advanced network infrastructures and fully exploit the opportunities offered by the connection to the physical work i.e the Internet of Things (IoT), powered by advanced computing capabilities and data infrastructure. The NGI and its link to IoT has to be at the service of people, industry and society, addressing present and specific societal challenges, combined with artificial intelligence (AI), secure transactions, sovereignty, edge computing, interactive technologies and social media, as depicted in Figure 2.1. Every technological design has to focus on making data and components easy to use and profitable in an open and democratic way to every single user.

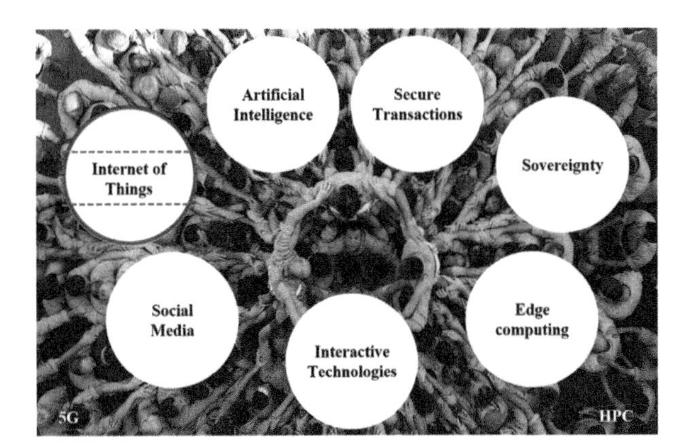

Figure 2.1 NGI Key Pillars.

IoT technologies and applications bring fundamental changes in individuals' and society's view of how technology and business work in the world, and it is therefore an essential element of the Next Generation Internet. IoT is seen today as a disruptive technology for enabling new opportunities and triggering new services and applications. However, collecting massive amounts of data in everyday life poses huge challenges for the user to keep control of his data in terms of managing access, sharing and protection. Additionally, one of Europe's greatest challenges is to keep the sovereignty of the underlying core infrastructure that computes and stores sensitive information, and protect IoT devices from misuse. The societal potential of IoT is extraordinary: better use of natural resources through smart farming, better food quality through devices enabling food traceability and control, better human health through devices linked to remote medicine and independent living, lower carbon emissions from autonomous driving and smart logistics, fewer accidents relying on connected driving, smart cities through smart use of massive data generated from a multitude of new sensors in a city.

The scope of this chapter is to review novel IoT concepts that have been gathered from groups of experts in different fora, including specific workshops organised by the European Commission, inputs the European Research Cluster on the Internet of Things (IERC), the Alliance for Internet of Things Innovation (AIOTI) and the IoT-European Platforms Initiative (IoT-EPI) cluster.

2.2 Key Technological Game Changers for IoT

Novel IoT architecture, platforms and solutions will emerge and will integrate new enabling technologies such as AI, secure Distributed Ledger Technologies (DLTs), or advanced communication networks, in order to meet new user requirements for performance, quality of services, trust and user control data. These IoT architectures, platforms and solutions will rely on the following game changers:

- Next generation IoT devices;
- Edge computing;
- Data-centric architectures;
- Community-driven business models; and
- A resilient and reliable infrastructure.

Towards Next Generation IoT devices. The IoT platform development will move in the next phase with the emergence of tactile interface based on human-centric sensing and actuating, augmented and virtual reality combined with new IoT end-point capabilities capturing contextual environment. Interactive and conversational IoT platforms will emerge with innovative user interfaces interfering with things and humans. These interactive platforms will enable real-time control, physical (haptic) experiences, interactive, context aware, event-driven IoT services with more intelligence at the edge. In supporting trust and security, information flows stay close to the user, decisions are taken at the point of interest, where data is collected and locally processed. For this to happen, the applications need to combine edge computing, IoT and mobile autonomous systems using AI technologies as functionality enablers.

Towards edge computing, shifting computing and data processing close to the source of data. The usual approach in most current IoT solutions is to execute data crunching in the cloud. In many scenarios this is the most suitable approach due to distributed nature of data collection. However, the value generated by many IoT devices decreases over time (e.g. for a thermostat control). With billions of IoT devices, it does not make sense to store all data in the cloud, but to limit data transfer to the cloud and store only information that is necessary to avoid data deluge. There are scenarios in which significant amount of data is collected at one location and the output of that local data processing is used to control a local process. In such cases, edge processing approach is desirable over processing in the cloud. For instance, this approach will require more computational capacity at device and gateway level to meet real-time requirements, preserve privacy and reduce the attack surface towards IoT devices by keeping most sensitive data local. This approach will imply a disruption from the vertical silo approach promoted by current commercial solutions, where all data are captured in cloud repositories and then fed back to the user. One of the most pertinent research tracks for edge computing will be to set the confidence level from information gathered from a cloud server, aggregated data from federated clouds and/or information retrieved from the internet. In the times of fake news, whilst experiencing novel possibilities of aggregating and manipulating data using AI, it poses unprecedented challenges for users and connected systems to set the appropriate confidence or criticality level of any external information. It remains a challenge for any future AI systems to make transparent how knowledge has been elicited that relies on trusted sources and algorithms. In contrast, edge computing novel architecture should support more decentralized decision and action support system available directly at the device level. In addition, edge computing solutions can create partial views on an environment to facilitate the decision-making process, perform data pruning, processing, anonymization, etc.

Towards data-centric architectures, dealing with the exploding volume of data generated across the different application fields and relying on AI techniques for pre-processing of data. Data storage and data flow will stress capabilities of the IoT platforms, mainly due to the large number of devices and objects, with the need of storing, processing and exchanging large amount of data in due time. Data storage is directly linked with security and privacy components, and data markets, including the availability for the regular citizen and not only corporations and stakeholders, other than with Data Sovereignty (subject to the laws of the country in which data is retrieved and located). The application of AI (mainly machine learning) across the whole IoT pipeline will have its roots in the cloud but will have to be deployed at the edge level, embedded in the things or the gateways to meet time constraints. An IoT data centre like the IBM Watson will be capable of re/defining experience and learning, detecting recurring patterns and systematic failures, in particular it will be able to adapt a holistic risk assessment of a system state in a complex environment. As said, critical functions have to be replicated and delegated to a local agent that ensures the functioning of a system even if it is offline. On the application/services side, AI-powered digital agents can act on behalf of the end users, interact with the most appropriate sensors and access the data related to the users' current activities. Up to an extent, these agents can act autonomously and proactively, for a seamless bridging of the real and digital world. Real-time intelligence provided by such lightweight agents would enable smart devices to have better understanding of their surroundings, the user's conditions and allow them to behave accordingly.

Towards community-driven business models ensuring security and privacy, building on DLTs. Novel business models and services increasingly built on social networks that are associated with daily life needs like mobility, shopping or home care, might be linked to a building, a quartier or city. We have seen success stories in the economy, like Uber, Airbnb or eBay, that have grown exponentially and that build on the fundamentals of a sharing economy. These peer-to-peer (P2P) marketplaces are driven by common interest and shared values.

In order to secure and enable growth of those P2P platforms, scale and secure technologies for authentication, authorization and accounting must evolve from isolated platforms to an ecosystem of connected platforms. DLT enables autonomy and ultimately, secures machine-to-machine (M2M) transactions without a central platform provider. Also, DLT can be a solution to manage the certificates for access to information from objects, including personal data, as well as smart contracts enabling new business models for P2P platform services. Things like money, loyalty points, intellectual property, certificates or even identity, can be sent across the globe, safely, (almost) instantly and without the need for a middle man/intermediary. Security and privacy mechanisms, based on blockchains or any other DLT, may provide new benefits and possibilities to the individual users to effectively and securely manage their personal data space, like authenticating the origin of the data and allowing the use of the data for specific stakeholders and applications, allowing the control of the re-selling of the data. The creation of micro-contracts and using cryptocurrencies may support the final benefit or revenues to the users. Traditional industrial sectors like energy, transport, or food chains may be transformed through P2P platform services, with an impact detrimental to today's business models. It remains a challenge and obligation not to ignore but to embrace P2P platforms that contribute to the growth of a community and demonstrate the opportunities of emerging technologies like DLT or blockchains for IoT platforms. DLT holds promise to mediate interactions in future decentralized IoT environments, but next-generation DLT solutions are needed to make this a reality. Current distributed ledgers seem not to be scalable and had difficulties to handle a high transaction load.

Towards resilient and reliable infrastructure. Future IoT services and applications will require infrastructures to support IoT device connectivity, data streaming and security with new requirements for service quality and reliability. Decentralized data governance and data security will be possible thanks to distributed architectures using DLT, where the control of personal data is significantly improved. But a trusted DLT platform will require beyond a protocol scalable, performant infrastructure and shared governance to establish trust and security. Another challenge for the infrastructure will be the treatment of IoT traffic which will be a major research and deployment issue, to increase availability, resilience and use of data coming from IoT. The emerging trends are related to distributed architectures, software defined technologies and new networking capabilities.

2.3 Interoperability

IoT environments are rather complex with heterogeneous physical devices supporting various communication protocols, while they are possibly connected to an intermediary gateway and then to their virtual representations (i.e. services) running on different platforms. Thus, it is possible to interact with a single IoT device in many ways using its varied interfaces and representations.

IoT platforms require interoperability on multiple levels, which means finding the characteristic functionalities of each layer and defining meta-protocols that can be mapped on the ones used in the platforms (i.e. on the level of syntactic interoperability, the characteristic functionality is resource access). A lot of work has been done in this field in particular in the IoT-EPI, which focuses mainly on architectures and semantic interoperability [2]. As an example, the INTER-IoT project [3] has defined an IoT multi-layer approach to provide semantic interoperability, as illustrated in Figure 2.2.

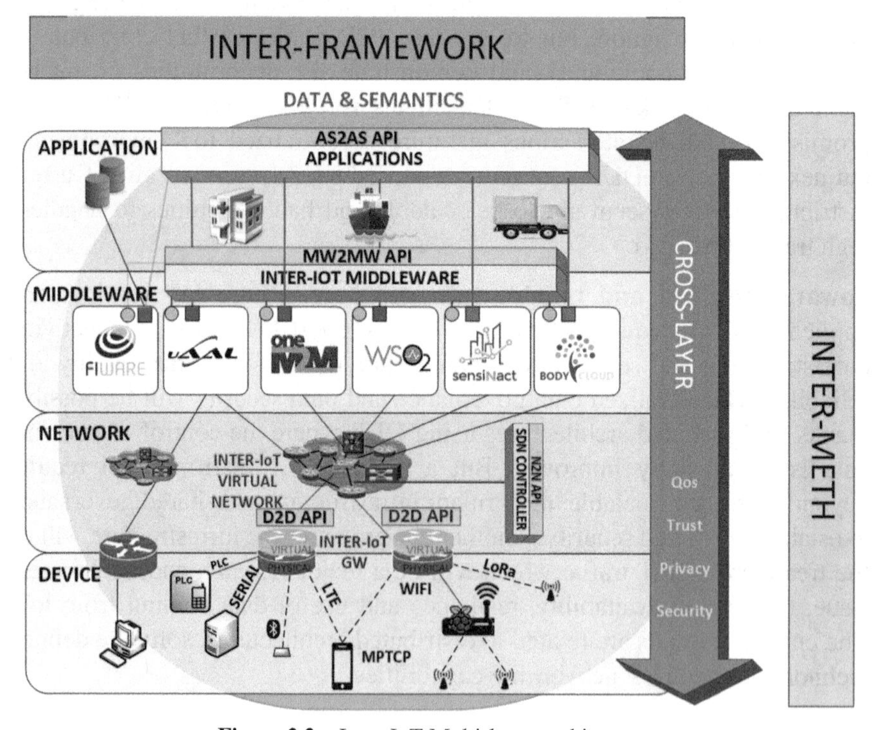

Figure 2.2 Inter-IoT Multi-layer architecture.

Nevertheless, research on a layer-oriented approach is still needed to address tighter interoperability at all layers of IoT systems (device, network, middleware, application, data and semantics) with a strong focus on guaranteeing trust, privacy and security aspects within this interoperability.

The demands of the future internet, including future IoT applications and services, will require a much larger object space, resource efficient implementation in devices, object interaction across so far siloed application spaces, as well as support for intelligent and trusted mechanisms for service provision. Standards have to support interoperability for any object to be seamlessly connected. New connected objects allow users to optimize functions in their daily life (to be safe, for entertainment and comfort, or daily activity support). This requires that objects seamlessly and securely connect, but that they are also identified due to their functionality. On semantic interoperability, despite several efforts to find common ontologies to be reused and different standardization efforts (e.g. SAREF, W3C or ETSI), in a real interoperability environment, new ontologies have to be defined, to address specific deployment. Efforts have to be devoted to semantic translation or alignment in order to provide an easy support for ontology matching between IoT platforms. Work needs to continue on common vocabularies, data models and semantic mapping techniques that could become the key technologies for semantic interoperability via common efforts on the abstract core model for IoT domains.

Under the new Focus Area on Digitisation in the Horizon 2020 Work Programme 2018–2020, the European Commission calls for a pilot on Interoperable Smart Homes and Grids under call DT-ICT-10-2018-19. IoT is expected to enable a seamless integration of home appliances with related home comfort and building automation services allowing to match user needs with the management of distributed energy across the grid. Through Digitisation of Energy, there will be much more assets connected to the grid, which are intelligently communicating with the grid. This comes with all kinds of complexities in terms of interoperability, but mainly due to a lot of different IoT platforms coming from different manufacturers and sectors, like building automation, heating, electrical vehicle charging, appliances, etc. The energy sectoral ecosystem finds itself in a transition period that entails the grid operator, the energy business and services, and the changing role of a consumer or prosumer. The interconnectivity of different systems and assets will become very powerful through IoT platforms if interoperability can be achieved across federated systems that enable the integration of data and novel services.

2.4 Boosting IoT Innovation and Deployment

In future IoT solutions, the importance of data will prevail and further grow. Measuring the economic value of data is a key challenge, focusing on the understanding of the economic value of the data instances and streams in different IoT infrastructure deployment use-cases. The openness of localized sensor data will provide new means to boost the IoT market. Providers of such data will experience new revenue streams. Moreover, new form of marketplaces will be created; that of local data marketplaces, which will also boost innovation. Especially when considering use cases like smart cities, smart transportation or smart grids, where sensing information is characterized by lot of heterogeneous and sensitive data sources, the real benefit from such kind of data markets is seized when data is shared across private, public and industrial value chains. Apart from technology enablers for data marketplaces like DLT, the European Commission favours communities and ecosystems that provide incentives for sharing data on any kind of assets or resources to create an added value through new services and applications (e.g. shared parking, car-sharing, P2P energy, etc.). It remains a challenge for public decision makers to adapt the regulatory framework for new data economy towards a Free Flow of Data, harmonization of data access across borders, data protection and portability in support of a Digital Single Market.

The IoT platform centric point of view will evolve to an ecosystem of platforms with IoT platforms, IoT nodes and sets of IoT things. Instead of IoT platform companies trying to lock-in their customers through closed system approaches, thus creating complex integration links, new common and open interoperation among all these structures will be needed. Ecosystem governance is necessary for controlling different degrees of interoperation and for managing the access to data and services across the whole ecosystem, especially for the use of personal data.

2.5 Conclusion

IoT is a key technology transversal to all sectors of activity and will be fundamental for the NGI initiative. The next generation of IoT will build on a new generation set of devices and systems that will make use of new infrastructure enhancements, better sensing and actuating capabilities, end-to-end semantic knowledge, more powerful computation capabilities on the edge, intrinsic adoption of AI from the edge to the backbone, and the ability to set-up new relationships (like smart contracts, context awareness or intelligent behaviour) among things, services and people, while respecting the human-centric concerns in terms of privacy, security, openness, sustainability and control of personal data.

The inputs collected from the relevant workshops and IoT stakeholder communities are key inspiring sources on the strategic directions needed to support future research, development and innovation of IoT in the context of the NGI initiative. These sources are major inputs to the elaboration of future research and innovation work programmes within Horizon 2020 and beyond.

References

[1] The Next Generation Internet initiative, online at: https://www.ngi.eu/
[2] IoT European Platforms Initiative – white paper on "Advancing IoT Platforms interoperability"
[3] INTER-IoT project, online at: http://www.inter-iot-project.eu/

3

The Next Generation Internet of Things – Hyperconnectivity and Embedded Intelligence at the Edge

Ovidiu Vermesan[1], Markus Eisenhauer[2], Martin Serrano[5], Patrick Guillemin[4], Harald Sundmaeker[3], Elias Z. Tragos[9], Javier Valiño[6], Bertrand Copigneaux[7], Mirko Presser[8], Annabeth Aagaard[8], Roy Bahr[1] and Emmanuel C. Darmois[10]

[1]SINTEF, Norway
[2]Fraunhofer FIT, Germany
[3]ATB Institute for Applied Systems Technology Bremen, Germany
[4]ETSI, France
[5]Insight Centre for Data Analytics, NUI Galway, Ireland
[6]Atos, Spain
[7]IDATE, France
[8]Aarhus University, Denmark
[9]Insight Centre for Data Analytics, University College Dublin, Ireland
[10]CommLedge, France

Abstract

The Internet of Things (IoT) and the Industrial Internet of Things (IIoT) are evolving towards the next generation of Tactile IoT/IIoT, which will bring together hyperconnectivity, edge computing, Distributed Ledger Technologies (DLTs) and Artificial Intelligence (AI). Future IoT applications will apply AI methods, such as machine learning (ML) and neural networks (NNs), to optimize the processing of information, as well as to integrate robotic devices, drones, autonomous vehicles, augmented and virtual reality (AR/VR), and digital assistants. These applications will engender new products, services and experiences that will offer many benefits to businesses, consumers and industries. A more human-centred perspective will allow us to maximise the effects of the next generation of IoT/IIoT technologies and applications as we move towards the integration of intelligent objects with social capabilities that need to address the interactions between autonomous systems and humans in a seamless way.

3.1 Next Generation Internet of Things

The IoT is enabled by heterogeneous technologies used to sense, collect, store, act, process, infer, transmit, create notifications of/for, manage and analyse data. The combination of emergent technologies for information processing and distributed security, e.g. AI, IoT, DLTs and blockchains, brings new challenges in addressing distributed IoT architectures and distributed security mechanisms that form the foundation of improved and, eventually, entirely new products and services.

New systems in the IoT that use smart solutions with embedded intelligence, connectivity and processing capabilities for edge devices rely on real-time analysis of information at the edge. These new IoT systems are moving away from centralized cloud-computing solutions towards distributed intelligent edge computing systems. Traditional centralized cloud computing solutions are perfect for non-real-time applications that require high data rates, huge amounts of storage and processing power, are not strict to very low latency, cost money and can be used for heavy data analytics and AI processing jobs. On the other hand, distributed edge solutions introduce computations at the edge of the network where information is generated and are perfect for real-time services, since they exhibit very low latency (in the order of milliseconds) and can be used for simple ultra-fast analytics jobs. The collection, storage and processing of data at the edge of the network in a distributed way contributes also to the increased privacy of the user data, since no personal information is stored in backbone centralized servers and each user retains the full control of his data.

IoT developments during recent years have been characterized by attributes that can be "labelled" the 6**A**s: **A**nything (any device), to be transferred from/to **A**nyone (anybody), located **A**ny place (anywhere), at **A**ny time (any context), using the most appropriate physical path from **A**ny path (any network) available between the sender and the recipient based on performance and/or economic considerations, to provide **A**ny service (any business). The IoT paradigm is evolving and entire IoT ecosystems are now built upon innervation elements known as the 6**C**s: **C**ollect (heterogeneity of devices of various complexities and intelligence, that enhance the real-time collection of data generated from the connections of devices and information), **C**onnect (ubiquitous distributed connections of heterogeneous devices and information, where the connections are the foundational component of the IoT), **C**ache (stored information in the distributed IoT computing/processing environment), **C**ompute (advanced processing and computation of data and information), **C**ognize (information analytics, insights, extractions, real-time AI processing and **C**reate (the creation of new interactions, services, experiences, business models and solutions). This is illustrated in Figure 3.1.

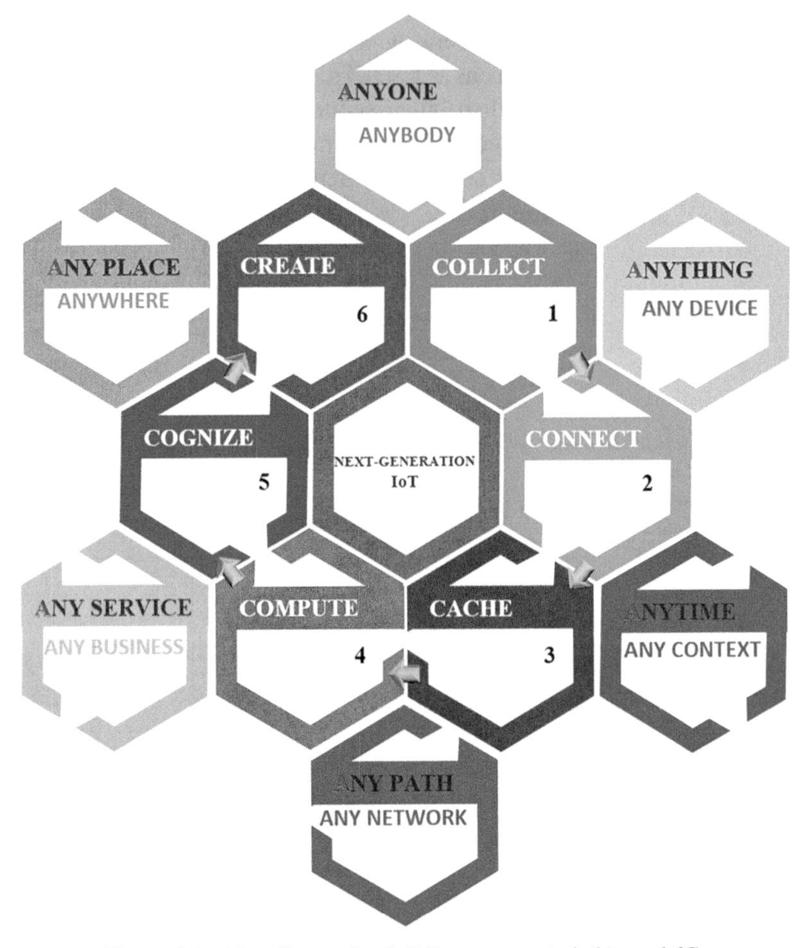

Figure 3.1 Next Generation IoT Hyperconnected: 6As and 6Cs.

The IoT transforms everyday physical objects in the surrounding environment into ecosystems of information that enrich people's lives [97]. The IoT not only influences the future Internet landscape, with implications for security and privacy (personal freedoms), but it could also help to reduce the digital divide. The increased dependence of AI and the IoT on the connectivity network, together with the severity of security challenges, increases their vulnerabilities in parallel. The ongoing and future success of the Internet as a driver for economic and social innovation is linked to how new technologies will respond to these threats. Combining AI with the IoT promises new opportunities, ranging from new services and breakthroughs in science to the augmentation of human intelligence and its convergence with the physical and digital world. The next generation of IoT-combining technologies as presented in Figure 3.3, such as AI, DLTs, hyperconnectivity, distributed edge computing, end-to-end distributed security and autonomous systems - robotics will require increased human-centred safeguards and prioritised ethical considerations in their design and deployment. Next generation IoT evolution is illustrated in Figure 3.2.

The IoT is bridging the gap between the virtual, digital and physical worlds by bringing together people, processes, data and things while generating knowledge through IoT applications and platforms. IoT achieves this addressing security, privacy and trust issues across these dimensions in an

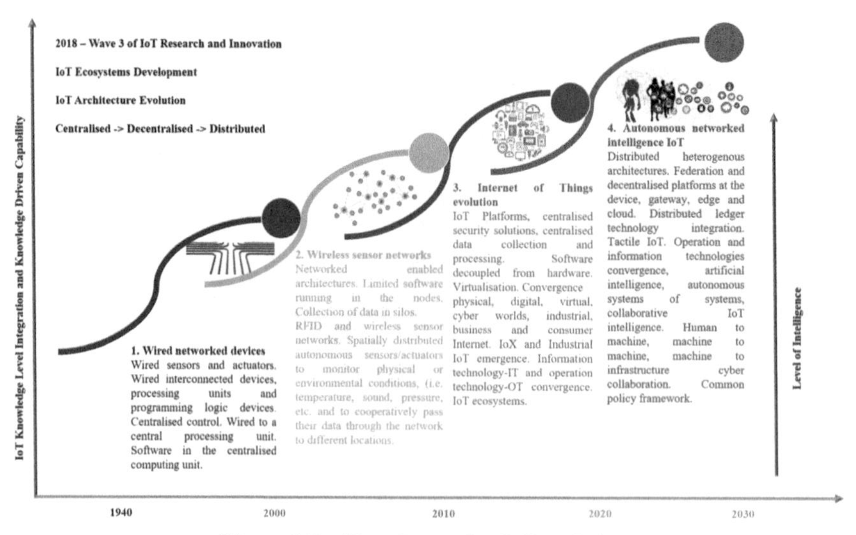

Figure 3.2 Next Generation IoT evolution.

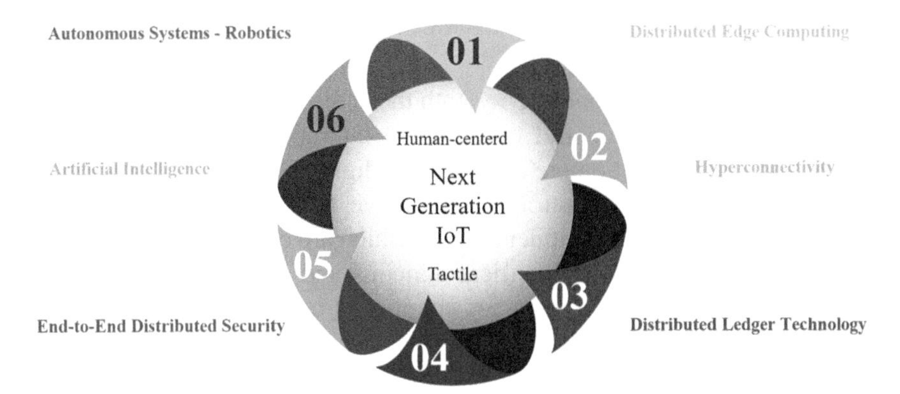

Figure 3.3 Next Generation IoT technology convergence.

era where technology, computing power, connectivity, network capacity and the number and types of smart devices are all expected to increase. In this context, IoT is driving the digital transformation.

As a global concept, the IoT requires a common high-level definition. The IoT is a paradigm involving multidisciplinary activities and has different meanings at different levels of abstraction through the information and knowledge value chain.

Considering the wide background and the number of required technologies, from sensing devices, communication subsystems, data aggregation and pre-processing to object instantiation and finally service provision, proposing an unambiguous definition of the "IoT" is non-trivial.

IoT is defined [60] as a dynamic global network infrastructure with self-configuring capabilities based on standard and interoperable communication protocols where physical and virtual 'things' have identities, physical attributes, and virtual personalities using intelligent interfaces for seamlessly integrating into the information network. In the IoT, 'things' are expected to become active participants in business, information and social processes where they are enabled to interact and communicate among themselves and with the environment by exchanging data and information 'sensed' about the environment, while reacting autonomously to the 'real/physical world' events and influencing it by running processes that trigger actions and create services with or without direct human intervention. Interfaces in the form of services facilitate interactions with these 'smart things' over the Internet, query and change their state and any information associated with them, considering security and privacy issues.

In the context of industry digitisation, IoT/IIoT brings together the primary characteristics of Next Generation Internet (NGI) technology, mobile systems and ubiquitous connectivity with those of industrial control systems, sensing, actuating and control capabilities. Interoperability, platform integration and standardisation are essential for digitising industry applications. IoT/IIoT and industrial control systems have three quality dimensions – integrity, availability and confidentiality – which are essential for implementing applications in industrial vertical domains and across different vertical domains. Whereas the IoT emerged as an add-on to the already existing Internet, it is important to consider the emergence of an NGI where the IoT is deeply embedded and no longer a mere add-on. IoT devices and systems that build on enhanced sensing/actuating, reasoning capabilities and computational power at the edge are already becoming a natural part of an integrated NGI rather than simple extensions of the Internet.

The IoT is promising in a hyperconnected world, where every object has the capability to sense its surrounding environment, transmit information, provide feedback or trigger an action through the application of AI processes in a distributed architecture with processing, intelligence and connectivity at the edge. It is becoming increasingly clear that the main benefit of IoT systems is the network effect, i.e., when different systems are integrated.

As many different systems become integrated, the IoT must face complex interoperability challenges before it can create real cross-domain services with seamless movements of devices and data. However, a lack of stable implementations and the variety of devices available undermine the promised interoperability. A standard solution for IoT interoperability could result in several implementations whose effectiveness would need to be verified and certified; current practices for interoperability testing require different vendors, developers and service providers to participate in physical events. The integration of hyperconnectivity, IoT/IIoT, AI, DLTs and edge computing requires the NGI to address these challenges. This implies the identification of the right business models and the proper governance framework, which support data movement across systems and identify liability in case of any issues, as well as an understanding of the means to overcome the current technical fragmentation in the IoT.

In many applications, the centralised services of cloud computing are being replaced with IoT edge-distributed solutions based on AI methods. With multi-access edge computing (MEC) and ubiquitous hyperconnectivity capabilities (5G and beyond), the IoT is now able to process large amounts of information, resulting from its connections, to be used for intelligent purposes by advanced AI algorithms, which can learn with less data and require fewer processing and memory resources.

The cognitive transformation of IoT applications also allows the use of optimised solutions for individual applications and the integration of immersive technologies, i.e., virtual reality (VR) and augmented reality (AR). Such concepts transform the way individuals and robots interact with one another and with IoT platform systems.

3.2 Next Generation IoT Strategic Research and Innovation

The Internet of Things European Research Cluster (IERC) concentrates the know-how regarding scientific production and research capacity for the Internet of Things in Europe; the IERC brings together EU-funded projects with the aim of defining a common vision for IoT technology and addressing European research challenges. The rationale is to leverage the large potential for IoT-based capabilities and promote the use of the results of existing projects to encourage the convergence of ongoing work; ultimately, the endpoints are to tackle the most important deployment issues, transfer research and knowledge to products and services, and apply these to real IoT applications.

The objectives of IERC are to provide information on research and innovation trends, and to present the state of the art in terms of IoT technology and societal analysis, to apply developments to IoT-funded projects and to market applications and EU policies. The final goal is to test and develop innovative and interoperable IoT solutions in areas of industrial and public interest. The IERC objectives are addressed as an IoT continuum of research, innovation, development, deployment, and adoption.

Every year, the IERC launches its Strategic Research and Innovation Agenda (SRIA), which is the outcome of discussions involving project representatives/coordinators, a collective group of experts from different stakeholders representing the different domains where IoT is relevant and industry representation that is not necessarily limited to IERC community participation. Such industry participation includes the Alliance for the Internet of Things Innovation (AIOTI), an industry-lead association representing the industrial European and non-European members.

Enabled by the activities of the IERC, IoT is bridging physical, digital, virtual, and human spheres through networks, connected processes, and data, and turning them into knowledge and action, so that everything is connected in a large, distributed network. New technological trends bring intelligence and cognition to IoT technologies, protocols, standards, architecture, data acquisition, and analysis, all with a societal, industrial, business, and/or human purpose in mind. The IoT technological trends are presented in the context of integration of hyperconnectivity, digital transformation, actionable data, information and knowledge.

The IERC works to provide a framework that supports the convergence of IoT architecture approaches; it will do so while considering the vertical definition of the architectural layers, end-to-end security, and horizontal interoperability.

The SRIA is developed with the support of a European-led community of interrelated projects and their stakeholders, all of whom are dedicated to the innovation, creation, development, and use of IoT technology.

Since the release of the first version of the SRIA, we have witnessed active research on several IoT topics. Updated releases of this SRIA build incrementally on previous versions [60, 62, 88] and highlight the main research topics associated with the development of IoT-enabling technologies, infrastructure, and applications [87].

The research activities include the IoT European Platforms Initiative (IoT-EPI) program that includes the research and innovation consortia that are working together to deliver an IoT extended into a web of platforms for connected devices and objects. The platforms support smart environments, businesses, services and persons with dynamic and adaptive configuration capabilities. The goal is to overcome the fragmentation of vertically-oriented closed systems, architectures and application areas and move towards open systems and platforms that support multiple applications. IoT-EPI is funded by the European Commission (EC) with EUR 50 million over three years (2016–2018) [67].

The research and innovation items addressed and discussed in the task forces of the IoT-EPI program, the IERC activity chains, and the AIOTI working groups form the basis of the IERC SRIA to address the roadmap of IoT technologies and applications; this is done in line with the major economic and societal challenges underscored by the EU 2020 Digital Agenda [87].

The IoT European Large-Scale Pilots Programme [68] includes the innovation consortia that are collaborating to foster the deployment of IoT solutions in Europe through integration of advanced IoT technologies across the value chain, demonstration of multiple IoT applications at scale and in a usage context, and as close as possible to operational conditions.

The programme projects are targeted and goal driven initiatives that propose IoT approaches to specific real-life industrial/societal challenges.

They are autonomous entities that involve stakeholders from supply side to demand side, and contain all the technological and innovation elements, the tasks related to the use, application and deployment as well as the development, testing and integration activities.

The scope of IoT European Large-Scale Pilots Programme is to foster the deployment of IoT solutions in Europe through integration of advanced IoT technologies across the value chain, demonstration of multiple IoT applications at scale and in a usage context, and as close as possible to operational conditions. Specific Pilot considerations include:

- Mapping of pilot architecture approaches with validated IoT reference architectures such as IoT-A enabling interoperability across use cases.
- Contribution to strategic activity groups that were defined during the LSP kick-off meeting to foster coherent implementation of the different LSPs.
- Contribution to clustering their results of horizontal nature (interoperability approach, standards, security and privacy approaches, business validation and sustainability, methodologies, metrics, etc.).

The IoT European Large-Scale Pilots Programme includes projects promoting the IoT innovation by means of market applications based on services' demand and impact in the European market, technology readiness and socioeconomic interests in European society. The IoT European Large-Scale Pilots Programme is funded by the European Commission (EC) with EUR 100 million over three years (2017–2019) [68].

The IoT is creating new opportunities and providing competitive advantages for businesses in both current and new markets. IoT-enabling technologies have changed the things that are connected to the Internet, especially with the emergence of Tactile Internet and mobile moments (i.e., the moments in which a person or an intelligent device pulls out a device to receive context-aware service in real-time). Such technology has been integrated into connected devices, which range from home appliances and automobiles to wearables and virtual assistants.

The IoT technologies and applications will bring fundamental changes in individuals' and society's views of how technology and business work in the world. A human-centred IoT environment requires tackling new technological trends and challenges. This has an important impact on the research activities that need to be accelerated without compromising the thoroughness, rigorous testing and needed time required for commercialisation.

A hyperconnected society is converging with a consumer-industrial-business Internet that is based on hyperconnected IoT environments. The latter require new IoT systems architectures that are integrated with network architecture (a knowledge-centric network for IoT), a system design and horizontal interoperable platforms that manage things that are digital, automated and connected, functioning in real-time, having remote access and being controlled based on Internet-enabled tools.

Research and development are tightly coupled. Thus, the IoT research topics should address technologies that bring benefits, value, context and efficient implementation in different use cases and examples across various applications and industries.

IoT devices require integrated electronic component solutions that contain sensors/actuators, processing and communication capabilities. These IoT devices make sensing ubiquitous at a very low cost, resulting in extremely strong price pressure on electronic component manufacturers.

The next generation IoT/IIoT developments, including human-centred approaches, are interlinked with the evolution of enabling technologies (AI, connectivity, security, etc.) that require strengthening trustworthiness with electronic identities, service and data/knowledge portability across applications and IoT platforms. This ensures an evolution towards distributed IoT architectures with better efficiency, scalability, end-to-end security, privacy and resilience. The virtualization of functions and rule-based policies will allow for free, fair flow of data and sharing of data and knowledge, while protecting the integrity and privacy of data. Vertical industry stakeholders will become more and more integrated in the connectivity-network value chain. Moreover, unified, heterogeneous and distributed applications, combining information and operation technologies (IT and OT), will expose the network to more diverse and specific demands.

Intelligent/cognitive connectivity networks provide multiple functionalities, including physical connectivity that supports transfer of information and adaptive features that adapt to user needs (context and content). These networks can efficiently exploit network-generated data and functionality in real-time and can be dynamically instantiated close to where data are generated and needed. The dynamically instantiated functions are based on intelligent algorithms that enable the network to adapt and evolve to meet changing requirements and scenarios and to provide context- and content-suitable services to users. The intelligence embedded in the network allows the functions of IoT platforms to be embedded within the network infrastructure and data, and the knowledge generated by the intelligent connectivity network and by the users/things can be used by the network itself. This knowledge can be taken advantage of in applications outside of the network.

The connectivity networks for next generation IoT/IIoT are transforming into intelligent platform infrastructures that will provide multiple functionalities and will be ubiquitous, pervasive and more integrated, further embedding telephone/cellular, Internet/data and knowledge networks.

Advanced technologies are required for the NGI to provide the energy-efficient, intelligent, scalable, high-capacity and high-connectivity performance required for the intelligent and dynamically adaptable infrastructure to provide digital services – experiences that can be developed and deployed by humans and things. In this context, the connectivity networks provide energy efficiency and high performance as well as the edge-network intelligence infrastructure using AI, Machine Learning (ML), Deep Learning (DL), Neural Networks (NNs) and other techniques for decentralised and automated network management, data analytics and shared contexts and knowledge.

Standardisation and solutions are needed for designing products to support multiple IoT standards or ecosystems and research on new standards and related APIs.

Summarizing, although huge efforts have been made within the IERC community for the design and development of IoT technologies, the continuously changing IoT landscape and the introduction of new requirements and technologies creates new challenges or raise the need to revisit existing well-acknowledged solutions. Thus, below is a list of the main open research challenges for the future of IoT:

- IoT architectures considering the requirements of distributed intelligence at the edge, cognition, artificial intelligence, context awareness, tactile applications, heterogeneous devices, end-to-end security, privacy, trust, safety and reliability.
- IoT systems architectures integrated with network architecture forming a knowledge-centric network for IoT.
- Intelligence and context awareness at the IoT edge, using advanced distributed predictive analytics.
- IoT applications that anticipate human and machine behaviours for social support.
- Tactile Internet of Things applications and supportive technologies.
- Augmented reality and virtual reality IoT applications.
- Autonomics in IoT towards the Internet of Autonomous Things.
- Inclusion of robotics in the IoT towards the Internet of Robotic Things.

- Artificial intelligence and machine learning mechanisms for automating IoT processes.
- Distributed IoT systems using securely interconnected and synchronized mobile edge IoT clouds.
- Stronger distributed and end-to-end holistic security solutions for IoT, preventing the exploitation of IoT devices for launching cyber-attacks, i.e., remotely controlling IoT devices for launching Distributed Denial of Service (DDoS) attacks.
- Stronger privacy solutions, considering the requirements of the new General Data Protection Regulation (GDPR) [80] for protecting the users' personal data from unauthorized access, employing protective measures (such as Privacy Enhancing Technologies – PETs) as closer to the user as possible.
- Cross-layer optimization of networking, analytics, security, communication and intelligence.
- IoT-specific heterogeneous networking technologies that consider the diverse requirements of IoT applications, mobile IoT devices, delay tolerant networks, energy consumption, bidirectional communication interfaces that dynamically change characteristics to adapt to application needs, dynamic spectrum access for wireless devices, and multi-radio IoT devices.
- Adaptation of software defined radio and software defined networking technologies in the IoT.

3.2.1 Digitisation

Digitisation is being utilised in many fields, and, as time passes, the influence of digital approaches and techniques is becoming more apparent in several industrial sectors. Buildings and cities are becoming smarter the larger the number of digital services they offer, vehicles are becoming self-driving, design processes are becoming highly efficient and objects and spaces can be visualised before being materialized thanks to the available digital information. Devices with embedded sensors featuring complex logic are scattered everywhere; they measure light, noise, sound, humidity and temperature and are empowered to communicate with each other to form IoT ecosystems.

A common element in all of these developments is that digitisation creates a great amount of information. A considerable part of this information reveals how objects work internally and as elements of more complex setups. Accordingly, many innovative technological installations offer creative solutions concerning how to collect and process this information and how to take necessary action.

The challenge with this information is related to how things interact with each other and with the environment while exhibiting behaviour that is often similar to human behaviour. This behaviour cannot be accurately handled by robots, drones, etc., so this is where technologies, such as swarm logic and AI, come into play.

Security-perceived threats almost always trigger interactive installations equipped to sense and react to surrounding parameters. Changes in these parameters can be visualised, increasing the chances of real threats being detected and asserted.

Thanks to advanced visualisation techniques, the threat landscape is better defined. While security used to be primarily about securing information, the landscape has widened considerably. The timely transfer of information, threat identification, isolation and correct and traceable actions all rely on security protection.

IoT ecosystems evolve, so too must security strategies, which have to account for the layered architecture, where all things, encryptions, communications and actions must be protected against a growing number of diverse attacks, whether via hardware, software or physical tampering.

The IoT system can be seen as a group of agents with non-coordinated individual actions that can collectively use local information to derive new knowledge as a basis for some global actions. The intelligence lies both in agents (AI) and in their interactions (collective intelligence). At the core of swarm logic is the sharing of information and interactions with each other and the surroundings to derive new information. However, this collective intelligence is prone to a number of attacks, especially related to malicious nodes sending false information to influence the decision-making system. Thus, reputation and trust management systems should be in place to be able to identify malicious or misbehaving system agents/nodes and remove them from the system until they behave normally again. These types of attacks can be easily identified and corrected at the edge of the network without having to move all the information to the cloud. Swarm agents can locate and isolate the threat and then converge towards a common point of processing. This is visualised by depicting the real-time state of the agent's movement.

Swarm-designed security is inspired by nature; hence, if IoT can uncover behaviour patterns (of birds, ants, etc.), it may also be capable of meeting security challenges with well-functioning solutions.

3.2.2 Tactile IoT/IIoT

The Tactile IoT/IIoT is a shift in the collaborative paradigm, adding human-centred perspective and sensing/actuating capabilities transported over the network to communications modalities, so that people and machines no longer need to be physically close to the systems they operate or interact with as they can be controlled remotely.

Tactile IoT/IIoT combines ultra-low latency with extremely high availability, reliability and security and enables humans and machines to interact with their environment, in real-time, using haptic interaction with visual feedback, while on the move and within a certain spatial communication range.

Faster Internet connections and increased bandwidth allow to increase the information garnered from onsite sensors within industrial IoT network. This requires new software and hardware for managing storing, analysing and accessing the extra data quickly and seamlessly through a Tactile IoT/IIoT applications. Hyperconnectivity is needed to take VR and AR to the next level for uniform video streaming and remote control/tactile Internet (low latency).

The Tactile IoT/IIoT provides the capabilities to enable the delivery of real-time control and physical (haptic) experiences remotely. The capabilities of the Tactile IoT/IIoT support the creation of a personal spatial safety zone, which is able to interact with nearby objects also connected to the Tactile IoT/IIoT. If applied to traffic, in the long term, this safety zone will be able to protect drivers, passengers and pedestrians. Autonomous vehicles could detect safety-critical situations and react instantly to avoid traffic accidents and warn other objects of impending danger. In production environments, occupational safety levels will improve as production machines or robots detect and avoid the risk of harm to people in their vicinity [45]. A representation of the Tactile Internet of Things Model is shown in Figure 3.4.

The Tactile IoT/IIoT is the next evolution that enables the control of the IoT/IIoT in real-time, with all human senses interacting with machines, by using various technologies both at the network and application level to enable and enhance the interaction in the cyberspace. At the edges, the Tactile IoT/IIoT will be enabled by the sensor/actuators and robotic "things". Content and data are transmitted over a 5G network, while intelligence is enabled close to the user experience through mobile edge computing. At the application level, automation, robotics, telepresence, AR, VR and AI will be integrated in various IoT/IIoT use cases.

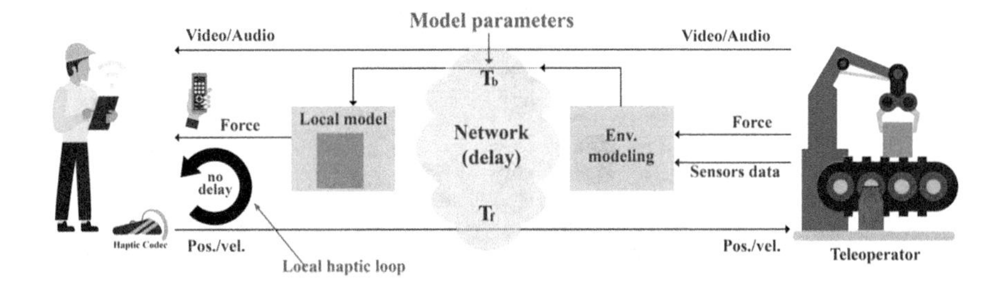

Stable haptic interaction for delays 10ms ... 200ms

Model errors / updates lead to reduced transparency

Figure 3.4 Tactile Internet of Things model.

Source: Adapted from Prof Eckehard Steinbach, TU Munich.

The Tactile IoT/IIoT provides a medium for remote physical interaction in real-time, which requires the exchange of closed-loop information between virtual and/or real objects (i.e., humans, machines and processes). The IEEE P1918.1 working group defines the Tactile Internet as a "network or network of networks for remotely accessing, perceiving, manipulating or controlling real or virtual objects or processes in perceived real-time by humans or machines" [44]. The domains of Tactile IoT are illustrated in Figure 3.5.

The Tactile Internet will benefit VR by providing the low-latency communication required to enable "Shared Haptic Virtual Environments", where several users are physically coupled via a VR simulation to perform tasks that require fine-motor skills. Haptic feedback is a prerequisite for high-fidelity interaction, allowing the user to perceive the objects in the VR not only audio-visually but also via the sense of touch. This allows for sensitive object manipulations as required in tele-surgery, micro-assembly or related applications demanding high levels of sensitivity and precision. When two users interact with the same object, a direct force coupling brought into existence by the VR and the users can feel one another's actions. High-fidelity interaction is only possible if the communication latency between the users and the VR is in the order of a few milliseconds. During these few milliseconds, the movements of the users need to be transmitted to the VR server, where the physical simulation is computed, and the result is returned to the users in the form of object status updates and haptic feedback. Typical update rates for the physical simulation and the display of haptic information are in the

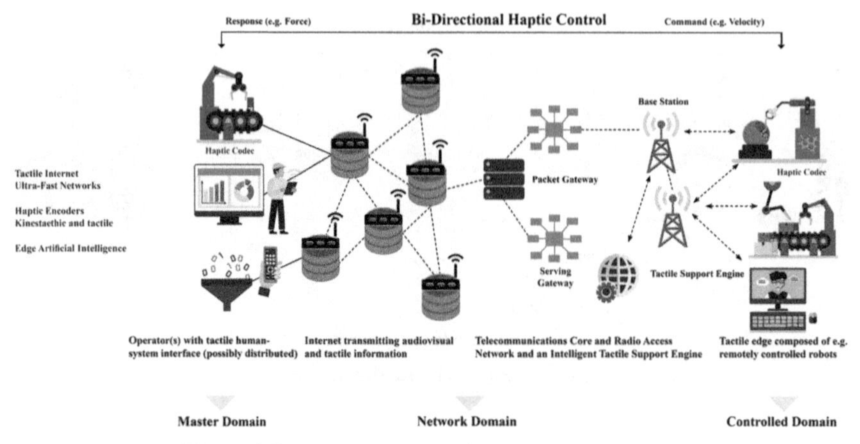

Figure 3.5 Tactile Internet of Things representation.

order of 1000 Hertz, which corresponds to an ideal round-trip communication latency of 1 millisecond (ms) [45].

The use of 5G wireless communications for Tactile IoT/IIoT requires latencies of 1 ms or less. The speed of light in fibre is about 200 km/s. Tactile IoT/IIoT which are distributed over distances larger than about 200 km will require a low-latency IoT core network [50].

Tactile Internet has to meet a number of design requirements such as very low end-to-end latency of 1 ms, high reliability for real-time response, data security, availability and dependability of systems without violating the very low latency requirement due to additional encryption delays. These key design objectives of the Tactile Internet can only be accomplished by keeping tactile applications local, close to the users, which calls for a distributed (i.e., decentralized) service platform architecture based on cloudlets and mobile edge computing. Furthermore, scalable procedures at all protocol layers are needed to reduce the end-to-end latency from sensors to actuators. Importantly, the Tactile Internet will set demanding requirements for future access networks in terms of latency, reliability, and also capacity (e.g., high data rates for video sensors) [51]. Tactile Internet of Things interactions are illustrated in Figure 3.6.

In the future, coworking with robots in IoT applications will favour geographical clusters of local production (“inshoring”) and will require human expertise in the coordination of the human-robot symbiosis with the purpose of inventing new jobs humans can hardly imagine or did not even know they

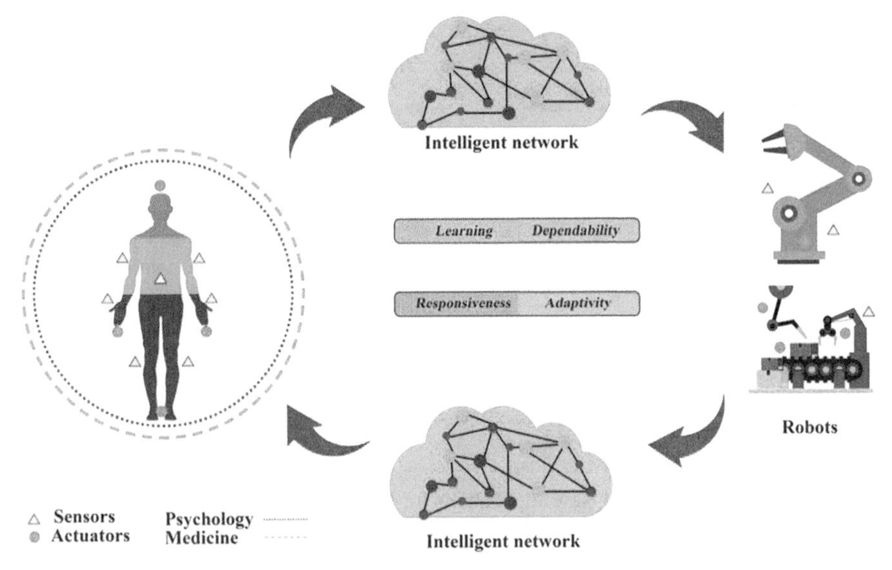

Figure 3.6 Tactile Internet of Things interactions.

Source: Adapted from 5G LAB.

wanted done. Fibre-wireless (FiWi) enabled Human-to-Robot (H2R) communications may be a stepping stone to merging mobile IoT/IIoT, and advanced robotics with automation of knowledge work and cloud technologies, which together represent the five technologies with the highest estimated potential economic impact in 2025 [51, 52].

As presented in Figure 3.7 current Internet cannot guarantee new application delivery constraints. In this context the future technological developments of 5G as the neutral next generation World Wide Wireless Internet by integrating new technologies with a holistic integrated approach combining IPv6-based, machine-to-machine, mobile IoT, mobile edge computing, software defined networks (SDN), network functions virtualisation (NFV), Fringe Internet, Tactile IoT/IIoT, based on seamless worldwide networking interoperability and spectrum harmonisation need to address and solve these constrains for the new applications.

3.2.3 Digital Twins for IoT

Digital twins are virtual representations of material assets. For the IoT, digital twins have never been trendier, as IoT vendors are using increasingly more advanced technology for their implementation, not least with an add-on marketing effect.

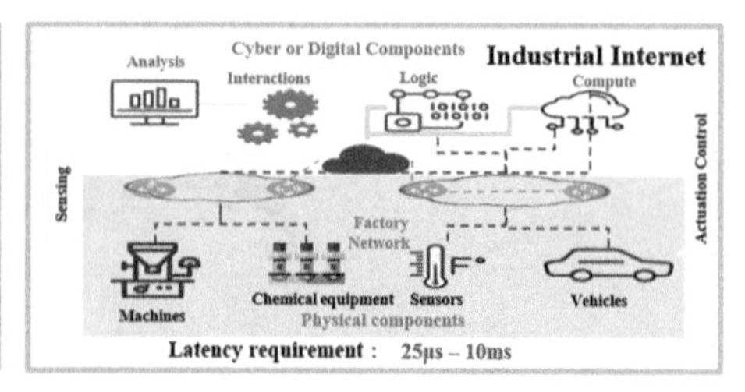

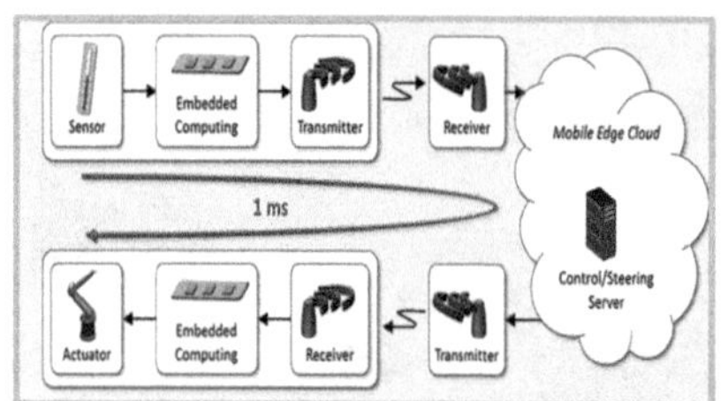

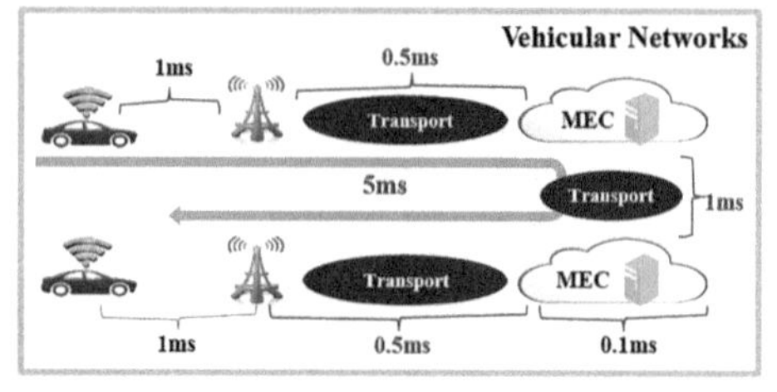

Figure 3.7 New applications for NGI and IoT/IIoT [99].

The current solutions provided by some of the key IoT platforms have mainly been for the representation of physical objects, while such features as simulation, manipulation and optimisation are still missing.

Thanks to technologies, such as blockchain, swarm logic and AI, digital twins now have these capabilities. In the pursuit of better security, digital twins can trigger and simulate threat scenarios in the digital world, as well as optimise the security strategy to handle such scenarios should they occur in the real world.

The digital twin, as a virtual representation of the IoT's physical object or system across its lifecycle, using real-time data to enable understanding, learning and reasoning, is a one element connecting the IoT and AI. The digital twin represents the virtual replica of the IoT physical device by acting like the real thing, which helps in detecting possible issues, testing new settings, simulating all kinds of scenarios, analysing different operational and behavioural scenarios and simulating various situations in a virtual or digital environment, while knowing that what is performed with that digital twin could also happen when it is done by the 'real' physical "thing". Digital twins as part of IoT technologies and applications are being expanded to more applications, use cases and industries, as well as combined with more technologies, such as speech capabilities, AR for an immersive experience and AI capabilities, enabling us to look inside the digital twin by removing the need to go and check the 'real' thing. A digital twin representation is shown in Figure 3.8.

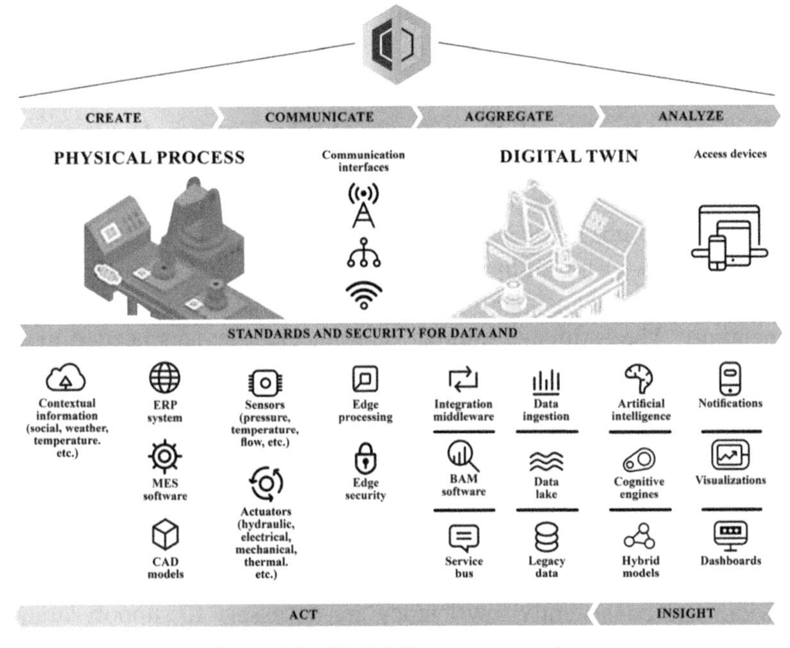

Figure 3.8 Digital Twin representation.

Source: Adapted Deloitte University Press.

Digital twins for IoT must possess at minimum the following attributes:

- Correctness – give a correct replication of the IoT ecosystem and its devices
- Completeness – updated vis a vis the functionality in the real-world system
- Soundness – exhibit only the functionality available in the real-world system
- Abstractness – free from details specific to particular implementations
- Expandability – adapt easily to emerging technologies and applications
- Scalability – must be able to operate at any scale
- Parameterised – accessible for analysis, design and implementation
- Reproducible – be able to replicate the same result for the same input as the real system.

The IoT's digital twins can expand the interface between man and machine through their virtual representation and advanced technologies on levels, such as AI and speech, which enable people and devices/machines to take actions based on operational data at the edge (provided by IoT devices and edge computing processing).

3.3 Future Internet of Things Enabling Technologies

3.3.1 Edge Computing

By 2023, the number of cellular IoT connections is forecast to reach 3.5 billion worldwide. The digitisation of assets, equipment, vehicles and processes in a factory means that the number of connected devices will increase exponentially. The estimated number of connected devices needed in a typical smart factory is 0.5 per square metre[1]. This calculation is based on potential use cases and assets that would benefit from a connection. This illustrates the distribution of cellular connectivity requirements (supporting the previously mentioned use cases) in a fully deployed smart factory. The share of each type of connected device[2] depends on whether the site has a low or high level of automation[3]. Evolving to a higher level of automation will increasingly lead to a higher share of 5G connected devices. Both high bandwidth and consistently low latency are necessary to support large data volumes and real-time critical data, as well as to ensure consistent and secure communication [20].

This requires change in IoT digital infrastructures. According to Gartner, for example, 80 percent of enterprises will have shut down their traditional data centre by 2025, versus 10 percent in 2018. Workload placement, which is driven by a variety of business needs, is the key driver of this infrastructure evolution. In this context, edge computing sits at the peak of Gartner's 2018 Hype Cycle for Cloud Computing and there is plenty of scope for false starts and disillusionment before standards and best practices are settled upon, and mainstream adoption can proceed. Edge computing delivers the decentralized complement to today's hyperscale cloud and legacy data centres. To maximize application potential and user experience, technology innovation leaders plan distributed computing solutions along a continuum from the core to the edge.

---

[1]Average number based on data from different manufacturing sites. In dense areas, the connection density could be up to one connected device per square metre.

[2]The exact distribution figures for a specific manufacturing site depend on the communication needs.

[3]The level of automation is a continuum from manual to fully automatic operations.

According to business-to-business (B2B) analysts MarketsandMarkets, the edge computing market will be worth $6.72 billion by 2022, up from an estimated $1.47 bn in 2017 – a Compound Annual Growth Rate (CAGR) of 35.4 per cent. Key driving factors are the advent of IoT and 5G networks, an increase in the number of “intelligent” applications and the growing load on cloud infrastructure. Among the vertical segments considered by MarketsandMarkets, Telecom and IT are expected to have the biggest market share during the 2017–2022 forecast period. That’s because enterprises faced with high network load and increasing demand for bandwidth will need to optimize and extend their Radio Access Network (RAN) to deliver an efficient Mobile (or Multi-access) Edge Computing (MEC) environment for their apps and services. The fastest-growing segment of the edge computing market during the forecast period, says MarketsandMarkets, is likely to be retail: high volumes of data generated by IoT sensors, cameras and beacons that feed into smart applications will be more efficiently collected, stored and processed at the network edge, rather than in the cloud or an on-premises data centre [19].

The use of intelligent edge devices requires reducing the amount of data sent to the cloud through quality filtering and aggregation, while the integration of more functions into intelligent devices and gateways closer to the edge reduces latency. By moving intelligence to the edge, local devices can generate value and optimise the processing of information and communication. This allows for protocol consolidation by controlling the various ways devices can communicate with each other. There are different edge computing paradigms, such as transparent computing, fog computing and mobile edge computing (MEC). MEC emerged in the context of 5G architectures and enables an open RAN as well as being able to host third party applications and content at the edge of the network. Fog computing, fog networking or fogging is a decentralized computing infrastructure in which data, processing, storage and applications are distributed in the most logical, efficient place between the data source and the cloud. Fog computing extends cloud computing and services to the edge of the network, bringing the advantages and power of the cloud closer to where information is created and acted upon. In a fog environment, intelligence is in the local area network. Information is transmitted from endpoints to a gateway, where it is then transmitted to sources for processing and return transmission. In edge computing, intelligence and power of the edge gateway or appliance are in devices such as programmable automation controllers. Edge computing allows the reduction of points of failure, as each edge device operates independently and determines which information to store locally and which to send to the cloud for further analysis. Fog computing is scalable and offers a view of the network as multiple data points feed information into it. Fog computing enables high-performance, interoperability and security in a multi-vendor computing-based ecosystem and is focusing on resource allocation at the service level, while transparent computing concentrates on logically splitting the software stack (including OS) from the underlying hardware platform to provide cross-platform and streamed services for a variety of devices. One more difference compared to MEC is the need to support exotic I/O and accelerator aware provisioning, real-time, embedded targets as well as real-time networks such as Time Sensitive Networks (TSN), e.g., IEEE 802.1. Another edge computing technology is represented by CMU's Cloudlet, which enables new classes of mobile applications that are both compute-intensive and latency-sensitive in an open ecosystem based on cloudlets. The Cloudlets have lately been transformed to Open Edge Computing[4] based on OpenStack[5]. Open Edge Computing has the vision that any edge node will offer computational and storage resources to any user in close proximity using a standardized mechanism. Edge computing technologies are characterized by openness, as operators open the networks to third parties to deploy applications and services, while their differences enable edge computing technologies to support broader IoT applications with various requirements.

The connectivity requirements of the manufacturing industry are matched by the capabilities of cellular networks. To enable smart manufacturing, there are different network deployment options depending on the case-by-case needs and the digitisation ambitions of the factory. One option is using virtualization and Dedicated Core Networks (DECOR) to map local private networks and virtual networks running within a mobile operator's public network. A 4G and 5G network with dedicated radio base stations and Evolved Packet Core in-a-box can be deployed on the premises to ensure that traffic stays local to the site. In this case, on-premises cellular network deployment with local data breakout ensures that critical production data do not leave the premises, using Quality of Service (QoS) mechanisms to fulfil use case requirements and optimize reliability and latency. Critical applications can be executed locally, independent of the macro network, using cellular network deployment with edge computing [20].

---

[4]http://openedgecomputing.org/

[5]https://www.openstack.org/

The Multi-access Edge Computing (MEC) standard is developed in the ETSI Industry Specification Group/ISG Multi-access Edge Computing (ETSI ISG MEC) [96]. The ETSI ISG MEC is the leading voice in standardization and industry alignment concerning MEC. It is a key building block in the evolution of mobile-broadband networks, complementing Network Function Virtualisation (NFV) and Software Defined Network (SDN), and is:

- A key enabler for IoT and mission-critical, vertical solutions
- Widely recognized as one of the key architectural concepts and technologies for 5G
- Able to enable many 5G use cases without a full 5G roll-out (i.e. with 4G networks)
- Enabling a myriad of new use cases across multiple sectors as well as innovative business opportunities.

The ETSI ISG MEC work on Phase 2 is extending the applicability of MEC technology and rendering MEC even more attractive to operators, vendors and application developers.

One example of deployment is the Cloud IoT Edge that extends Google Cloud's data processing and machine learning to edge devices (e.g., robotic arms, wind turbines, oil rigs, etc.) so they can act on the data from their sensors in real-time and predict outcomes locally. Cloud IoT Edge can run on Android Things or Linux-based operating systems. It is composed of two runtime components, Edge IoT Core and Edge ML, and takes advantage of Google's purpose-built hardware accelerator ASIC chip, Edge TPU™. The Edge TPU is a purpose-built small-footprint ASIC chip designed to run TensorFlow Lite machine-learning models on edge devices. Cloud IoT Edge is the software stack that extends Google's cloud services to IoT gateways and edge devices. Cloud IoT Edge a runtime component for gateway-class devices (with at least one CPU) to store, translate, process and extract intelligence from edge data, while interoperating with the rest of Google's Cloud IoT platform (see Figure 3.9) [21].

Computing at the edge of the mobile network defines IoT-enabled customer experiences and requires a resilient and robust underlying network infrastructure to drive business success. IoT assets and devices are connected via mobile infrastructure and cloud services are provided to IoT platforms to deliver real-time and context-based services. Edge computing uses the power of local computing and different types of devices to provide intelligent services. Data storage, computing and control can be separated and distributed among the connected edge devices (servers, micro servers, gateways, IoT nodes, etc.). Edge computing advantages, such as improved scalability, local

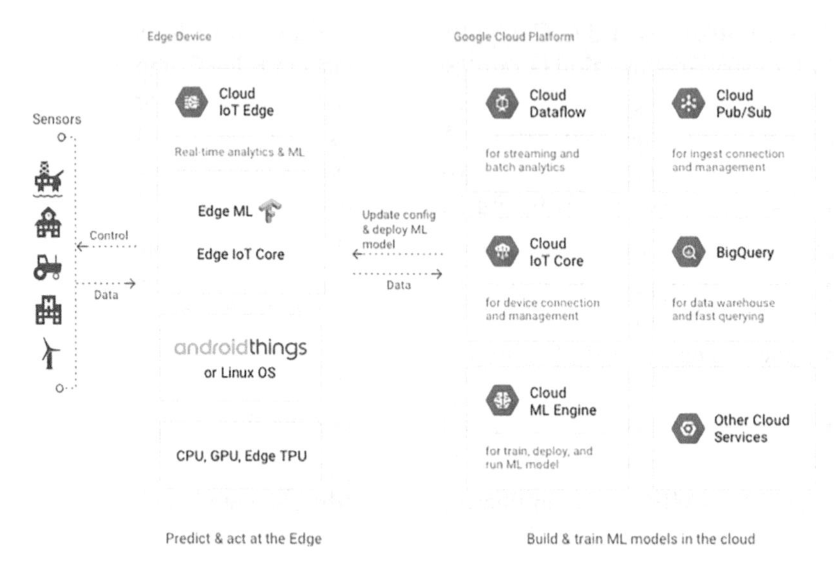

Figure 3.9 How Cloud IoT Edge works [21].

processing, contextual computing and analytics, make it well suited to IoT application requirements. Edge computing technologies like MEC – offering low latency, proximity, high bandwidth, real-time insight into radio network information and location awareness – enable the development of many new types of IoT applications and services for industrial sectors. Augmented Reality (AR) mobile applications have inherent collaborative properties in terms of data collection in the uplink, computing at the edge and data delivery in the downlink [17].

AR information requires low latency and a high rate of data processing in order to provide correct information depending on the location of the device. The processing of information can be performed on a local MEC server instead of a centralized server to provide the user experience required. IoT devices generate additional messaging on telecommunication networks and require gateways to aggregate messages and ensure low latency and security. An architecture used for leveraging MEC to collect, classify and analyse the IoT data streams is presented in [18]. The MEC server manages different protocols and distribution of messages and processes the analytics. The MEC environment supports the creation of new value chains and new type of ecosystems, which provide new opportunities for mobile operators and application and content providers.

Information transmission costs and latency limitations of mobile connectivity pose challenges to many IoT applications that rely on cloud computing. Mobile edge computing enables IoT applications to deliver real-time and context-based mobile moments to users of IoT solutions, while managing the cost base for mobile infrastructure. The benefits are improved performance, deployment of intelligence and analytics at the edge, reduced overload of the communication networks, low latency, compliance, satisfaction of concerns related to data privacy and data security and reduced operational costs. Several challenges listed below, however, have to be addressed when considering edge-computing implementations [91]:

- Mobile edge computing provides real-time network and context information, including location, while giving application developers and business leaders access to cloud computing capabilities and a cloud service environment that is closer to their actual users.
- Mobile edge computing implementation and integration pose the challenge of providing a distributed architecture with improved robustness, reliability and local intelligence, as well as processing that enables the autonomous execution of processes, rules and algorithms.
- Mobile edge computing is an important network infrastructure component for blockchain. The continuous replication of "blocks" via devices on this distributed data centre poses a tremendous technological challenge. Mobile edge computing reveals one opportunity to address this challenge.
- The need to optimize and reduce connectivity, data migration and bandwidths costs associated with sending data to the cloud, while implementing local intelligence, processing and distributed storage.
- Edge computing solutions for avoiding intermittent connectivity, low bandwidth and/or high latency at the network edge considering the increased numbers of smart edge devices running software for machine learning or AI software
- Optimization of the communication with nodes in the intervening edge computing infrastructure.

Regarding future IoT applications, it is expected that more of the network intelligence will reside closer to the source. This will push for the rise of edge cloud/fog and MEC-distributed architectures, as most data will be too noisy, latency-sensitive or expensive to be transferred to the cloud. Edge computing technologies for IoT require developers to address issues such as unstable and intermittent data transmission via wireless and mobile links, efficient distribution and management of data storage and computing, edge computing interfacing with the cloud computing to provide scalable services and, finally, mechanisms to secure IoT applications. The edge computing model requires a distributed architecture and needs to support various interactions and communication approaches to be used broader in consumer/business/industrial domains. To do this, it needs to provide peer-to-peer networking, edge-device collaboration (self-organizing, self-aware, self-healing, etc.), distributed queries across data stored in edge devices as well as in the cloud and temporary storage locations, distributed data management, (e.g., for defining where, what, when and how long, in relation to data storage) and information governance (e.g., information quality, discovery, usability, privacy, security, etc.). In this context, the research challenges in this area are:

- Open distributed edge computing architectures and implementations for IoT and IIoT (IT/OT convergence for IoT applications as traditionally the operational technologies (OT) used to manage and automate industrial equipment are placed at the edge of the network, while information technologies (IT) are more centralized).
- Integrated IoT distributed architecture for IT/OT integration to be used with new business models needed for interpreting or contextualizing IoT data for decision-making, while leveraging integrated data and standard processes to drive outcomes.
- Modelling and performance analysis for edge computing in IoT.
- Built-in end-to-end distributed security at every level of the architecture, in addition to mechanisms for monitoring and managing computing and networking endpoints for IoT systems.
- Heterogeneous wireless communication and networking in edge computing for IoT to handle multiple connectivity solutions using different protocols. Providing different orchestration solutions (e.g., operating both vertically and horizontally with vertical orchestrators to handle services in a specific domain, while horizontal orchestrators manage services across different domains providing integration among them) for edge computing to implement a platform to support both IT and OT activities in IIoT.
- Orchestration techniques for providing compute resources in separate islands, where it is possible to process information and provide services at the local level for a period of time without a coordinate computation and communication.
- Resource allocation and energy efficiency in edge computing for IoT.

- QoS and quality of experience (QoE) provisioning in edge computing for IoT.
- Trustworthiness distributed end-to-end security and privacy issues in edge computing for IoT.
- Federation and cross-platform service supply in transparent computing for IoT.

3.3.2 Artificial Intelligence

Artificial intelligence concerns activity devoted to making machines intelligent, with intelligence understood as a quality that enables an entity to function appropriately and with foresight in its environment [43].

Intelligent IoT devices are considered intelligent machines, while the collective attributes of a machine (i.e., computer, robot or other device) capable of performing functions, such as learning, decision-making or other intelligent human behaviours, are defined as AI. IoT-based sensor data generated in healthcare, bioinformatics, information sciences and policy- and decision- making in governments and enterprises can be processed using methods that rely on AI to provide new data insights and generate new types of knowledge. The benefits of both AI and the IoT can be expanded when the technologies are combined, both on the edge devices' end and core servers' end. AI machine-learning methods can obtain insights from the data to analyse and predict the future connections of IoT devices in advance.

AI is playing a starring role in the IoT because of its ability to quickly bring insights from data. ML offers the ability to automatically identify patterns and detect anomalies in the data that smart sensors and devices generate: information such as temperature, pressure, humidity, air quality, vibration and sound.

Companies are finding that machine learning can provide significant advantages over traditional business intelligence tools for analysing IoT data, including being able to make operational predictions up to 20 times sooner and with greater accuracy than threshold-based monitoring systems [25].

AI techniques extend machine learning strategies that can be applied to intelligent IoT devices for complex decisions based on detecting patterns, self-learning, self-healing, context-awareness and autonomous decision-making. These will involve and affect the future implementations of digital twin models and continuous learning with roles in autonomous vehicles applications, the IoRT and predictive maintenance.

Democratized AI, defined as the possibility to put the AI techniques under the reach of everyone, is one of five trends, along with digitalized

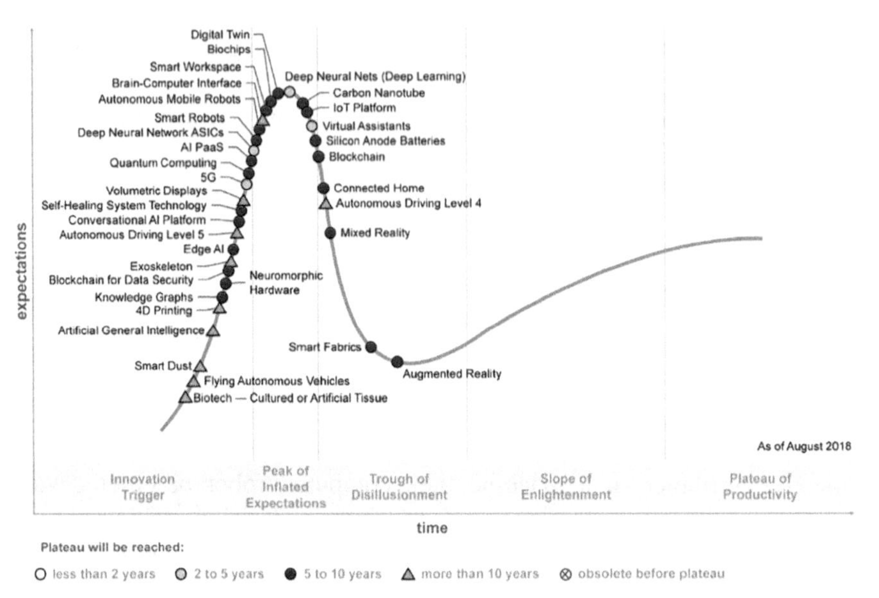

Figure 3.10 Gartner's Hype Cycle for emerging technologies 2018.

ecosystems, do-it-yourself biohacking, transparently immersive experiences and ubiquitous infrastructure, that is driving Gartner's latest Hype Cycle for emerging technologies (see Figure 3.10) [42], derived from 35 individual technologies.

The five trends blur the lines between human and machine with the AI group containing technologies such as AI platform as a service (PaaS), artificial general intelligence, autonomous driving (Levels 4 and 5), autonomous mobile robots, conversational AI platform, deep neural nets, flying autonomous vehicles, smart robots and virtual assistants.

The technologies enabling the next generation IoT are included under all five areas and comprise AI, edge AI, autonomous systems, blockchain, digital twins, augmented reality (AR), 5G, neuromorphic hardware and IoT platforms. The ubiquitous infrastructures of edge computing and the always-on, always-available, limitless infrastructure environment are enabling technologies that form the basis for the next generation IoT landscape.

When combined, AI and IoT transform both the Internet, the global economy and societal interactions. Within the next decade, it is expected that AI and machine learning to be embedded in various forms of technology that incorporate information exchange, analysis and knowledge.

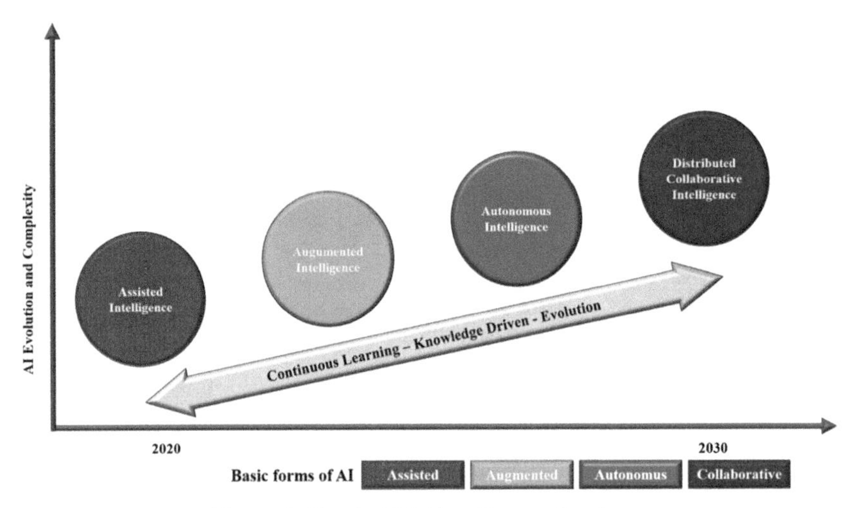

Figure 3.11 Artificial Intelligence Roadmap.

The opportunities created range from new services and breakthroughs in science, to the augmentation of human and machine intelligence and their convergence with the digital, virtual and cyber worlds. The future challenges related to the delegation of decision-making to machines and IoT autonomous systems, lack of transparency and whether technological change will outpace the development of governance and policy norms need to be addressed and solutions must be provided.

The evolution of basic forms of AI from assisted, augmented, autonomous to collaborative is illustrated in Figure 3.11.

In this context, the development of software and IoT devices capable of making ethical judgements as part of autonomous collaborative systems is emerging. As IoT autonomous systems are developing and combined with the ubiquity of AI in applications, such as the Internet of Vehicles for driverless vehicles, artificial ethical agents could become a legal necessity.

The combined developments in AI and the IoT enable new ways of interacting with connected objects through voice or gesture, while AR and virtual reality (VR) are powered by data generated by the IoT. Sensor/actuator technologies, the IoT, AI and increased connectivity bandwidth (ubiquitous, reliable and secure connectivity) are pushing the development of the Tactile IoT based on the convergence of these technologies where the lines between the digital and the physical blur.

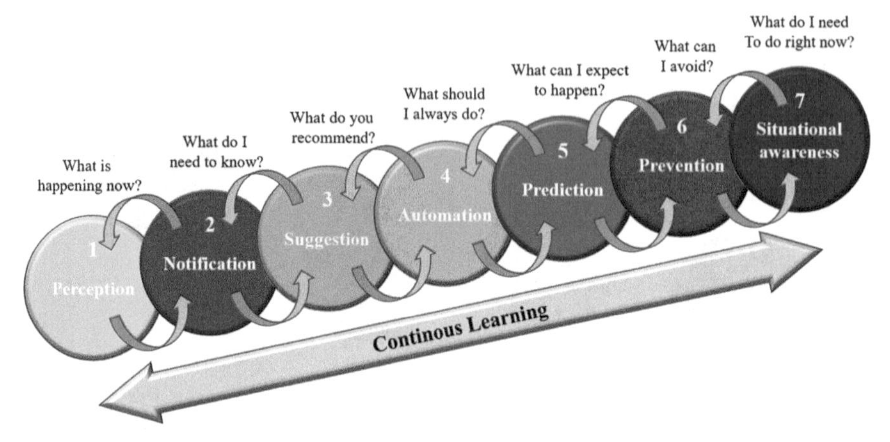

Figure 3.12 Outcomes of Artificial Intelligence.

Source: Constellation Research.

The disruptive nature of AI comes from the speed, precision, and capacity of augmenting humanity. When AI is defined through seven outcomes as presented in Figure 3.12, the business value of AI projects gain meaning and can easily show business value through a spectrum of outcomes [54, 55]:

- Perception describes what is happening now.
- Notification is a way of providing answers to questions through alerts, workflows, reminders and other signals that help deliver additional information through combined manual input and machine learning.
- Suggestion recommends action. This is built on past behaviours and modifications over time that are based on weighted attributes, decision management and machine learning.
- Automation repeats recurrent actions. It is leveraged as machine learning matures over time and tuning takes place.
- Prediction informs what to expect. It builds on deep learning and neural networks to anticipate and test for behaviours.
- Prevention helps avoid negative outcomes. It applies cognitive reckoning to identify potential threats.
- Situational awareness explains what must be known immediately. It resembles mimicking human capabilities in decision making.

AI methods to search for information in data and for learning from past and predict the future is illustrated in Figure 3.13.

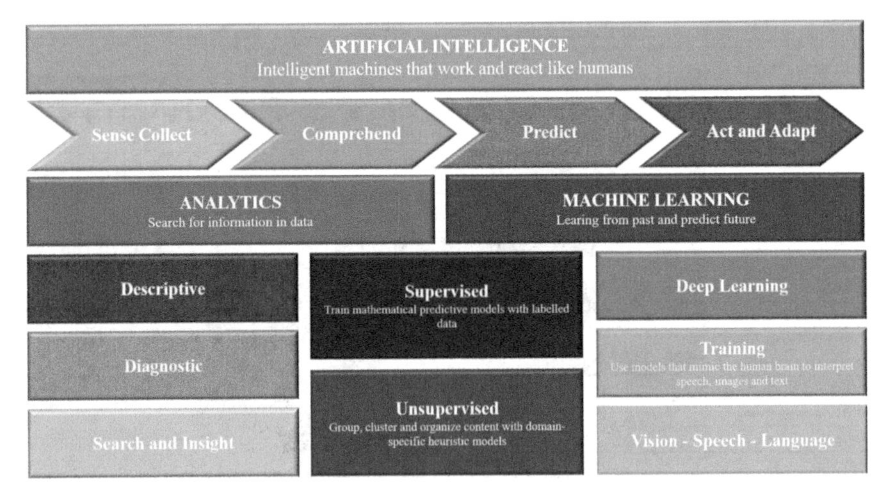

Figure 3.13 Artificial Intelligence methods.

Companies face a difficult task when deciding which opportunities to pursue, among the hundreds available, but they can narrow their options through a structured approach. The first step involves picking an industry and identifying the potential for disruption within the industry, which is estimated by looking at the number of AI use cases, start-up equity funding, and the total economic impact of AI, defined as the extent to which solutions reduced costs, increased productivity, or otherwise benefited the bottom line in a retrospective analysis of various applications. The greater the economic benefit, the more likely that customers will pay for an AI solution. Figure 3.14 shows the data compiled for 17 industries for AI-related metrics [46].

AI is a promising technological innovation, raising already high expectations for 2025. The IoT is the source of data for AI and machine learning applications, as fleets of connected IoT devices, autonomous vehicles and robots need to be automated to allow them to react to environmental conditions in real-time. By 2021, AI will support more than 80% of emerging technologies, while, in the following year, it will support more than 80% of enterprise IoT projects, according to Gartner. By 2020, it will create 2.3 million jobs, although 50% of organizations will lack the relevant AI and data talent.

While software has been a predominant factor in most corporate and investor interest for many years, hardware has become important again with the growth of AI. The cloud continues to be an option for various applications, not least due to its scale advantage, and the choice between cloud or edge

AI Demand and Trends	Market size	Pain points		Willingness to pay
	Global industry size $ trillion	AI use cases #	Start-up equity raised $ billion	Average AI economic impact %
Public/Social sector	25 +	50 +	1.0 +	5 - 10
Retail	10 - 15	50 +	0.5 - 1.0	5 - 10
Healthcare	5 - 10	50 +	1.0 +	15 - 20
Banking	15 - 25	50 +	1.0 +	< 5
Industrials	5 - 10	50 +	0.5 - 1.0	10 - 15
Basic materials	5 - 10	10 - 30	< 0.5	15 - 20
Consumer package goods	15 - 25	10 -30	0.5 - 1.0	5 - 10
Automotive and assembly	5 - 10	10 - 30	0.5 - 1.0	10 - 15
Telecom	< 5	30 - 50	< 0.5	20 +
Oil and gas	5 - 10	30 - 50	< 0.5	< 5
Chemicals and agriculture	5 - 10	10 - 30	< 0.5	5 - 10
Pharmaceuticals and medical	< 5	10 - 30	< 0.5	20 +
Transport and logostics	5 - 10	30 - 50	< 0.5	5 - 10
Insurance	< 5	30 - 50	< 0.5	15 - 20
Media and entertainment	< 5	10 - 30	< 0.5	15 - 20
Travel	< 5	10 - 30	< 0.5	5 - 10
Technology	< 5	10 - 30	< 0.5	10 - 15

Figure 3.14 AI dependency on market size, pain points, and willingness to pay across different industries.

Source: Adapted from McKinsey & Company, [46].

solutions will depend on the IoT use cases and applications. Regarding cloud hardware, the market remains fragmented. The hardware preference of customers and suppliers vary for application-specific integrated circuit (ASIC) technology and graphics processing units (GPUs).

The low latency connectivity at the edge is critical, driving the current development and growing role for inference at the edge. ASICs — with their superior performance per watt — provide a more optimized user experience, including lower power consumption and higher processing, for many applications. Enterprise edge is covered by several technologies, such as field programmable gate arrays, GPUs and ASIC technology.

The ML and DL technology stack is divided into nine layers [46], across services, training, platform, interface, and hardware as presented in Figure 3.15.

Despite rather old technological foundations, in recent years, machine learning has brought about important progress for applications such as computer vision or natural language processing.

It has also recently attracted sizeable investments with an explosion in VC money and a growing focus (through buy outs and investments) amongst Internet companies. The key AI innovations are presented Figure 3.16.

Technology stack and layers			Definitions	Examples
Services	Solution and use case	9	Solution to problems using trained deep-learning model.	Autonomous vehicles (visual recognition).
Training	Data types	8	Data presented to AI system based on a specific application given data.	Labelled versus unlabelled.
	Methods	7	Techniques for optimizing the model weights for the specific application given data.	Unsupervised, supervised, reinforcement.
Platform	Architecture	6	Structures approach to extract features from data given the specific problem.	Convolutional neural network, recurrent neural network.
	Algorithm	5	A set of rules that gradually modifies the weights of neural network to achieve optimal inference, as defined by the training method.	Back propagation, evolutionary, contrasted divergence.
	Framework	4	SW packages to define architecture and invoke algorithms on the HW through the interface.	Caffe, Torch, Theano.
Interface		3	Classes within framework that determine and facilitate communication between SW and underlying HW.	Compute unified device architecture, open computing language.
Hardware	Head node	2	HW unit that orchestrates and coordinates computations among accelerators.	Central processing units.
	Accelerator	1	Silicon chip designed to perform highly parallel operations required by AI.	Training: GPUs, FPGAs, and ASICs. Inference: CPUs, GPUs, ASICs, and FPGAs.

CPU - Central processing unit; GPU - Graphic processing unit; FPGA - Field-programmable gate arrays; ASIC - Application-specific integrated circuit

Figure 3.15 Machine Learning (ML) and Deep Learning (DL) technology multi-layered stack.

Source: Adapted from McKinsey & Company, [46].

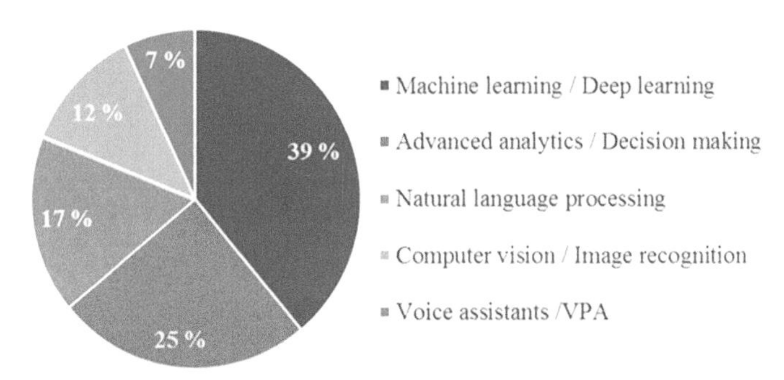

Figure 3.16 Key AI innovations according to the IDATE Technology 2025 survey.

Source: IDATE DigiWorld.

The most anticipated AI applications for 2025 move beyond the current focus on language and vision by targeting advanced data analytics capacities and enabling decision-making applications.

Unprecedented abilities in Data Analytics

If computers are starting to catch up with humans in their ability to detect objects in images, applying deep learning to a field where algorithms are already ahead of most humans, such as data analytics, promises potentially momentous breakthroughs.

Applying deep learning to data analytics enables complex pattern recognition and prediction. This is especially noteworthy in the case of "unsupervised training" machine learning, that is, when the algorithm is fed with unstructured data and tries to spot interesting patterns on its own.

Several industries offer the strongest opportunities for AI: public sector, banking, retail, and automotive as presented in Figure 3.17. While the public sector's prominence may seem surprising in an age where governments are cutting budgets, many officials see the value of AI in improving efficiency and efficacy, and they are willing to provide funding. As they plan their AI strategies, suppliers may focus their investments on potential consumers of AI solutions who are willing to be the first domino [46].

An important domain concerning the application of deep learning data analytics is the health sector. Using deep learning approaches can help in health record data analysis to improve diagnostics, risk analysis and preventive medication.

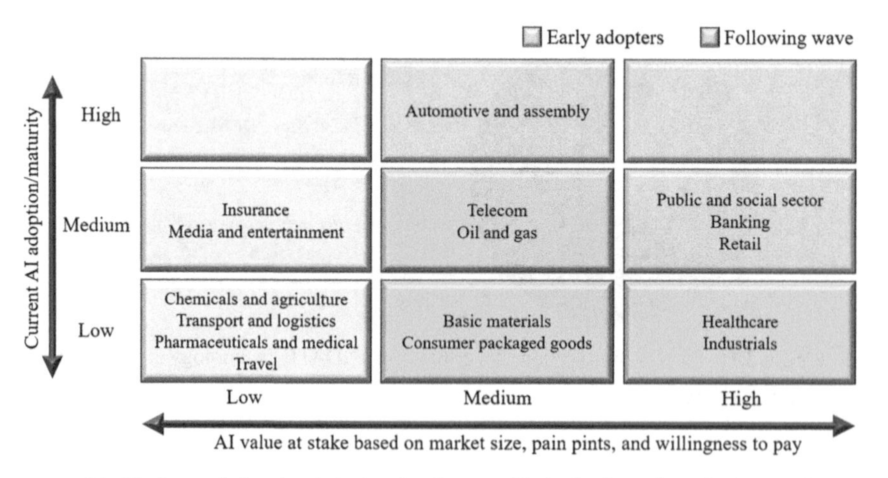

Figure 3.17 AI adoption/maturity vs. value at stake.

Source: Adapted from McKinsey & Company, [46].

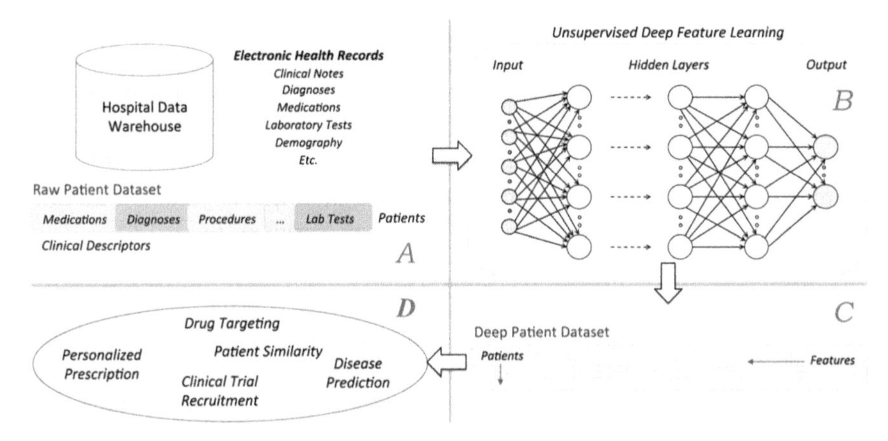

Figure 3.18 Use of unstructured deep learning in the analysis of hospital patient data. *Source*: Nature/Mount Sinai Hospital.

Examples of health applications include reconstructing brain circuits, predicting the activity of potential drug molecules or predicting the effects of mutations in non-coding DNA on gene expressions.

The Mount Sinai Hospital (see Figure 3.18) recently highlighted the potential for using unstructured deep learning in the secondary use of electronic health records in order to predict health status, as well as to help prevent disease or disability.

The interest in deep learning approaches is especially strong in the case of traditional data analysis methodologies, which have been applied with limited success, as it can improve prediction results.

The IoT as Key Data provider for AI

Access to relevant data sets (often derived from vertical industries) in order to train the deep learning algorithms will be critical. The development of the IoT can play a critical role in providing access to the relevant data sets for training future AI including the digital twin models.

Shrinking computer chips and improved manufacturing techniques have led to cheaper and more powerful sensors. As the number of sensors employed in IoT ecosystems increases rapidly, so do the amounts of raw data produced, in turn calling for new computational models to handle this by employing intelligence at the edge for information processing. The new computational models need to cope not only with larger quantities but also with increased complexity of the raw data in terms of their syntax and semantics. Machine learning and deep learning are able to collect and process data from billions of sensors.

Algorithmic developments in AI are coupled with increased data resources and computational demands that are served mainly today by cloud infrastructures. New developments to address AI algorithms for processing data at the edge are underway. Despite the rapid advances of AI in vision, speech recognition, natural language processing and dialog, there is still room for improvement when developing end-to-end intelligent systems that must encapsulate multiple competencies and deliver services in real-time using limited resources. In this direction, the developments are focusing on designing and delivering embedded and hierarchical AI solutions in the combined IoT/IIoT, edge and cloud computing environments that provide real-time decisions, using less data and computational resources, while orchestrating the access to each type of resource in a way that enhances the accuracy and performance of the models. The distributed AI concept builds on top of a hierarchy where low-level, context-agnostic models, which run on the IoT and on IoT devices, can dynamically feed higher-order models running on higher-capacity resources, so as to better capture the context knowledge. Due to the adoption of AI methods and techniques making use of a wide range of available resources (IoT/IIoT/edge/cloud), IoT applications are now able to offer a trade-off between accuracy and performance, depending on the overall requirements.

Complex IoT applications allow distributed intelligence embedded in limited resource sensors at the edge of the network, providing an effective demonstrator of the potential of AI as a service model. There is a need to identify AI and machine learning (ML) methodologies for temporal data to build distributed learning systems that can scale from the IoT/IIoT to edge and cloud resources.

The development of connected vehicles, smart cities and connected health are especially likely to provide access to the massive amounts of data required by deep learning approaches. However, other domains of IoT deployments, such as the industrial Internet are increasingly generating data sets that could fit the deep learning approach. The number of data points that manufacturing facilities generate and the ability to find correlations and pattern and recommend decisions among these IoT data could be addressed by deep learning or other AI methods approaches.

AI and IoT/IIoT requirements for complex integrated systems

In complex IoT applications, the same conditions will seldom apply twice to the same situation, even when they involve the same process; the context and the operation conditions in a specific environment, such as the stress and fatigue of the equipment, always contribute to the creation of an entirely new set of input values for the model. To cope with this expected richness of the model, feedback and training data need to be requested continuously and by a multitude of manufacturing equipment, resulting in a default global (and hence, cross-border) AI model. Deep learning approaches can aggregate the underlying model inputs and gradually build the necessary user and context profiles, but the models (low and higher order) need to continuously adapt to the influx of new data.

A multi-parametric model of manufacturing equipment profiles requires the federation of a multitude of underlying models and data (functional, behavioural, environment, operational, informational, etc.), orchestrated in a distributed and decentralized fashion so as to ensure that the evaluation is happening close to the data sources (ensuring real-time reactions) in an efficient and collaborative fashion.

Thus, the applications across industrial sectors integrating AI and IoT need to address a multitude of requirements in order to fulfil the integration of functional and non-functional attributes for such complex systems. The requirements for complex IoT/IIoT systems that have embedded artificial intelligence (AI) techniques and methods can be summarized as presented in Figure 3.19 and the following list:

Explainability: Enables human users to understand the decisions made by AI systems and the rationale behind them. This ability will make it easier to track down eventual failures and assess decisions' strengths and weaknesses. Ultimately, this will increase the trust in the systems' decisions. This ability will have to integrate with human-computer interface techniques which are able to track complex reasoning processes.

Availability: Enables IoT applications to provide data and resources in a timely manner for a set percentage of time (i.e., the uptime) as well as retain their core functionality, even if the system has undergone a security attack. Industrial IoT applications may target mission-critical tasks along the production line; system outages will therefore have direct economic impact. In the near future, it is also envisaged that IoT systems, due to embedded AI, will be able to perform autonomously via online learning over their lifetime and remove even the downtime needed for maintenance. AI systems should be available in terms of integration into new applications and process steps.

Trustworthiness: Enables IoT systems to be trusted, only allowing authenticated devices or services that can be uniquely identified to participate in the decision-making processes of the system. This makes it possible to report

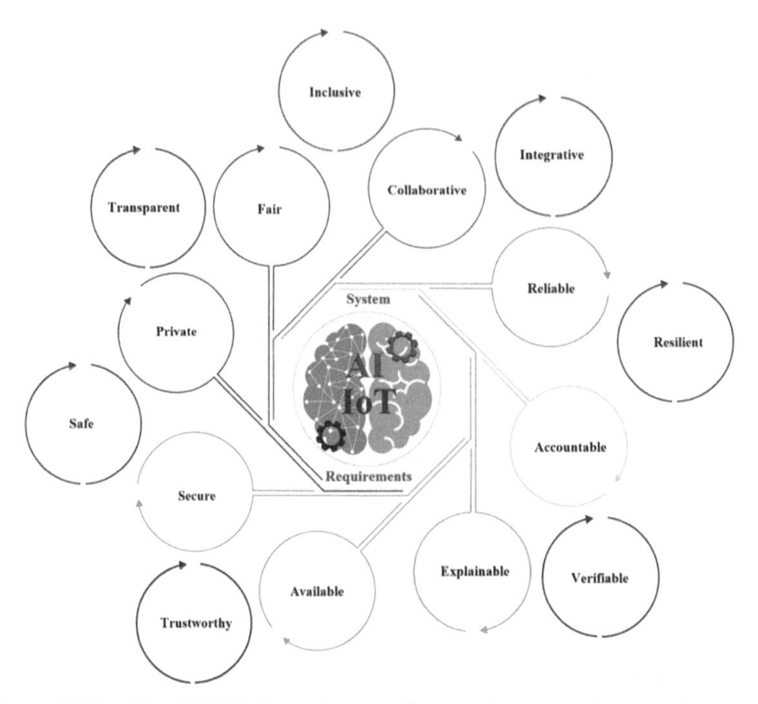

Figure 3.19 AI and IoT/IIoT requirements for complex system integrated systems.

the source of vulnerabilities and inconsistencies. As more and more AI-enabled systems become connected through the IoT, trustworthiness becomes an indispensable requirement. Precisely due to the AI, trustworthiness will become multi-dimensional, far beyond verifying identity. Consequently, trust will no longer be 'true' or 'false', but rather about degrees of trustworthiness that will control the access levels of devices/users to critical services.

Security: Enables systems to guarantee distributed end-to-end security, which is essential to ensure robustness against all types of attack vectors in the IoT. This includes securing the AI system itself as well as securing communication between edge computing IoT devices with encryption and authentication mechanisms against attacks with manipulated input data.

Safety: Enables systems to protect persons and objects during operation. AI systems that operate physically next to and collaboratively with humans through robots or other machines must not exhibit random or unpredictable behaviour. Safety by design is essential, entailing compliance with relevant safety standards. Importantly, the employed AI and IoT systems must be robust against implausible data and operate with extremely low latency to quickly and appropriately react to unforeseen events (i.e., to prevent accidents).

Privacy: Enables IoT and AI-based systems that operate on mission- and business-critical data to keep this data private. This entails both limiting access to and placing restrictions on certain types of information with the goal of preventing unauthorized access (confidentiality) as well as protecting data from being modified or corrupted without detection. Such data must therefore be processed locally at the edge and only leverage data available within privacy limits (smart data).

Transparency: Enables IoT and AI-based systems to provide insight into devices and processes in situations such as auditing, inspections to assess vulnerabilities, or when security breaches arise. This may be supported by digital twins that represent the complete system state at any point in time.

AI methods for data visualization can further enhance transparency and contribute to making the systems state easier to understand.

Fairness: Enables IoT systems which embed AI technologies to support or automate decision processes while adhering to the same fairness and compliance standards as humans.

Inclusiveness: Enables AI-based IoT systems to allow human intervention even in the most automated decision and communication processes. This is essential to avoid the formation of isolated non-AI capable sub-systems within a process, production system or supply chain.

Collaboration: Enables AI-based IoT systems to self-organize around a common goal; for example, in the presence of a threat, as well as to collaborate with humans, both physically (e.g., human-robot collaboration) and by exchanging information (human-machine interfaces). Collaboration is an emergent property of complex interactions and dynamics, increasingly present in industry. Industry-grade AI will not be concentrated on a single device or system. Instead, many different AI-enabled subsystems will be distributed (distributed AI) across IoT nodes, embedded devices and other edge devices (embedded AI).

Integration: Enables IoT-embedding AI systems to exhibit an open and flexible perspective by consolidating insights from all existing systems and processes. Bridging possible gaps is a key prerequisite of the establishing AI methods in the industry according to a sustainable roadmap.

Reliability: Enables IoT systems to operate without systems outages and regular human intervention. Reliability is essential for productivity and is a key prerequisite for AI systems that are put into continuous operation with short maintenance time in mission-critical production environments.

Resiliency: Enables IoT with embedded AI to always operate in stable states, including to return to such states after failures. Resilience is essential for their safe support for our digital economy. In the future, they should even be able to detect failure and initiate measures for compensating it.

Accountability: Enables IoT systems with embedded AI systems that support or even replace human decisions to be accountable to their customers, partners and regulators. Normally, accountability features will be integrated "by design" and will be available via the supplier of these systems.

Verifiability: Enables IoT and AI-based systems to demonstrate the functionality and properties they are supposed to have. AI systems for industrial applications must fulfil the same standards as legacy systems and will be applied to safety-, mission- and business-critical tasks. This requires that AI embedded systems can be validated (to reach correct results), verified (verifiable AI) and certified (certifiable AI) for the targeted applications.

The research challenges for implementing AI at the edge of networks for IoT applications are as follows:

- Mechanisms for collecting and aggregating data and information and developing edge models that generate insights from the data available in real-time by providing methods and techniques to train models in the edge environment with appropriately distributed storage capabilities
- 'AI-friendly' processors to address the AI workloads for IoT applications requiring AI computationally intensive capabilities; research and development concerning architectural concepts to shift central control to the edge and the use of modified graphics processor units, hybrid processors and AI-based processors, embedding accelerators and neural networks for processing specific AI algorithms
- New energy- and resource-efficient methods for image recognition and geospatial processing using AI at the edge, based on machine learning and other AI techniques
- Edge computing implementation based on neuromorphic computing and in-memory computing to process unstructured data, such as images or video, used in IoT applications

- Edge computing implementations based on distributed approaches for IoT computing systems at the edge
- Distributed IoT end-to-end security for AI-based solutions that process data at the edge using a group of edge nodes to work together on a particular task, thereby ensuring that no security holes or attacks are possible
- AI for smart data storage in edge-based IoT
- AI for software-defined networking in edge-based IoT
- Swarm intelligence algorithms for edge-based IoT/IIoT
- Machine learning, deep learning and multi-agent systems for edge-based IoT/IIoT
- Cognitive aspects of AI in edge-based IoT/IIoT
- Neural networks for AI in edge-based IoT/IIoT
- Distributed heterogeneous memory systems design for AI in edge-based IoT/IIoT

3.3.3 Networks and Communication

It is predicted that the adoption of low-power short-range networks for wireless IoT connectivity will increase through 2025 and will coexist with wide-area IoT networks [82], while 5G networks will deliver 1,000 to 5,000 times more capacity than 3G and 4G networks today. IoT technologies are extending known business models, leading to the proliferation of different ones as companies push beyond the data, analytics and intelligence boundaries. IoT devices will be contributing to and strongly driving this development. Changes will first be embedded in given communication standards and networks and subsequently in the communication and network structures defined by these standards.

5G and the IoT promise new capabilities and use cases, which are set to impact not only consumer services but also many industries embarking on their digital transformations. New massive IoT cellular technologies, such as NB-IoT and Cat-M1, are taking off and driving growth in the number of cellular IoT connections, with a CAGR of 30 percent expected between 2017 and 2023. These complementary technologies support diverse LPWAN use cases over the same underlying LTE network [20].

3.3.3.1 Network technology – hyperconnectivity beyond 5G

The development of critical communication capabilities will be an essential enabler for the development of the IoT. It will enable IoT use cases to go beyond data collection and respond to complex scenarios requiring precise actuation, automation and mission critical communications.

The next generation technological enhancements to telecommunication networks, brought about by 5G, will allow new connectivity to become the catalyst for next generation IoT services by creating innovations such as advanced modulation schemes for wireless access, network slicing capabilities, automated network application lifecycle management, software- defined networking and network function virtualization, as well as providing support for edge- and cloud-optimized distributed network applications.

The requirements of critical IoT communications are numerous and diverse, ranging from the increased reliability and resilience of the communication network, to ultra-low latencies and high capacity, while also integrating the context of the mission with the ability to respond to strict energy efficiency constraints or to cover large outdoor areas, deep indoor environments or vehicles moving at high speeds. Bandwidth and delay for services enabled by legacy networks and 5G are presented in Figure 3.20.

Starting with LTE Advanced, cellular communication standards have begun responding to these requirements by developing new technologies.

These first developments are opening new possibilities, especially for public safety operations, but they are still limited in scope. They notably lack the ability to provide the ultra-low latencies required by many critical use cases.

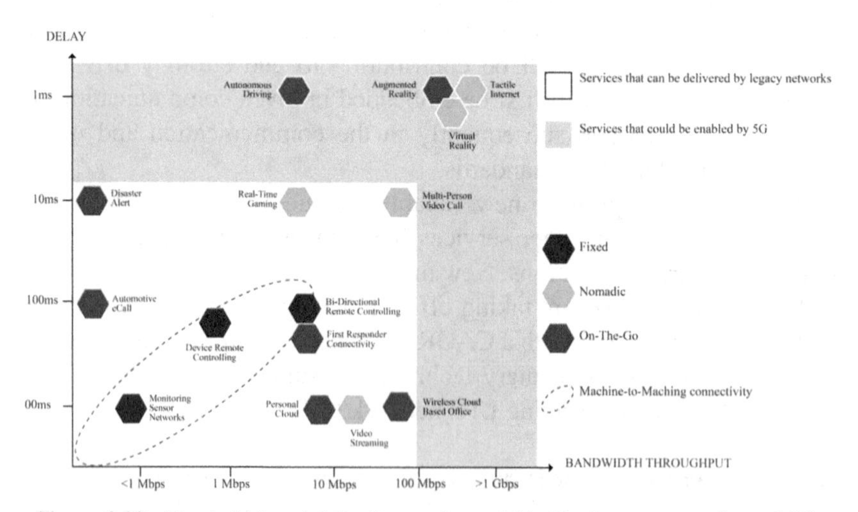

Figure 3.20 Bandwidth and delay for services enabled by legacy networks and 5G.
Source: Adapted from [95].

Table 3.1 Key critical IoT communications requirements

Requirements	Details
Reliability	High availability of the network Low packet losses
Resilience	Ability to function in degraded conditions Low convergence time
Energy efficiency	Projected lifespan of equipment batteries
Low latencies	End-to-end latencies of communication systems under 10 ms and sometimes inferior (under 5 ms or even under 1 ms).
Coverage	Coverage of very a large area (rural) Deep indoor coverage Coverage of moving vehicles Ability to deploy and use private networks
Security	Authentication of communications Encryption of communications Attack detections
Capacity	Ability of the network to operate with a very large number of users

Source: IDATE.

The development of 5G is seen as central to enabling critical IoT communications; indeed, many of the planned 5G features (from network slicing and massive multi-user multiple-input, multiple-output (MIMO) to new messaging services, cellular vehicles- to-everything or improved relay capabilities) represent highly important advances for critical scenarios, including reliable, low latencies. The critical IoT communications requirements are presented in Table 3.1.

End-to-end (E2E) network slicing is a foundation to support diversified 5G services and is key to 5G network architecture evolution. Based on Network Functions Virtualisation (NFV) and Software Defined Network (SDN), physical infrastructure of the future network architecture consists of sites and three-layer data centres (DCs). Sites support multiple modes (such as 5G, LTE, and Wi-Fi) in the form of macro, micro, and pico base stations to implement the RAN real-time function. These functions have high requirements for computing capability and real-time performance and require the inclusion of specific dedicated hardware. Three-layer cloud DC consists of computing and storage resources. The bottom layer is the central office DC, which is closest in relative proximity to the base station side. The second layer is the local DC, and the upper layer is the regional DC, with each layer of arranged DCs connected through transport networks. According to diversified service requirements, networks generate corresponding network topologies and a series of network function sets (network slices) for each corresponding service type using NFV on a unified physical infrastructure. Each network slice is derived from a unified physical network infrastructure, which greatly reduces subsequent operators' network construction costs. Network slices feature a logical arrangement and are separated as individual structures, which allows for heavily customizable service functions and independent operation and management [47].

Advanced ML and AI techniques can also be used for optimizing the connectivity of future mobile heterogeneous IoT devices to allow them to support efficiently a number of diverse services. ML techniques can be used to identify the optimal radio technology in mobile IoT devices considering the load and the services they support. Additionally, in future cognitive radio-based IoT devices, ML techniques can be used to optimize the spectrum channel and width, the devices will use and create a self-organizing network of cooperating devices to improve spectrum utilization [92–94].

As illustrated in Figure 3.21, Enhanced Mobile Broad Band (eMBB), Ultra Reliable Low Latency Communications (uRLLC), and Machine Type Communications (mMTC) are independently supported on a single physical infrastructure. eMBB slicing has high requirements for bandwidth to deploy cache in the mobile cloud engine of a local DC, which provides high-speed services located in close proximity to users, reducing bandwidth requirements of backbone networks. uRLLC slicing has strict latency requirements in application scenarios of self-driving, assistant driving, and

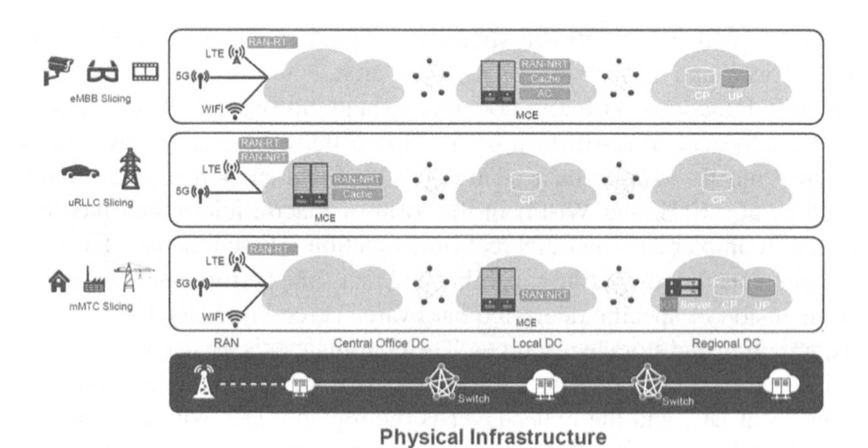

Figure 3.21 End-to-End Network Slicing for Multiple Industries Based on One Physical Infrastructure [47].

remote management. RAN Real-Time and non-Real-Time processing function units must be deployed on the site side providing a beneficial location preferably based in close proximity to users. Vehicle-to-Everything (V2X) server and service gateways must be deployed in the mobile cloud engine of the central office DC, with only control-plane functions deployed in the local and regional DCs. mMTC slicing involves a small amount of network data interaction and a low frequency of signalling interaction in most MTC scenarios. This consequently allows the mobile cloud engine to be deployed in the local DC, and other additional functions and application servers can be deployed in the regional DC, which releases central office resources and reduces operating expenses [47].

Highly dependent upon both the creation of new technologies and the deployment of new communication networks (requiring both important investments all along the value chain), critical IoT capabilities are unlikely to be largely available before 2025.

The 5G spectrum high bands are expected to be deployed include the 28 GHz band, as well as the 26 GHz, 37 GHz and 39 GHz bands. The 28 GHz band may be used in certain countries by the end of 2018 or early 2019, while the other high bands are estimated to be available in late 2019. Low bands below 1 GHz are of interest due to their favourable radio wave propagation characteristics, as they provide coverage in remote areas and into buildings. A new band in the 600 MHz range is expected to be made available by the end of 2018 for 5G services. Part of the mid-bands between 1 GHz and 7 GHz are expected to be allocated in several countries. Mid-bands within the 3.3 GHz to 5 GHz range will likely be made available around 2020 and are seen as important spectrum resources for terrestrial 5G access networks. The midbands are particularly beneficial as they offer a favourable "middle ground" between propagation characteristics (coverage) and bandwidth (capacity). There are several spectrum bands already in use by service providers. In general, all the current 3GPP bands including low bands (600 MHz, 700 MHz, 800 MHz, 850 MHz and 900 MHz) and mid-bands (1.5 GHz, 1.7 GHz, 1.8 GHz, 1.9 GHz, 2.1 GHz, 2.3 GHz and 2.6 GHz) are being considered for 5G services in the future. These bands, and composite arrangements of these bands, will be central to delivering 5G coverage and capacity for enhanced mobile broadband, IoT, industrial automation and mission-critical business cases, as well as for Public Protection and Disaster Relief (PPDR) services. In addition, 3 GPP has recently started a separate Study Item to investigate the feasibility of using the 6.5 GHz band (5,925 MHz to 7,125 MHz) for 5G services [20]. Frequency ranges being studied

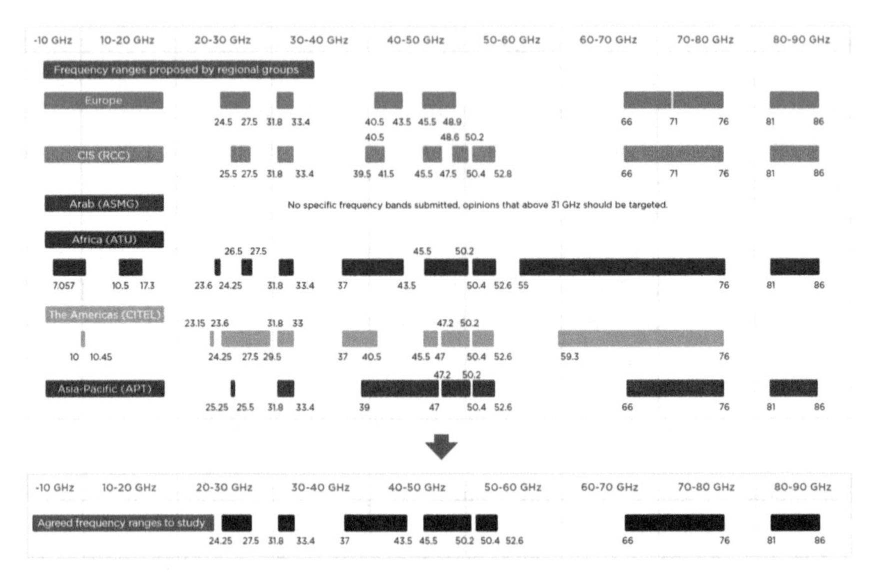

Figure 3.22 Frequency ranges being studied for identification at World Radio Communication Conference 2019 [95].

for identification at World Radio Communication Conference 2019 [95] are presented in Figure 3.22.

IoT applications, based on AR and VR, will revolutionize customer experience in gaming, retail shopping and other customer-centric applications. Consumer experience will be enhanced by high data rates, while extremely low latencies will be achieved.

However, these developments are of interest to many industries, including the automotive, manufacturing, health, energy and public service sectors. There are several factors that could impact commercial adoption of Network Slicing as presented in Figure 3.23. The adoption of Network Slicing influences the IoT applications and the selection of connectivity solutions. In this context, industry activities to standardise Network Slicing should focus on minimising the complexity of the technical solution so that adoption can be made relatively easy, the IoT use cases need to be defined to drive economies of scale and reduce unitary cost of deployment and the operators need to make the cost of deploying Network Slicing marginal to the broader investment case for 5G [49].

The development of a critical IoT is mainly a business-to-business (B2B) and business-to-business-to-consumer (B2B2C) demand, and strongly driven

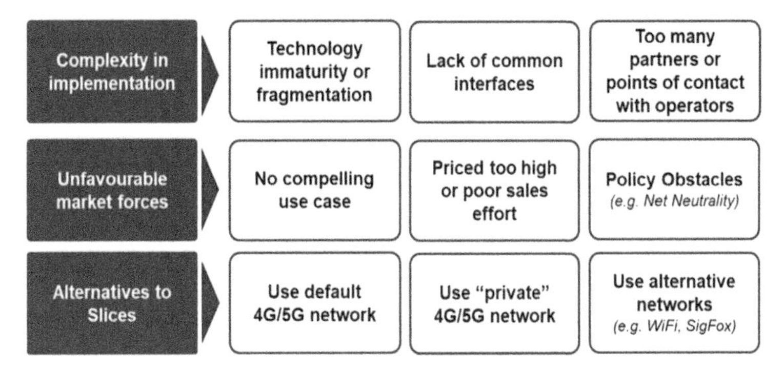

Figure 3.23 Factors that could impact commercial adoption of Network Slicing [49].

by the digital transformation of vertical industries and expanding the traditional cellular technology development (which relies heavily on consumer brands).

The global market is thus limited in volume, with market estimates according to IDATE of about 60 million units by 2030; but this will be compensated by high average revenue per units (ARPUs) as the technology will respond to a critical demand in many industries in terms of generating important cost reductions and new revenue opportunities.

The leading market in volume will be the automotive sector, in which the development of the most advanced autonomous cars will use critical IoT capabilities to perform tasks, such as complex intersection control, dynamic area management, and cooperative cruise control and platooning. The automotive industry is already strongly involved in the standardization process of 5G with the set-up of the 5G Automotive Association (5GAA).

Other verticals of importance include connected health, in which critical IoT capabilities promise the generalization of teleoperations and robotics surgery. Manufacturing will also be strongly impacted, as critical IoT capabilities are among the building blocks of the smart factory (enabling advanced automation and remote control). The key requirements for critical IoT communications in different industrial sectors are presented in Table 3.2.

The 5G use cases can be realized to provide solutions in the B2C, B2B, B2B2X and IoT market segments. In B2C market, operators offer the services such as high definition video (TV, movies, streaming live sports) or home security solutions directly to the end consumers. In B2B market, operators offer the services such as mobility solutions and Cloud services to the businesses (SMEs, large corporations) where the services are typically consumed

Table 3.2 Vertical industrial sectors – key requirements for critical IoT communications

Verticals	Critical IoT Scenarios	Demand Strength	Key Requirements
Automotive	Automated cars	+++	Latency, reliability, coverage (large scale and mobility), point-to-point communication (V2V, V2I)
Health	Robotics	++	Latency, reliability, energy efficiency.
Industrial IoT	Automation, time-critical automation, remote control	++	Latency, reliability, coverage (deep indoor) point-to-point communication, Energy efficiency and local (private) deployments
Energy	Fault prevention and alert, grid backhaul network	++	Latency, reliability, point-to-point communication, large-scale coverage
Public safety	Mission-critical communications	++	Reliability, coverage, resilience, energy efficiency.
Agriculture, forestry, environment	Automation	+	Latency, reliability, energy efficiency, coverage of rural areas

Source: IDATE.

by the employees of the business. In B2B2X market, the services such as 'In stadium' high definition video service are offered to businesses like stadium operators and they in turn offer the service to their premium customers. In IoT market, operators can leverage the low latency, high reliability, high bandwidth and massive connections capabilities to offer several vertical industry use cases like connected vehicles, smart utilities and remote surgery types of applications by participating in the industry specific ecosystems and innovating new business models. Figure 3.24 illustrates the 5G applications market potential and readiness matrix presenting the connectivity and value-added services opportunities in different sectors [98].

International Mobile Telecommunications system requirements for the year 2020 mapped to 5G use cases [95] are presented in Figure 3.27. These promising prospects are attracting many actors to define their future role in the critical IoT market. The capabilities of 5G will indeed lead to more complex value chains with more actors providing connectivity and bundling

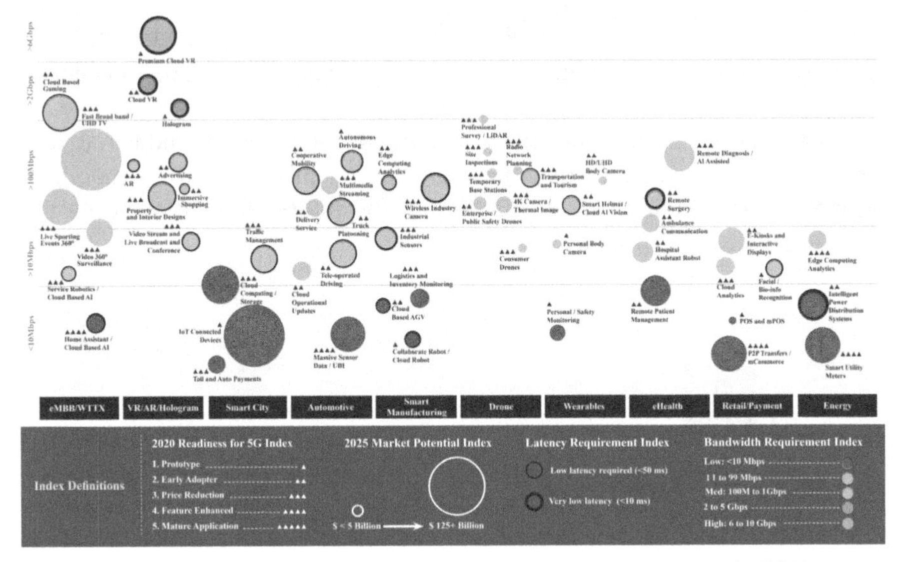

Figure 3.24 5G Applications Market Potential and Readiness Matrix [100].

connectivity with vertical specific services. It is seen by telecommunication companies as an opportunity to diversify and offer vertical specific services, but the rest of the value chain is also eager to benefit from new revenue streams: from equipment providers betting on small cell networks, to over-the-top (OTT) players looking over unlicensed networks or pure vertical players integrating connectivity in their new services. Figure 3.25 illustrates the use of 5G connectivity in different industrial application areas.

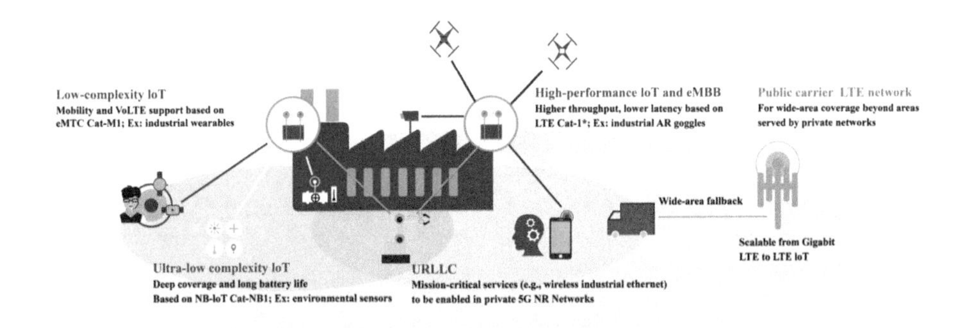

Figure 3.25 5G use in different industrial application areas.

Industrial sectors will depend on smart wireless technologies like 5G and LTE advanced for efficient automation of equipment, predictive maintenance, safety, process tracking, smart packing, shipping, logistics and energy management.

Smart sensor technology offers unlimited solutions for industrial IoT for smarter, safe, cost effective and energy efficient industrial operation. A number of key requirements for factory of the future automation scenarios for connectivity are presented in Figure 3.26.

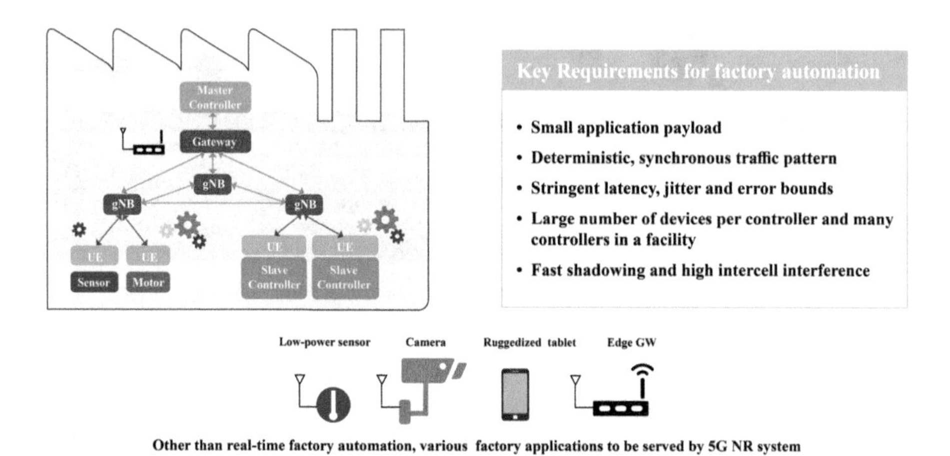

Figure 3.26 Key requirements for connectivity for factory of the future automation.

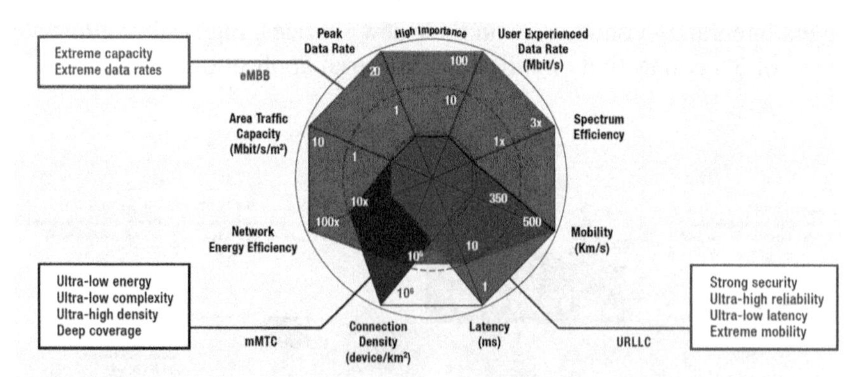

Figure 3.27 International Mobile Telecommunications system requirements for the year 2020 (IMT-2020) mapped to 5G use cases [95].

5G communications could be considered a disruptive element enabling the vision of a truly global IoT, given that one of the key features of 5G is the focus on the integration of heterogeneous access technologies, including satellite communication systems. Satellites could play an important role in providing ubiquitous coverage and reliability in remote areas and enabling new IoT services. IoT devices are not equipped with satellite connectivity, while IoT protocols are not designed with satellite requirements in mind. Thus, cross-layer optimization is required to allow the collection of IoT data from satellites, load balance and the offloading of terrestrial networks, in turn enabling smooth integration of IoT and satellite networks.

5G offers a more reliable network and will deliver a secure network for the IIoT by integrating security into the core network architecture. Industrial facilities will be among the major users of private 5G networks.

Future networks have to address the interference between different cells and radiation and develop new management models to control roaming, while exploiting the coexistence of different cells and radio access technologies.

New management protocols controlling the user assignment with regard to cells and technology will have to be deployed in the mobile core network for better efficiency in accessing the network resource. Satellite communications need to be considered as a potential radio access technology, especially in remote areas. With the emergence of safety applications, minimizing latency and the various protocol translations will benefit end-to-end latency. Densification of the mobile network strongly challenges the connection with the core network. Future networks should however implement cloud utilization mechanisms in order to maximize efficiency in terms of latency, security, energy efficiency and accessibility.

In this context, there is a need for higher network flexibility, which combines cloud technologies with software-defined networks and network function virtualization, which will enable network flexibility to integrate new applications and configure network resources to an adequate degree (sharing computing resources, splitting data traffic, security rules, QoS parameters, mobility etc.).

The evolution and pervasiveness of present communication technologies have the potential to grow to unprecedented levels in the near future by including the Web of Things (WoT) into the developing IoT. Network users will be humans, machines and things, and groups of them.

3.3.3.2 Communication technology

Global connection growth is mainly driven by IoT devices, both on the consumer side (e.g., smart home) and on the enterprise/B2B side (e.g., connected machinery). The number of IoT devices that are active is expected to grow to 10 billion by 2020 and 22 billion by 2025. Figure 3.28 presents the global number of connected IoT devices categorised by the communication/protocol technology [27].

These trends require the extension of the spectrum in the 10–100 GHz range and unlicensed band and technologies, such as WiGig or 802.11ad, which are mature enough for massive deployment and can be used for cell backhaul, point-to-point or point-to-multipoint communication.

Modular integrated connectivity creates a scalable mobile platform (modems for 2G/3G/4GLTE), enabling high-speed data and voice and various onboard selected LoRa, Sigfox, On Ramp Wireless, NWave/Weightless SIG, 802.11 Wi-Fi/Wi-Fi Aware, Bluetooth, ZigBee, 6LowPAN, Z-Wave, EnOcean, Thread, wMBus protocols with the simultaneous use of multiple ISM radio bands (i.e., 169/433/868/902 MHz, 2.4 GHz and 5 GHz). Connectivity modules are based on integrated circuits (ICs), reference designs and feature-rich software stacks created according to a flexible modular concept, which properly addresses various application domains.

The load of the network will differ, with some models using the unbalanced load of the ad hoc network from the core network point of view, and

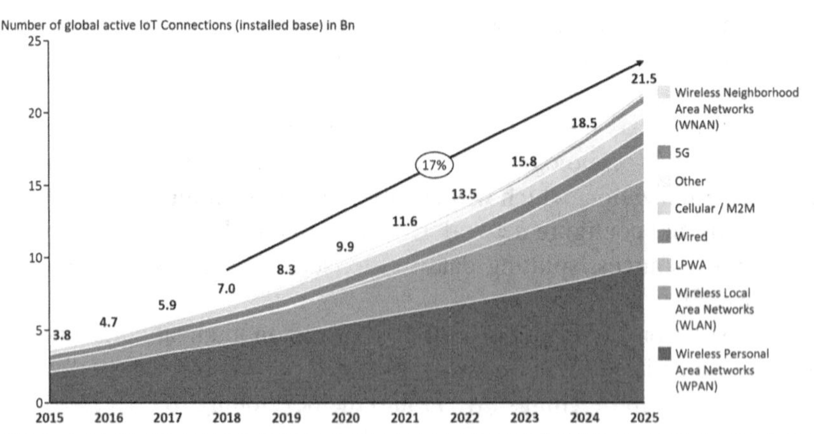

Figure 3.28 Global number of connected IoT devices [27].

others using network-based solutions by balancing the topology from the core network point of view. In this case, the identified network requirements to be supported are the calculation of the optimal ad hoc network topology, by using monitoring information, and the notification of appropriate actions.

Wireless Personal Area Networks (WPANs)

The highest number of IoT devices is connected through short-range technology (WPAN), which typically does not exceed 100 m in the maximum range. These include Bluetooth-connected devices, such as headsets, as well as ZigBee- and Z-Wave-connected devices, which can mostly be found in smart homes, e.g., for connecting smoke alarms or thermostats [27]. Zigbee 3.0 is a networking solution used on top of IEEE 802.15.4 radio technology, which includes Wi-Fi and IP Internet capability. Zigbee 3.0 has meshing capability and is used as an IoT connectivity solution for a range of smart home and industrial applications, including lighting, security, thermostats and remote controls. It is secure and supports battery-free devices, meshing, low latency and energy harvesting (e.g., motion, light, piezo, Peltier). Zigbee 3.0 also includes Zigbee Green Power, which was developed as an ultra-low-power wireless standard to support energy-harvesting IoT devices and is effective for IoT devices that are only sometimes on the network (i.e., when they have power), enabling them to go on and off the network securely, so they can be off most of the time.

6LowPAN is a network protocol that defines encapsulation and header compression mechanisms. The standard has the freedom of a frequency band and a physical layer and can also be used across multiple communications platforms, including Ethernet, Wi-Fi, 802.15.4 and sub-1 GHz ISM. The protocol is implementing open IP standards including TCP, UDP, HTTP, COAP, MQTT and web sockets, and offers end-to-end addressable nodes, allowing a router to connect the network to IPs. 6LowPAN is a mesh network that is robust, scalable and self-healing. Mesh router devices can route data destined for other IoT devices, while hosts are able to sleep for long periods of time.

Wireless Local Area Networks (WLANs)

Another large category comprises WLANs, which offer a range of connectivity up to 1 km. Wi-Fi is the most common standard in this category and experiencing significant growth, mostly through the use of home assistants, smart TVs and smart speakers, but also increasingly through use in industrial settings such as factories (although it continues to play a minor role in those settings compared to other technologies) [27]. With the introduction of Wi-Fi 6 (802.11ax standard), the connectivity performance is enhanced for the use of IoT devices and businesses and operators running large-scale deployments.

Wi-Fi 6 brings more capabilities to support next generation connectivity uses. Wi-Fi 6 offers faster speeds for all devices on the 2.4 GHz and 5 GHz spectra, with a raw throughput speed boost of as much as 37%. Wi-Fi 6 IoT devices can shut down Wi-Fi connections most of the time using the Target Wake Time feature and connect only briefly as scheduled in order to transmit data they have gathered since the last time this was performed, thus extending battery life. Wi-Fi 6 uses orthogonal frequency-division multiple access (OFDMA) to improve the efficiency of multi-user multiple-input, multiple-output (MIMO) streams. MIMO works both on the uplink and on the downlink and can simultaneously receive data from different devices on different channels (maximum of eight in Wi-Fi 6) at once. The Target Wake Time feature improves sleep and wake efficiency, reduces power consumption and decreases congestion on crowded networks. In Wi-Fi 6, the theoretical maximum bandwidth of a single stream is 3.5 Gbit/s, and up to four streams can be delivered to a single device, which means a maximum of up to 14 Gbit/s.

Low-Power Wide Area Networks (LPWANs)

A large chunk of the future growth in the number of IoT devices is expected to come from LPWANs. By 2025, it is expected that more than two billion devices will be connected through LPWANs. The technology, which promises extremely high battery life and a maximum communication range of over 20 kilometres, is used by the three main competing standards, LoRa, Sigfox and NB-IoT, which are currently being rolled out worldwide with more than 25 million devices already connected to date, the majority of which are smart meters [27]. Another research report predicts that there will be 2.7 billion LPWAN IoT connections by 2029 [28]. LPWANs operate in the unlicensed industrial, scientific and medical (ISM) spectrum at 900 MHZ, 2.4 GHz and 5 GHz.

LoRa is the standard protocol of the LoRa Alliance (open, non-profit association established in 2015 with more than 500 members). LoRa has a bandwidth of 250 kHz and 125 kHz and a maximum data rate of 50 Kbps, enabling bidirectional communication albeit not simultaneously, and has a maximum payload of 243 bytes. The range is up to 5 km in urban areas and 20 km in rural areas depending on the application. There are around 50 million LoRa-based end nodes and 70,000 LoRa gateways that have already been deployed worldwide [29]. LoRa networks use gateway devices to work and manage the network for connecting IoT devices.

Sigfox operates and commercializes its own proprietary communications technology, which is an ultra-narrowband (100 Hz) with a maximum data rate of 100 bps. It also operates in the unlicensed ISM spectrum and its small payload (maximum 12 bytes) means it can offer greater coverage geographically, reaching up to 10 km in urban areas and 40 km in rural areas. Sigfox offers bidirectional connections, with the downlink from base stations to end IoT devices occurring following uplink communication. The daily uplink messages are limited to 140.

Weightless (Weightless-N, Weightless-W and Weightless-P) operates in the unlicensed spectrum and is an open standard (Weightless SIG) designed to operate in a variety of bands, with all the unlicensed sub-GHz ISM bands, while featuring a 100 kbps maximum data rate on uplink and downlink. Weightless can handle 2,769 end points per base station on standard smart meter set-ups (200 bytes uploaded every 15 min).

NB-IoT coexists with GSM and LTE and is a 3GPP LTE standard, based on licensed cellular networks providing a 200 kHz bandwidth and a 200 kbps maximum data rate, which offers bidirectional communication, albeit not simultaneously, and has unlimited messages with a maximum payload 1,600 bytes. The global shipments of NB-IoT devices will have a compound annual growth rate of 41.8% from 106.9 million units in 2018 to 613.2 million in 2023 [30].

LTE-M is another 3GPP providing extended coverage by using an installed LTE base with the same spectrum, radios and base stations. It is implemented as a 4G technology with an important role in 5G. The uplink/downlink transfers 1 Mbps and, due to low latency and full duplex operation, can carry voice traffic. LTE-M supports more demanding IoT mobile devices, which require real-time data transfer (e.g., transport, wearable), while NB-IoT supports more IoT static sensors and devices.

Cellular and non-cellular LPWA network connections will grow globally at a 53% CAGR until 2023, driven by market growth in smart meters and asset trackers. In 2017, smart meters and asset trackers contributed to almost three quarters of all LPWA network connections, dominated by non-cellular LPWA network technologies. By 2023, non-cellular LPWA will cede its market-share dominance to NB-IoT and LTE-M, as cellular LPWA moves to capture over 55% of LPWA connections.

Private LPWA networks, built to address a single vertical application or an individual enterprise, have been popular choices for over a decade and accounted for 93% of LPWA connections in 2017. LoRa and other non-cellular LPWA technologies have benefited from the decreasing cost of ICs, low implementation costs and flexibility of private networks, which can be tailored to meet specific enterprise IoT applications. As the geographic footprint of public networks rapidly expands, cellular and non-cellular public networks will capture over 70% of LPWA connections by 2023 [31].

Future IoT devices will require network agnostic solutions that integrate mobile, NB-IoT, LoRa, Sigfox, Weightless etc. and high-speed wireless networks (Wi-Fi), particularly for applications spanning multiple jurisdictions.

LPWA networks have several features that make them particularly attractive for IoT devices and applications, which require low mobility and low levels of data transfer:

- Low power consumption that enable devices to last up to 10 years on a single charge
- Optimized data transfer that supports small, intermittent blocks of data
- Low device unit costs
- Few base stations required to provide coverage
- Easy installation of the network
- Dedicated network authentication
- Optimized for low throughput, long or short distance
- Sufficient indoor penetration and coverage

Wired

Few people think of wired connections when they think of the IoT. In many settings, a wired device connection is still the most reliable option that provides very high data rates at very low cost, albeit without much mobility. Particularly in industrial settings, fieldbus and Ethernet technologies use wired connections to a large extent, and it is expected that they will continue to do so in the future [27]. Sensor/actuator units that are installed within a building automation system can use wired networking technologies like Ethernet. Power Line Communication (PLC) is a hard-wired solution that uses existing electrical wiring instead of dedicated network cables and for industrial applications has significant advances. According to the frequency bands allocated for operation PLC systems can be divided into narrowband PLC (NBPLC), and broadband PLC (BPLC). NBPLC refers to low bandwidth communication, utilising the frequency band below 500 kHz and providing data rates of tens of kpbs, while BPLC utilises a wider frequency band, typically between 2 MHz and 30 MHz, and allows for data rates of hundreds of Mbps. BPLC is recommended for smart home applications requiring high-speed data transfer applications like Internet, HDTV, and audio, while the use of NBPLC systems is more appropriate for remote data acquisition, automatic measuring systems, renewable energy generation, advanced metering, street lighting, plug-in electric vehicles, etc. BPLC has higher speed, that reduce the data collection period and ensures a real-time remote-control command, but its stability and reliability are still determined by the quality of power lines. Another way to classify PLC is as PLC over AC lines and PLC over DC lines. Most companies are currently providing AC-PLC solutions, PLC in DC lines also has applications for distributed energy generation, and transportation (electronic controls in airplanes, automobiles and trains).

Cellular / M2M

2G, 3G and 4G technology, for a long time, were the only option for remote device connectivity. As LPWA and also 5G gain momentum, it is expected that these legacy cellular standards will cede their share to new technologies as they present a more lucrative opportunity to many end users [27].

5G

5G is under development and the technology, which promises a new era of connectivity through its massive bandwidth and extremely low latency, is now heavily promoted by governments, particularly China. The Chinese government views 5G adoption as a competitive asset in the quest to move the equilibrium of technological innovation from the US and Europe towards China. In the US, the first pre-standard 5G networks will provide fixed wireless access (FWA) services to residential and small business users by the end of this year. While many more use cases will be targeted once the final standard is ratified in 2020, we should already see first adopters next year and expect quick growth from there [27].

5G includes two of the tree scenarios, massive machine type communications (mMTC) and ultra-reliable and low latency communications (URLLC), which support IoT applications for industries with available, ultra-low latency links for next generation IoT services.

The Internet as network technology is focusing on the internet working among underlay technologies in order to provide end-to-end services. Telecoms/communication and Internet/computer communication are converging via telephone/cellular and Internet/data networks. The TCP/IP paradigm started as an overlay of network technologies, while TCP/IP is nowadays integrated in pre-existent network infrastructures and starting to include transport and application functionality in the network as well.

5G is including the wired/core section of the network, as well as LANs, to architecturally integrate cloud/fog/edge systems, based on software network functions, providing differentiated service support. The next generation Internet and 5G are converging into a fully integrated interoperable network, where people and IoT physical, digital and virtual devices interact in real-time. 5G connectivity is one important element in building real-time interactive systems and implementing a tactile IoT/IIoT in order to provide the communication infrastructure with low latency, very short transit time, high availability, reliability and security. LTE evolution will continue, while LTE and 5G will co-exist in upcoming years. The availability of device hardware and attractive service pricing will influence the adoption of 5G for various IoT applications across different industrial sectors.

5G monetization is a critical success factor for the deployment of 5G and monetization models must be supported by different pricing models. Some of the monetization models to be considered are [98]:

- Monetize network, infrastructure and business services by leveraging the network and infrastructure capabilities:
  - Operators provide services such as Network as a Service, Information broadcasting, Cloud services with high QoS that attract premium in B2B segments.
  - Operators become platform providers for a variety of microservices, assuring low latency where needed by providing them on the edge. With this model, application developers and vendors could simply define how they want their applications to perform and let the connectivity provider make it happen.
  - Operators enable the developers to monetize their applications that connect to many millions of devices and in turn will be able to secure revenues from developers and the end users of these devices.
  - Leveraging the infrastructure, vertical industrial solutions can be offered by establishing ecosystems with complex partnerships and revenue sharing models. Key for success is to make the economics work for both the operators and other participants in the ecosystem to bring useful solutions to the market quickly.
- Monetize value: 5G creates opportunity for operators to monetize the 'value' created by services with revenue sharing type of models rather

than being simply the connectivity provider. Applications such as translation services, home automation can be monetized by putting compute functionality on the ultra-low latency edge networks and thus putting it on the Cloud without compromising on latency.

- Advertisements in the new digital services like high definition content offers new monetization opportunities for operators.
- Data monetization: Operators have access to customer data – customer priorities and interactions, network data – usage, pattern, massive number of device related data. This massive amount of data along with data analytics creates new opportunities for operators to monetize insights.

Wireless Neighbourhood Area Networks (WNANs)

WNANs sit between WLAN and long-range technologies, such as cellular, in terms of communication range. Typical proponents of this technology include mesh networks such as Wi-SUN or JupiterMesh. In some cases, the technology is used as an alternative to LPWA/cellular (e.g., in utilities' field area networks) and in other cases such as a complementary element (e.g., for deep indoor metering where nothing else reaches) [27].

Wi-SUN is an open standards-based field area network (FAN) used for the IoT and can support applications such as advanced metering infrastructure, distribution automation, intelligent transport and traffic systems, street lighting, and smart home automation. The suite of IoT technologies is based on IEEE 802.15.4, TCP/IP and related standard protocols with a bandwidth of up to 300 kbps, a low latency of 20 ms, power efficiency (less than 2 mA when resting; 8 mA when listening), resilience, scalability (networks to 5,000 devices; 10 million end points worldwide) and using security mechanisms based on public key certificates, AES, HMAC, dynamic key refresh and hardened crypto. The PHY layer is based on IEEE 802.15.4g, which provides bidirectional communication. The network layer is IPv6 with 6LoWPAN adaptation supporting star and mesh topologies, as well as hybrid star/mesh deployments.

These different types of networks are needed to address IoT products, services and techniques so as to improve the grade of service (GoS), quality of service and quality of experience (QoE) for end users. Customization-based solutions are addressing the IIoT while moving to a managed wide-area communications system and ecosystem collaboration.

Intelligent gateways will be needed at lower cost to simplify the infrastructure complexity for end consumers, enterprises and industrial environments. Multifunctional, multiprotocol processing gateways are likely to be deployed for IoT devices and combined with Internet protocols and different communication protocols.

These different approaches show that device interoperability and open standards are key considerations in the design and development of internetworked IoT systems.

Ensuring the security, reliability, resilience and stability of Internet applications and services is critical to promoting the concept of a trusted IoT, based on the features and security provided by devices at various levels of the digital value chain.

3.3.4 Distributed Ledger Technology/Blockchain Technology

A distributed ledger is a record of transactions or data that is maintained in a decentralized form across different systems, locations, organizations or devices. It allows data or funds to be effectively sent between parties in the form of peer-to-peer transfers without relying on any centralized authority to broker the transfer. A distributed consensus mechanism allows members of the network (nodes) to establish a common "truth". There are different mechanisms for this: in the case of Bitcoin and other "cryptocurrencies", a computationally complex "proof-of-work" algorithm is used to protect the integrity of the network against change to the public "blockchain" by making it impractical for malevolent players to alter the chain. Whilst Bitcoin operates on a public blockchain, there is also the possibility to operate distributed ledgers privately where network participants are provided with relevant permissions to either read or write to (i.e., append) the ledger [33, 34].

Blockchain is a technological disruption in secured infrastructures. It is based on a combination of encrypted algorithms and duplicated data storage on a network of computers. Used as a secured infrastructure, it can meet the demand for security from various industries.

From a technological perspective, blockchain is a data storage infrastructure technology. It makes it possible to store data securely (each entry is authenticated, irreversible and duplicated), with decentralized control: there is no central authority that controls the information on the chain. This is achieved using encryption technologies (hash function and asymmetric cryptography) and a computer network of independent nodes.

Blockchain technologies were initially designed to be used with the Bitcoin cryptocurrency, where they were employed to create a reliable ledger of all financial transactions. But blockchain technology is also developing in ways that are opening new prospects:

- The use of blockchains as a ledger of transactions (the initial use case)
- The use of blockchains to accurately archive and date important pieces of information
- The introduction of smart contracts: automated conditional transactions that are executed without human intervention or the involvement of a trusted third party
- The advent of decentralized applications: applications that use the blockchain as their execution infrastructure, without a centralized IT platform

The information architecture used by Bitcoin technology provides a source for the development, contextualization, exchange and distributed security of data needed for the IoT.

Blockchains for the IoT transform the way business transactions are conducted globally within a trustworthy environment to automate and encode business transactions while preserving enterprise-level privacy and security for all parties in the transaction. Blockchain solutions are, for instance, being developed to identify IoT objects and to sign automatic and decentralized contracts between connected devices.

The benefits of blockchains for the IoT are providing mechanisms for building trust between stakeholders in an IoT application and IoT devices with blockchain cryptography, to reduce the risk of collusion and tampering, to facilitate cost reductions by removing the overheads associated with middlemen and intermediaries, and to accelerate transactions by reducing the settlement time from days to almost real-time. Considerations in the application of distributed ledgers for the IoT include addressing the storage space, the computing power of the devices, security, communication power, transaction confirmation time, consensus mechanisms, congestions, costs/fees and price volatility.

IoT applications using distributed ledger technologies (DLTs) must evaluate several attributes regarding the implementation of use cases, which must take into account the retention in the distributed ledger, multiparty sharing needs, the trade-off between retrieval and flexibility performance for the ledger database features and the trade-off in real-time, as there is a time lapse between the moment when data or transactions are generated and when the consensus mechanism confirms that the information is part of the ledger. The evolution of the blockchain is illustrated in Figure 3.29.

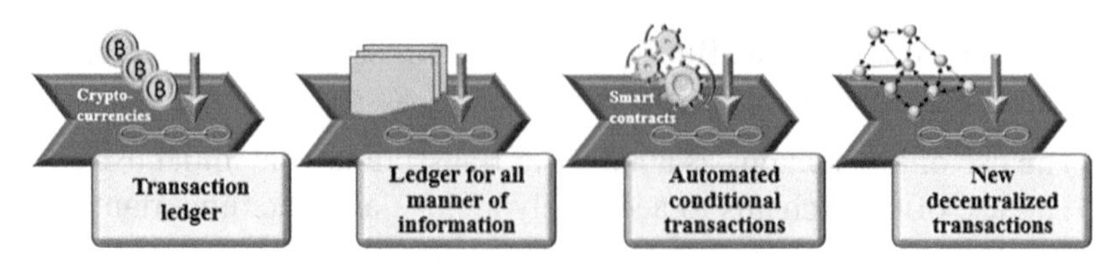

Figure 3.29 Evolution of the blockchain.

Source: Adapted from IDATE DigiWorld, Blockchain, October 2016.

IoT devices will be used in building blockchain-based solutions to support applications aimed at improving operational efficiency, transforming the user experience and adopting new business models in a secure, private and decentralised manner, so that all stakeholders benefit. This is especially the case for blockchain applications that can track and control property: from asset management applications (IoT devices being used to track assets along the logistics chain) to a radical transformation of business relationships, transitioning to a world where any property or object can easily be rented out to another user securely and without the need to interact directly with the user (the user signs a smart rental contract, which, once the payment has been made, gives him/her access to the lock for a set period of time).

The IoT can make use of blockchain-based computing platforms (e.g., iExec, Golem, Sonm, Hypernet, Ripple) or Hyperledger, which is an open source Linux Foundation platform. Recently, the Enterprise Ethereum Alliance, a blockchain standards organization, and Hyperledger announced that they have joined each other's groups [22, 23]. Other solutions offered by Ripple, BigchainDB and Sovrin exist [37–39].

The Hyperledger platform [22, 24] connects data from the IoT via specific adapters in order to integrate a variety of existing sensors and protocols, as well as integrate and connect transactions that are related to these sensors with blockchain systems that might belong to different stakeholders. The platform allows for the use of cognitive artificial intelligence (AI) components to infer new insights from these combined data. Further research is needed to define how to optimally combine blockchains, cognitive AI and the IoT for various industry domains.

Combinations of blockchain technology and the IoT into an IoT-driven blockchain, as used in the aviation industry, are presented in [40, 41] (see Figures 3.30 and 3.31).

The combination of blockchains and the IoT provides several benefits in supply chains such as: tracking objects as they travel along the export/import

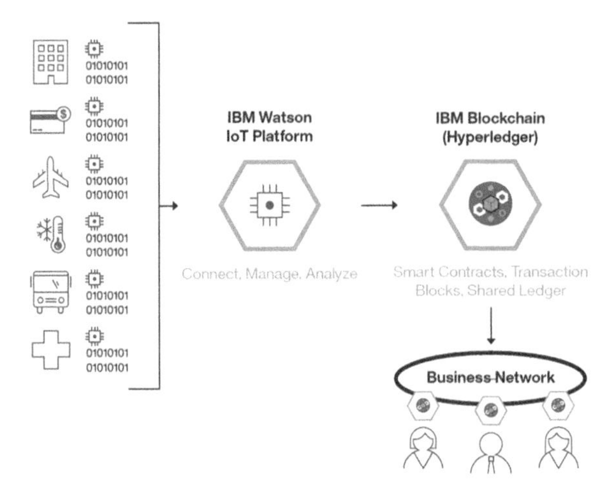

Figure 3.30 Combining blockchain technology and the IoT with the use of IBM Watson and blockchain platforms [41].

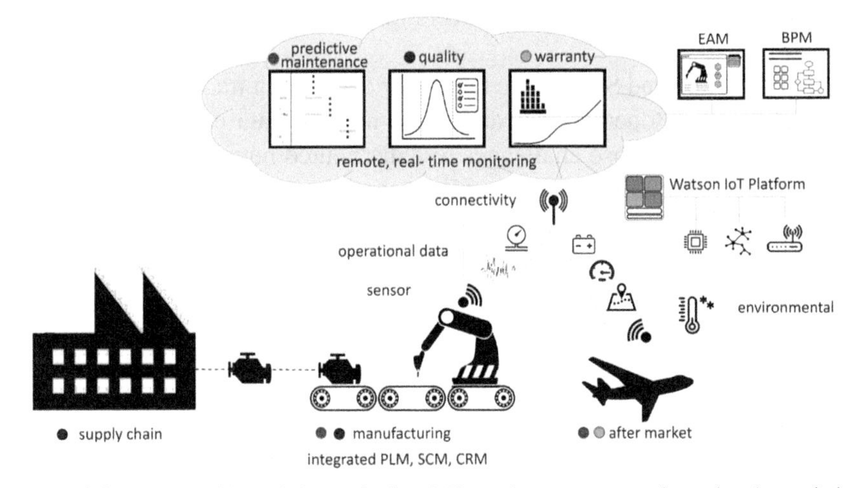

Figure 3.31 Using blockchain and the IoT to improve operations in the aviation industry [40].

supply chain, while enforcing shipping and lines of credit contracts and expediting incremental payments; maintaining an indelible history of parts and end assembly through supply chains, potentially including critical events that affect life or scheduled maintenance; providing decentralized edge computing to securely run computing workloads, such as analytics, on edge devices owned by third parties; interconnecting IoT devices by allowing distributed devices to request and pay for services through distributed role management and micropayments, as well as regulatory compliance, in order to track equipment or process history in an indelible record, and enabling easy sharing of this information with regulatory agencies or insurers [41].

IOTA is a next generation blockchain focused on use in the IoT as a "ledger of things". IOTA uses a revised distributed ledger design known as a "tangle", which aims to be massively scalable as well as avoid the cost of replicating all data to all nodes [35, 36].

Hypernet is proposing new architecture and has implemented a new programming model beneath the blockchain layer to handle distributed computation problems, which require inter process communication. Hypernet is based on the principle of distributed average consensus (DAC), and the combination with blockchains allows for the efficient distribution of compute jobs, while effectively managing processing units in dropping on and off the network. The platform creates a secure backbone, where buyers and providers of computational power can engage, based on trust. The on-chain (scheduled) and off-chain (DAC) technology layers of Hypernet fit together, with both driven by consensus.

Golem, iExec and Sonm have built their concept on traditional computing architectures developed specifically to be used in data centres. These data centre architectures pose challenges to a distributed network used in the IoT as the amount of network communication and data transfer overhead is very high and the architectures do not tolerate computers randomly dropping in and out of the network. As data centre architectures are optimised for one particular topology, they cannot be used on a distributed network, as the network topology is unknown.

Blockchain-based systems require devices in the blockchain to have the resources run the blockchain software and process blockchain data. However, distributed ledgers are open, with devices connected to a distributed ledger known as "nodes". Each "block" within the ledger has a maximum size of 1 MB and IoT devices used to hold a full copy of the ledger need to have processing and storage capabilities necessary to hold at least a few "full nodes" containing the complete ledger.

IoT security issues are relevant when using blockchain technology, as there is a need for proper security credentials to view a transaction and IoT device commissioning and secure key management are challenging issues in the case of IoT devices.

Addressing and solving the limitations of blockchain technology in the future could allow for the integration of blockchain-based platforms for the

IoT. Furthermore, considering that a blockchain contains the transaction and can also contain the contract, the IoT device can process financial information, buying/selling data from/to another IoT device or system, which could produce a transactional system less prone to the problems of resilience.

The blockchain model has several limitations for the use with IoT devices as blockchain processing tasks are computationally intensive and timeconsuming, while IoT devices have limited processing and storage resources to directly participate in a blockchain.

Lower-end IoT edge devices with limited storage space, communications bandwidth and processing power are not especially suitable to support resource-intensive distributed ledgers but can utilize the services of a distributed ledger network (e.g., by using an API). IoT gateway devices, such as an IoT home gateway, could potentially support blockchains (e.g., Raspberry Pi as a "full node"). The requirement for large amounts of disk storage adds cost and complexity, making it more likely that this would be reserved for higher-end gateway products. Lower-end IoT gateways and connections with limited bandwidth are more likely to access the distributed ledger network using an API. High-end IoT edge nodes, such as industrial controllers, smart building controllers and enterprise systems should be able to run capable distributed ledger solutions. Maintaining a local copy of the distributed ledger provides for local high-performance access to the data held on the ledger, as well as continuity in the case that connectivity to the Internet may be disrupted. IoT mobile edge computing nodes (i.e., deployed in the carrier network) can be used to build new distributed ledger solutions typically offered by telecoms operators to enterprise customers as a permissioned distributed ledger [33].

The research challenges for implementing DLTs and blockchains at the edge of networks for IoT applications are as follows:

- Techniques for increased scalability, as DLTs and blockchains do no scale as required by IoT applications for use in a distributed system.
- Solutions for dealing with the required processing power, as IoT devices do not have the processing and storage capabilities required to perform encryption for all the objects involved in a blockchain-based ecosystem. Connecting large numbers of IoT devices requires large volumes and very low cost, while the majority of these IoT devices are not capable of running the required encryption algorithms at the desired speed.
- Techniques to speed up the process of validating the transactions for IoT devices.

- Storage capabilities (e.g., internal flash memory or external NOR or NAND flash) to be used to store transactions and device IDs, as well as the ledger on the nodes as the ledger increases in size as time passes.
- Addressing the complexities of the convergence of DLTs, blockchains and IoT technologies and providing simpler implementations at the system level.
- Interoperability issues when combining data sources from different applications, while considering the lack of data model standards for industrial vertical markets.
- Legal and compliance issues for hybrid transactions management across different industrial sectors.
- Security, privacy and trust of blockchain and decentralized schemes.
- Performance optimization of blockchain and decentralized schemes.
- Lightweight protocols and algorithms based on blockchains.
- Blockchain-based lightweight data structures for IoT data.
- Blockchain-based IoT security solutions.
- Blockchains in 5G.
- Blockchains in edge and cloud computing.

3.4 Emerging IoT Security Technologies

IoT-based businesses, applications and services are scaling up and going through various digital transformations in order to deliver value for money and remain competitive. In this context, they are becoming increasingly vulnerable to disruption from denial-of-service attacks, identity theft, data tampering and other threats.

Emerging distributed end-to-end security technologies enhance the ability of an IoT ecosystem and its devices to exhibit complex behaviour independently or collectively in the presence of threats, in a pursuit to achieving end-to-end security. By using such technologies as blockchain, swarm logic and AI, IoT can offer security by design and end-to-end security solutions never implemented before. Techniques such as simulation and optimisation allow for the integration of security early in the design, where a diversity of security breach scenarios can be tested and guarded before they occur in real life.

Interoperability, scalability and security are three of the most essential attributes of IoT environments and ecosystems, which are absent or not fully addressed in today's architectures. Several technologies have succeeded in offering sound and complete solutions to these matters, although not without challenges still remaining. A new 3D IoT layered architecture capturing the

Figure 3.32 3D IoT Layered Architecture.

IoT systems functions and cross-cutting functions is presented in Figure 3.32. Among them, blockchain technology has been developed for scale and with interoperability in mind; hence, it is often generalized as DLT. Its security mechanism, based on public ledger and consensus, is applied across the stack and the network, whether this is centralized, decentralized or distributed. Nevertheless, in spite of all security advancements, guarding against single scenarios of fraud, hacking and other breaches still remains a challenge.

Different IoT topologies require different security configurations and strategies, and this is especially true at the edges, where devices can be diverse, less traditional, small and possibly out of reach for security updates. The edges are therefore vulnerable, providing entry points for malicious attacks, which are difficult to track and therefore easily propagate throughout the whole IoT ecosystem.

As edge devices are often unsophisticated devices, it may be difficult to build security into the design. However, it is here that swarm technology may come to the rescue. Edge devices may form clusters, where they collaborate and share resources and functions in the presence of perceived danger. Each edge device, now belonging to a cluster, will exhibit collective intelligence and be able to evolve and adapt to new requirements and threat situations. Swarm technology helps to identify the threat and define its landscape.

In the evolving IoT market, security goes beyond securing the information exchanged among the IoT nodes. The entire operation of an IoT ecosystem depends on protection at all levels, from single devices to communications. Moreover, devices must exhibit a high level of resilience against a growing range of attacks, including hardware, software and physical tampering.

Security is therefore critical to IoT technologies and applications, and end-to-end security is essential to enabling the implementation of trustworthy IoT solutions for all stakeholders in IoT ecosystems and IoT value networks to enable the development, deployment and maintenance of systems in IoT applications and provide a common framework to enable the growth of IoT value network solutions.

The standard security services that are valid for the Internet framework and technology, such as authentication, confidentiality, integrity, non- repudiation, access control and availability, should be extended to also apply to IoT technologies but adapted with their particularities and constraints in mind.

Identification: is the act of allowing a device or service to be specifically and uniquely identified without ambiguity. This may take the form of RFID tag identifiers, IP addresses, global unique identifiers, functional or capability identifiers, or data source identifiers.

Authentication: is the act of confirming the truth of an attribute of an entity or a single piece of data by using passwords, PINs, smart cards, digital certificates, or biometrics to sign in. In contrast with Identification, Authentication is the process of actually confirming the Identity of a device or confirming that data arriving or leaving are genuine and have not been tampered with or forged.

Authorization: is the function of specifying access rights to resources and ensuring that any request for data or control of a system is managed within these policies. Authorization mechanisms tend to be centralized, which may be a challenge in IoT systems that tend to be increasingly decentralized, without an authority involved. Whatever the degree of democratized authorization, where more entities can grant permissions, the authorization system must be consistent, persistent and attack resistant.

Availability: has two definitions within the IoT domain. Firstly, as with mainstream Information Assurance, the system must provide data and resources in a timely manner for a set percentage of the time (e.g. 99.99% uptime availability). Secondly, in the IoT it is critical that many devices are available or retain their critical functionality, even if the system has undergone an attack.

Confidentiality: is a set functionality that limits access or places restrictions on certain types of information, with the goal of preventing unauthorized access. Confidentiality is usually achieved through encryption and cryptographic mechanisms and is essential within an IoT ecosystem where a large amount of information is exchanged among the nodes.

Integrity: is a critical measure in information assurance and is defined as providing consistency or a lack of corruption within the IoT system. It requires the final information received to correspond with the original information sent and that data cannot be modified without detection. Malicious modification of the information exchanged may disrupt the correct functioning of an entire IoT ecosystem.

Non-repudiation: is an aspect of authentication that enables systems to have a high level of mathematical confidence that data, including identifiers, are genuine. This ensures that either a transmitting or receiving party cannot later deny that the request occurred (cannot later "repudiate") and provides data integrity around the system. This is of particular importance in terms of tracking illegal activities within an IoT system, as it allows for accountability to be enforced. Whether Non-repudiation needs to be enforced under certain circumstances will depend on the particular applications.

A Root of Trust: is an immutable boot process within an IoT system based on unique identifiers, cryptographic keys and on-chip memory, to protect the device from being compromised at the most fundamental level. The Chain of Trust extends the Root of Trust into subsequent applications and use cases. Given that IoT systems rely on a large number of devices that collect and process information, it is paramount to ensure their credibility so that they are honest and leverage correct outputs.

Secure Update: enables IoT systems and devices to install new firmware from authorized sources without the firmware being compromised. Software updates are critical processes and are susceptible to a number of threats and attacks. During an update, the device receives the firmware wirelessly and installs it, removing the previous version. However, to reassure that the process is being done properly and securely, the sender of the firmware should be verified as trusted, the firmware should be validated as not compromised, the initial security keys should be protected, etc. Additionally, depending on the services that the device offers, the downtime during a firmware update may need to be kept at a minimum. If not properly protected, devices may be open to manipulation, typically through the installation of malicious code on a device.

3.5 IoT/IIoT Technology Market Developments

IoT/IIoT components, communication, systems, platforms, solutions applications and services markets are developing steadily, posing new challenges for research and innovation concerning IoT technologies addressing next generation developments.

The IoT chip market is expected to register a CAGR of over 13.68% during the forecast period of 2018–2023. The report profiles end user segments (such as healthcare, building automation and automotive segments) in the IoT chip market in various regions. Chipsets designed for IoT systems have unique factors including the need for optimal energy efficiency. The network effect is clearly evident as the impact of increasingly interconnected IoT systems will cause an acceleration in overall demand for chipsets due to the interdependency of platforms, gateways and devices [26].

The number of connected devices that are in use worldwide now exceeds 17 billion, with the number of IoT devices at seven billion (not including smartphones, tablets, laptops or fixed-line phones). Global connection growth is mainly driven by IoT devices – both on the consumer side (e.g., smart home) and on the enterprise/B2B side (e.g., connected machinery). The number of IoT devices that are active is expected to grow to 10 billion by 2020 and 22 billion by 2025 (see Figure 3.33). The global market for IoT (end user spending on IoT solutions) is expected to grow by 37% from 2017 to $151 billion. Due to the market acceleration regarding the IoT, those estimates have been revised upwards and it is now expected that the total market will reach $1,567 billion by 2025. Software and platforms are expected to continue to drive the market as more data are moved to the cloud, new IoT applications are brought to market, and analytics continue to gain in importance [27].

3.5.1 Digital Business Model Innovation and IoT as a Driver

The growing digitisation of businesses as well as societies has facilitated an increase in the amount of data made available and to be adopted and explored in the development of businesses. Digitisation is creating a second economy that is vast, automatic and invisible – thereby bringing the biggest change since the Industrial Revolution [1]. Data has become massive and has

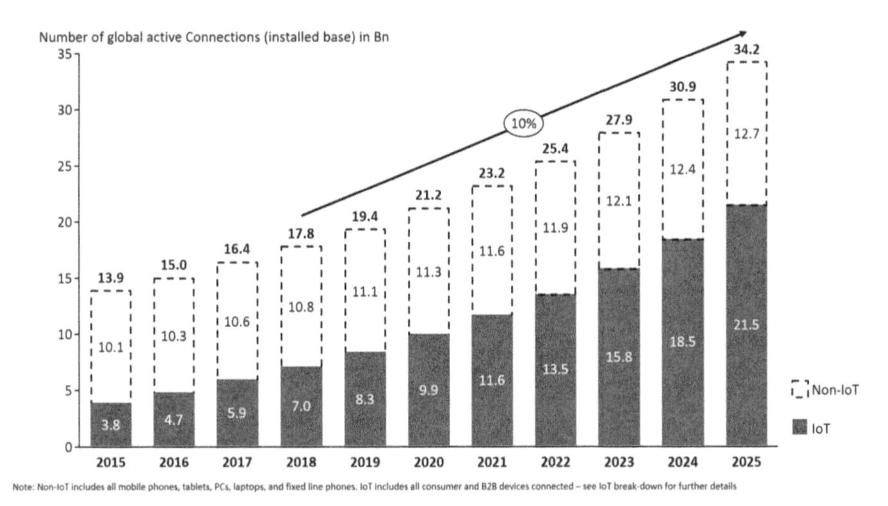

Figure 3.33 Total number of active device connections worldwide [27].

moved from static data to real-time data streams created by the IoT based on a large number of transactions of millions of sensors and devices across the ecosystems of many organizations, even now moving from the central paradigm of the cloud more and more towards the edge in a distributed manner [2]. Some studies estimate an increase in annually created, replicated and consumed data from around 1,200 exabytes in 2010 to 40,000 in 2020 [3], with a growing proportion of data generated and consumed by machines [3]. In businesses, IoT data can be applied, for instance, to target customers more effectively; make better pricing decisions; predict failures; and optimise the use of assets, production or logistics. To fully exploit IoT in business we need to understand how businesses integrate technology [2].

3.5.1.1 Business models and business model innovation

Business models are intended to make sense of how businesses work. Business models are abstracted in different ways in the literature. Business models are discussed in [6] as a narrative that describes the customer, customer value, revenue collection of the model and the delivery of this value. Another level of abstraction is presented in [11]. In this reference the business model is described as an archetype of 55 different business model building blocks that can be combined in various ways to accommodate the business model in which the business operates. The most popular and most adopted breakthrough on another level of abstraction is the graphical framework. The most

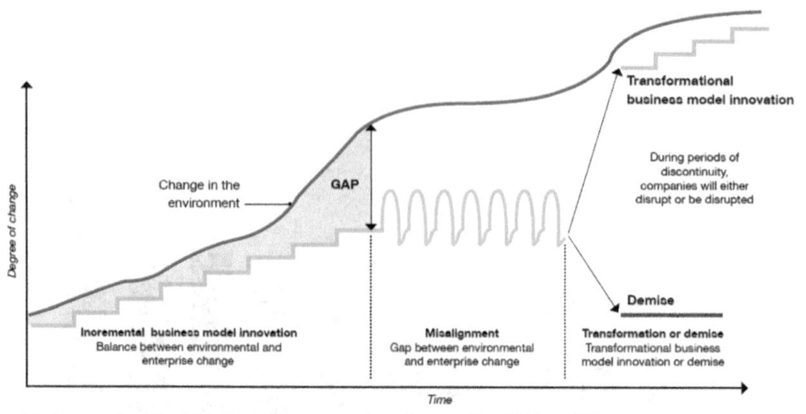

Figure 3.34 The importance of Business Model Innovation with respect to external changes in the environment.

widely adopted graphical framework is the Business Model Canvas presented in [10]. The business model literature points to the fact that the technological development of the Internet and recent developments of ICT has boosted the usage of the business model concept and innovation in general. According to [9] in the context of innovation, the term Business Models is used to either commercialize new technology or ideas and as a source of innovation to the business model itself, that can lead to a competitive advantage.

The use and importance of business model innovation is stated in [12] and illustrated in Figure 3.34:

- Business Model Innovation will continue to become increasingly important for ensuring sustained competitiveness of both large and small businesses.
- Business Model Innovation is expected to serve as a facilitator of new market exploration and an important source of competitive advantage.
- Continuous adaptation of business models is imperative to ensure organizational fit with the environment.

However, there are significant research gaps combining technology trends with business model innovation:

- Empirical evidence remains patchy and often builds on observations of businesses that have successfully implemented new business models without knowing how the innovation has been created in the first place.

- Business Model Innovation frameworks are technology agnostic and often abstract the complexity of ecosystems, technology development/operations and organisations challenges.

3.5.1.2 The use of IoT for digital business development

IoT and digital technologies are central for digital business development and the disruptive business innovation tendencies of this decade and probably also decades to come. Consequently, Nambisan et al. [4] conceptualise digital innovation as "the creation of (and consequent change in) market offerings, business processes, or models that result from the use of digital technologies" and therefore, digital innovation management refers to the "practices, processes, and principles that underlie the effective orchestration of digital innovation". Thus, we need to examine how different company types, industries and sectors apply digital technologies to design digital businesses and digital business models.

3.5.1.3 The design and implementation processes of digital business development

Through digitisation of the business functions through IoT, data can be provided to enhance and develop each of these functions and thus the entire value chain. In practice this is demonstrated in the dramatic change of the marketing functions focus on online, social media and mobile marketing and less of a focus on traditional advertising, thus creating stronger interactions and continuous data collections with customers through social networks. Through the online environment, assortment and pricing decisions is made easier and much more flexible. Logistics and logistics streams are key to competitive delivery and services, and the marketing and logistic functions therefore need to cooperate more effectively in order to deliver superior customer value, and at a lower and more competitive cost [2]. With standards to represent different forms of data (text, numbers, pictures and video) facilitating communication via Bluetooth and the internet has led to the evolution of new products and services, and thus data has become a commodity. Thus, we need to explore the specific processes that go into the adoption and implementation of digital business development viewed from both a business and technology angle.

The effect and business opportunities of digitisation across ecosystems.

Digitisation affects entire ecosystems, their business models and the underlying business functions of a company's value chain. With intelligent devices becoming interconnected in IoT, new developments have created associated infrastructure and an expanding knowledge base, and these innovative combinations are being reflected in enterprises' "digital" business models [5]. Thus, we need to examine which metric and measurement systems are applied and require development to better assess the tangible as well as intangible value creation and capture of digital business development.

How technology, organisation and business interact is still poorly understood. Most studies have focussed on successful businesses and have been conducted within a discipline like business, innovation, technology or organisation. The literature provides some patchy evidence which shows Business Model Innovation (BMI) as a cross-disciplinary activity, connected with technology. It is also clear that IoT is a strong driver for business development and digitisation in industry, with many new applications and services emerging and driving new business models. We call for more concerted efforts to link business and technology at the applied research level and to devise new methods of studying IoT.

Acknowledgments

The IoT European Research Cluster - European Research Cluster on the Internet of Things (IERC) maintains its Strategic Research and Innovation Agenda (SRIA), considering its experiences and the results from the on-going exchange among European and international experts.

The present document builds on the 2010, 2011, 2012, 2013, 2014, 2015, 2016 and 2017 Strategic Research and Innovation Agendas.

The IoT European Research Cluster SRIA is part of a continuous IoT community dialogue supported by the EC DG Connect – Communications Networks, Content and Technology, E4 - Internet of Things Unit for the European and international IoT stakeholders. The result is a lively document that is updated every year with expert feedback from on-going and future projects financed by the EC. Many colleagues have assisted over the last few years with their views on the IoT Strategic Research and Innovation agenda document. Their contributions are gratefully acknowledged.

List of Contributors over the Years of IERC Activities:

Abdur Rahim Biswas, IT, CREATE-NET, WAZIUP
Alessandro Bassi, FR, Bassi Consulting, IoT-A, INTER-IoT
Alexander Gluhak, UK, Digital Catapult, UNIFY-IoT
Amados Daffe, SN/KE/US, Coders4Africa, WAZIUP

Antonio Kung, FR, Trialog, CREATE-IoT
Antonio Skarmeta, ES, University of Murcia, IoT6
Arkady Zaslavsky, AU, CSIRO, bIoTope
Arne Bröring, DE, Siemens, BIG-IoT
Arthur van der Wees, Arthurs Legal, CREATE-IoT
Bruno Almeida, PT, UNPARALLEL Innovation, FIESTA-IoT, ARMOUR, WAZIUP
Carlos E. Palau, ES, Universitat Politècnica de Valencia, INTER-IoT
Charalampos Doukas, IT, CREATE-NET, AGILE
Christoph Grimm, DE, University of Kaiserslautern, VICINITY
Claudio Pastrone, IT, ISMB, ebbits, ALMANAC
Congduc Pham, FR, Université de Pau et des Pays de l'Adour, WAZIUP
Dimitra Stefanatou, Arthurs Legal, CREATE-IoT
Elias Tragos, IE, Insight Centre for Data Analytics, UCD and FORTH-ICS, RERUM, FIESTA-IoT
Emmanuel C. Darmois, FR, COMMLEDGE, CREATE-IoT
Eneko Olivares, ES, Universitat Politècnica de Valencia, INTER-IoT
Fabrice Clari, FR, inno TSD, UNIFY-IoT
Franck Le Gall, FR, Easy Global Market, WISE IoT, FIESTA-IoT, FESTIVAL
Frank Boesenberg, DE, Silicon Saxony Management, UNIFY-IoT
François Carrez, UK, University of Surrey, FIESTA-IoT
Friedbert Berens, LU, FB Consulting S.à r.l, BUTLER
Gabriel Marão, BR, Perception, Brazilian IoT Forum
Gert Guri, IT, HIT, UNIFY-IoT
Gianmarco Baldini, IT, EC, JRC
Giorgio Micheletti, IT, IDC, CREATE-IoT
Giovanni Di Orio, PT, UNINOVA, ProaSense, MANTIS
Harald Sundmaeker, DE, ATB GmbH, SmartAgriFood, CuteLoop
Henri Barthel, BE, GS1 Global
Ivana Podnar, HR, University of Zagreb, symbIoTe
JaeSeung Song, KR, Sejong University, WISE IoT
Jan Höller, SE, EAB
Jelena Mitic DE, Siemens, BIG-IoT
Jens-Matthias Bohli, DE, NEC
John Soldatos, GR, Athens Information Technology, FIESTA-IoT
José Amazonas, BR, Universidade de São Paulo, Brazilian IoT Forum
Jose-Antonio, Jimenez Holgado, ES, TID

Jun Li, CN, China Academy of Information and Communications Technology, EU-China Expert Group
Kary Främling, FI, Aalto University, bIoTope
Klaus Moessner, UK, UNIS, IoT.est, iKaaS
Kostas Kalaboukas, GR, SingularLogic, EURIDICE
Latif Ladid, LU, UL, IPv6 Forum
Levent Gürgen, FR, CEA-Leti, FESTIVAL, ClouT
Luis Muñoz, ES, Universidad De Cantabria
Manfred Hauswirth, IE, DERI, OpenIoT, VITAL
Marco Carugi, IT, ITU-T, ZTE
Marilyn Arndt, FR, Orange
Markus Eisenhauer, DE, Fraunhofer-FIT, HYDRA, ebbits
Martin Bauer, DE, NEC, IoT-A
Martin Serrano, IE, DERI, OpenIoT, VITAL, FIESTA-IoT
Martino Maggio, IT, Engineering - Ingegneria Informatica Spa, FESTIVAL, ClouT
Maurizio Spirito, IT, Istituto Superiore Mario Boella, ebbits, ALMANAC, UNIFY-IoT
Maarten Botterman, NL, GNKS, SMART-ACTION
Ousmane Thiare, SN, Université Gaston Berger, WAZIUP
Pasquale Annicchino, CH, Archimedes Solutions, CREATE-IoT
Payam Barnaghi, UK, UNIS, IoT.est
Philippe Cousin, FR, FR, Easy Global Market, WISE IoT, FIESTA-IoT, EU-China Expert Group
Philippe Moretto, FR, ENCADRE, UNIFY-IoT, ESPRESSO, Sat4m2m
Raffaele Giaffreda, IT, CNET, iCore
Ross Little, ES, Atos, CREATE-IoT
Roy Bahr, NO, SINTEF, UNIFY-IoT, CREATE-IoT
Sébastien Ziegler, CH, Mandat International, IoT6
Sergio Gusmeroli, IT, Engineering, POLIMI, OSMOSE, BeInCPPS
Sergio Kofuji, BR, Universidade de São Paulo, Brazilian IoT Forum
Sergios Soursos, GR, Intracom SA Telecom Solutions, symbIoTe
Sonia Compans, FR, ETSI, CREATE-IoT
Sophie Vallet Chevillard, FR, inno TSD, UNIFY-IoT
Srdjan Krco, RS, DunavNET, IoT-I, SOCIOTAL, TagItSmart
Steffen Lohmann, DE, Fraunhofer IAIS, Be-IoT
Sylvain Kubler, LU, University of Luxembourg, bIoTope
Takuro Yonezawa, JP, Keio University, ClouT
Toyokazu Akiyama, JP, Kyoto Sangyo University, FESTIVAL

Veronica Barchetti, IT, HIT, UNIFY-IoT
Veronica Gutierrez Polidura, ES, Universidad De Cantabria
Xiaohui Yu, CN, China Academy of Information and Communications Technology, EU-China Expert Group

Contributing Projects and Initiatives

SmartAgriFood, EAR-IT, ALMANAC, CITYPULSE, COSMOS, CLOUT, RERUM, SMARTIE, SMART-ACTION, SOCIOTAL, VITAL, BIG IoT, VICINITY, INTER-IoT, symbIoTe, TAGITSMART, bIoTope, AGILE, Be-IoT, UNIFY-IoT, ARMOUR, FIESTA, ACTIVAGE, AUTOPILOT, CREATE-IoT, IoF2020, MONICA, SYNCHRONICITY, U4IoT, BRAIN-IoT, ENACT, IoTCrawler, SecureIoT, SOFIE, CHARIOT, SEMIoTICS, SerIoT.

References

[1] Arthur, W. B. (2011). "The second economy". McKinsey Quarterly, Vol. 4, No. 1, pp. 90–99.
[2] Aagaard, A. (2018). "Digital business models – driving transformation and innovation". Palgrave MacMillan.
[3] Gantz, J., and Reinsel, D. (2012). "The digital universe in 2020: big data, bigger digital shadows, and biggest growth in the Far East", International Data Corporation, Framingham.
[4] Nambisan, S., Lyytinen, K., Majchrzak, A., and Song, M. (2017). "Digital innovation management: reinventing innovation management research in a digital world". MIS Quarterly, Vol. 41, No. 1, pp. 223–238.
[5] Kiel, D., Arnold, C., Collisi, M., and Voigt, K. I. (2016). "The impact of the industrial Internet of Things on established business models". In Proceedings of the 25th International Association for Management of Technology (IAMOT) Conference, Orlando, Florida, USA, May 15–19.
[6] Magretta, J. (2002). "Why business models matter". *Harvard Business Review*, Vol. 80, No. 5, pp. 86–92.
[7] Amit, R., and Zott, C. (2012). Creating value through business model innovation. *MIT Sloan Management Review*, Vol. 53, No. 3, p. 41.
[8] Afuah, A., and Tucci, C. L. (2001). *Internet business models and strategies*. New York: McGraw-Hill, p. 358.

[9] Dodgson, M., Gann, D. M., and Phillips, N. (Eds.). (2013). *The Oxford handbook of innovation management.* OUP Oxford.

[10] Osterwalder, A., and Pigneur, Y. (2010). *Business model generation: a handbook for visionaries, game changers, and challengers*. Wiley.

[11] Gassmann, H., Frankenberger, K., and Csik, M. (2014). "The St. Gallen business model navigator". Working paper: University of St. Gallen: ITEM-HSG.

[12] Knab, S., and Rohrbeck, R. (2014). "Why intended business model innovation fails to deliver: insights from a longitudinal study in the German smart energy market." Proceedings of the R&D Management Conference, Stuttgart, Germany, June 3–6, 2014.

[13] Vermesan, O., and Friess, P. (Eds.). (2016). Digitising the Industry Internet of Things Connecting the Physical, Digital and Virtual Worlds, ISBN: 978-87-93379-81-7, River Publishers, Gistrup.

[14] Vermesan, O., and Friess, P. (Eds.). (2015). Building the Hyperconnected Society – IoT Research and Innovation Value Chains, Ecosystems and Markets, ISBN: 978-87-93237-99-5, River Publishers, Gistrup.

[15] Outlier Ventures Research, Blockchain-Enabled Convergence - Understanding The Web 3.0 Economy, Online: https://gallery.mailchimp.com/65ae955d98e06dbd6fc737bf7/files/Blockchain_Enabled_Convergence.01.pdf

[16] What is a blockchain? https://www2.deloitte.com/content/dam/Deloitte/ch/Documents/innovation/ch-en-innovation-deloitte-what-is-blockchain-2016.pdf

[17] Lin, S., Cheng, H. F., Li, W., Huang, Z., Hui, P., and Peylo, C. (2017). Ubii: physical world interaction through augmented reality, IEEE Trans. Mob. Comput. Vol. 16. pp. 872–885.

[18] Sun, X., and Ansari, N. (2016). EdgeIoT: mobile edge computing for the Internet of Things, IEEE Commun. Vol. 54, pp. 22–29.

[19] Edge Computing Market, MarketsandMarkets Report, 2017, Online: https://www.marketsandmarkets.com/Market-Reports/edge-computing-market-133384090.html

[20] Ericsson Mobility Report, June 2018, Online: https://www.ericsson.com/assets/local/mobility-report/documents/2018/ericsson-mobility-report-june-2018.pdf

[21] Cloud IoT Edge, Online: https://cloud.google.com/iot-edge/

[22] Hyperledger, Online: https://www.hyperledger.org/

[23] Enterprise Ethereum Alliance (EEA), Online: https://entethalliance.org/

[24] Hyperledger Fabric, Online: https://www.ibm.com/blockchain/hyperledger/fabric-support
[25] Herr, G., Lyon, J., and Gillen, S. (2016). "Industrial intelligence: cognitive analytics in action," presentation at EMEA Users Conference, Berlin.
[26] Research and Markets, Online: https://www.researchandmarkets.com/
[27] State of the IoT 2018: number of IoT devices now at 7B – Market accelerating, IoT analytics, Online: https://iot-analytics.com/state-of-the-iot-update-q1-q2-2018-number-of-iot-devices-now-7b/
[28] IDTechEx, Comparison of Low Power Wide Area Networks (LPWAN) for IoT 2018–2019, Online: https://www.idtechex.com/research/articles/comparison-of-low-power-wide-area-networks-lpwan-for-iot-2018-2019-00014777.asp
[29] SEMTECH, Online: https://www.semtech.com/
[30] Ryberg, T., BERG INSIGHT, NB-IoT networks are here, now it's time to make business, Online: https://www.iot-now.com/2018/07/04/85156-nb-iot-networks-now-time-make-business/
[31] NB-IoT, CAT-M, SIGFOX and LoRa Battle for Dominance Drives Global LPWA Network Connections to Pass 1 Billion By 2023, ABIresearch, 2018, Online: https://www.abiresearch.com/press/nb-iot-cat-m-sigfox-and-lora-battle-dominance-drives-global-lpwa-network-connections-pass-1-billion-2023/
[32] Wi-SUN FAN Overview, Online: https://tools.ietf.org/id/draft-heile-lpwan-wisun-overview-00.html
[33] Opportunities and Use Cases for Distributed Ledgers in IoT, GSMA 2018, Online: https://www.gsma.com/iot/wp-content/uploads/2018/09/Opportunities-and-Use-Cases-for-Distributed-Ledgers-in-IoT-f.pdf
[34] Nakamoto, S., Bitcoin: A Peer-to-Peer Electronic Cash System, Online: https://bitcoin.org/bitcoin.pdf
[35] Popov, S., (2018). The Tangle, Online: https://assets.ctfassets.net/r1dr6vzfxhev/2t4uxvsIqk0EUau6g2sw0g/45eae33637ca92f85dd9f4a3a218e1ec/iota1_4_3.pdf
[36] IOTA, Online: https://iota.org
[37] Ripple, Online: https://ripple.com
[38] Sovrin, Online: https://sovrin.org
[39] BigchainDB, Online: https://www.bigchaindb.com
[40] Gutierrez, C. (2017). Boeing Improves Operations with Blockchain and the Internet of Things, Online: https://www.altoros.com/blog/boeing-improves-operations-with-blockchain-and-the-internet-of-things/

[41] Gutierrez, C., and Khizhniak, A. (2017). Improving Supply Chain and Manufacturing with IoT-Driven Blockchains, Online: https://www.altoros.com/blog/ibm-aims-to-improve-manufacturing-and-supply-chain-by-coupling-iot-and-blockchain/

[42] Panetta, K. (2018). 5 Trends Emerge in the Gartner Hype Cycle for Emerging Technologies, Online: https://www.gartner.com/smarterwithgartner/5-trends-emerge-in-gartner-hype-cycle-for-emerging-technologies-2018/

[43] Nilsson, N. J. (2010). The Quest for Artificial Intelligence: A History of Ideas and Achievements, Cambridge, UK Cambridge University Press.

[44] IEEE, Tactile Internet Emerging Technologies Subcommittee, Online: http://ti.committees.comsoc.org/

[45] The Tactile Internet ITU-T Technology Watch Report ITU-T, 2014, Online: https://www.itu.int/dms_pub/itu-t/oth/23/01/T23010000230001PDFE.pdf

[46] Batra, G., Queirolo, A., and Santhanam, N. (2018). McKinsey & Company, Artificial intelligence: The time to act is now, Online: https://www.mckinsey.com/industries/advanced-electronics/our-insights/artificial-intelligence-the-time-to-act-is-now

[47] 5G Network Architecture A High-Level Perspective, (2016). White Paper, HUAWEI Technologies Co., Ltd., Online: https://www-file.huawei.com/-/media/CORPORATE/PDF/mbb/5g_nework_architecture_whitepaper_en.pdf?la=en&source=corp_comm

[48] 5G Security Architecture White Paper, (2017). HUAWEI Technologies Co., Ltd., 2017, Online: https://www-file.huawei.com/-/media/CORPORATE/PDF/white%20paper/5g_security_architecture_white_paper_en-v2.pdf?la=en&source=corp_comm

[49] Network Slicing Use Case Requirements, GSMA, April 2018, Online: https://www.gsma.com/futurenetworks/wp-content/uploads/2018/07/Network-Slicing-Use-Case-Requirements-fixed.pdf

[50] Szymanski, T. H. (2016), Securing the Industrial-Tactile Internet of Things with Deterministic Silicon Photonics Switches, IEEE Access, Vol. 4, pp. 8236–8249.

[51] Maier, M., Chowdhury, M., Prasad Rimal, B., and Pham Van, D. (2016). The Tactile Internet: Vision, Recent Progress, and Open Challenges, IEEE Communications Magazine, Vol. 54, No. 5, pp. 138–145.

[52] Maier, M. (2014). "FiWi Access Networks: Future Research Challenges and Moonshot Perspectives," Proc. IEEE Int'l. Conf. Commun.

(ICC), Workshop on Fiber-Wireless Integrated Technologies, Systems and Networks, Sydney, Australia, pp. 371–375.

[53] Chowdhury, M., and Maier, M. (2017). Collaborative Computing for Advanced Tactile Internet Human-to-Robot (H2R) Communications in Integrated FiWi Multirobot Infrastructures, IEEE Internet of Things Journal, Vol. 4, No. 6, pp. 2142–2158.

[54] Spectrum of Seven Outcomes for AI, Online: https://www.constellationr.com/

[55] Wang, R. (2016). Monday's Musings: Understand The Spectrum Of Seven Artificial Intelligence Outcomes, Online: http://blog.softwareinsider.org/2016/09/18/mondays-musings-understand-spectrum-seven-artificial-intelligence-outcomes/

[56] Chan, R., Consensus Mechanisms used in Blockchain. https://www.linkedin.com/pulse/consensus-mechanisms-used-blockchain-ronald-chan

[57] Ekblaw, A. et al. (2016). A Case Study for Blockchain in Healthcare: "MedRec" prototype for electronic health records and medical research data, https://www.healthit.gov/sites/default/files/5-56-onc_blockchainchallenge_mitwhitepaper.pdf

[58] Crosby, M. et al. (2015). BlockChain Technology. Beyond Bitcoin. Berkeley, University of California, http://scet.berkeley.edu/wp-content/uploads/BlockchainPaper.pdf

[59] Samman, G. S. (2016). How Transactions Are Validated On A Distributed Ledger, https://www.linkedin.com/pulse/how-transactions-validated-distributed-ledger-george-samuel-samman

[60] ITU-T, Internet of Things Global Standards Initiative, http://www.itu.int/en/ITU-T/gsi/iot/Pages/default.aspx

[61] International Telecommunication Union – ITU-T Y.2060 – (06/2012) – Next Generation Networks – Frameworks and functional architecture models – Overview of the Internet of things

[62] Vermesan, O., Friess, P., Guillemin, P., Sundmaeker, H. et al., (2013). "Internet of Things Strategic Research and Innovation Agenda", Chapter 2 in Internet of Things – Converging Technologies for Smart Environments and Integrated Ecosystems, River Publishers, ISBN: 978- 87-92982-73-5

[63] Yole Développement, Technologies & Sensors for the Internet of Things, Businesses & Market Trends 2014–2024, 2014, Online: http://www.yole.fr/iso_upload/Samples/Yole_IoT_June_2014_Sample.pdf

[64] Parks Associates, Monthly Wi-Fi usage increased by 40% in U.S. smartphone households, Online: https://www.parksassociates.com/blog/article/pr-06192017

[65] Gluhak, A., Vermesan, O., Bahr, R., Clari, F., Macchia, T., Delgado, M. T., Hoeer, A. Boesenberg, F., Senigalliesi, M., and Barchetti, V. (2016). "Report on IoT platform activities", 2016, Online: http://www.internet-of-things-research.eu/pdf/D03_01_WP03_H2020_UNIFY-IoT_Final.pdf

[66] McKinsey & Company, Automotive revolution – perspective towards 2030. How the convergence of disruptive technology-driven trends could transform the auto industry, 2016.

[67] IoT Platforms Initiative, Online: https:// www.iot-epi.eu/

[68] IoT European Large-Scale Pilots Programme, Online: https://european-iot-pilots.eu/

[69] Où porterons-nous les objets connectés demain?, Online: http://lamontreconnectee.net/les-montres-connectees/porterons-objets-connectes-demain/

[70] Moore, S. (2016). Gartner survey shows wearable devices need to be more useful, Online: http://www.gartner.com/newsroom/id/3537117

[71] Digital Economy Collaboration Group (ODEC), Online: http://archive.oii.ox.ac.uk/odec/

[72] Maidment, D. (2014). Advanced Architectures and Technologies for the Development of Wearable Devices, White paper, Online: https://www.arm.com/files/pdf/Advanced-Architectures-and-Technologies-for-the-Development-of-Wearable.pdf Accenture. Are you ready to be an Insurer of Things?, Online: https://www.accenture.com/_acnmedia/Accenture/Conversion-Assets/DotCom/ Documents/Global/PDF/Strategy_7/Accenture-Strategy-Connected-Insurer-of-Things.pdf#zoom=50

[73] Connect building systems to the IoT, Online: http://www.electronics-know-how.com/article/1985/connect-building-systems-to-the-iot

[74] Kejriwal, S., and Mahajan, S. (2016). Smart buildings: How IoT technology aims to add value for real estate companies The Internet of Things in the CRE industry, Deloitte University Press, Online: https://www2.deloitte.com/content/dam/Deloitte/nl/Documents/real-estate/deloitte-nl-fsi-real-estate-smart-buildings-how-iot-technology-aims- to-add-value-for-real-estate-companies.pdf

[75] ORGALIME Position Paper, 2016, Online: http://www.orgalime.org/sites/default/files/position-papers/Orgalime%20Comments_EED_EPBD_Review%20Policy%20Options_4%20May%202016.pdf

[76] Hagerman, J. (2014). U.S. Department of Energy, Buildings-to-grid technical opportunities, https://energy.gov/sites/prod/files/2014/03/f14/B2G_Tech_Opps–Intro_and_Vision.pdf

[77] Ravens, S., and Lawrence, M. (2017). Defining the Digital Future of Utilities – Grid Intelligence for the Energy Cloud in 2030, Navigant Research White Paper, Online: https://www.navigantresearch.com/research/defining-the-digital-future-of-utilities
[78] Roland Berger Strategy Consultants, Autonomous Driving, (2014). Online: https://www.rolandberger.com/publications/publication_pdf/roland_berger_tab_autonomous_driving.pdf
[79] Roland Berger Strategy Consultants, (2017). Automotive Disruption Radar – Tracking disruption signals in the automotive industry, Online: https://www.rolandberger.com/publications/publication_pdf/roland_berger_disruption_radar.pdf
[80] The EU General Data Protection Regulation (GDPR) (Regulation (EU) 2016/679).
[81] RERUM, EU FP7 project, www.ict-rerum.eu
[82] Gartner Identifies the Top 10 Internet of Things Technologies for 2017 and 2018, Online: http://www.gartner.com/newsroom/id/3221818
[83] A Look at Smart Clothing for 2015, Online: http://www.wearable-technologies.com/2015/03/a-look-at-smartclothing-for-2015/
[84] Best Smart Clothing – A Look at Smart Fabrics, (2016), Online: http://www.appcessories.co.uk/best-smart-clothing-a-look-at-smart-fabrics/
[85] Brunkhorst, C. (2015). "Connected cars, autonomous driving, next generation manufacturing – Challenges for Trade Unions", Presentation at IndustriAll auto meeting Toronto 14th Oct. 2015, Online: http://www.industriall-union.org/worlds-auto-unions-meet-in-toronto
[86] Market research group Canalys, Online: http://www.canalys.com/
[87] Digital Agenda for Europe, European Commission, Digital Agenda 2010–2020 for Europe, Online: http://ec.europa.eu/information_society/digital-agenda/index_en.htm
[88] Vermesan, O., Friess, P., Woysch, G., Guillemin, P., Gusmeroli, S. et al., "Europe's IoT Stategic Research Agenda 2012", Chapter 2 in The Internet of Things 2012 New Horizons, Halifax, UK, 2012, ISBN: 978-0-9553707-9-3
[89] Vermesan, O. et al., (2011). "Internet of Energy – Connecting Energy Anywhere Anytime" in Advanced Microsystems for Automotive Applications 2011: Smart Systems for Electric, Safe and Networked Mobility, Springer, Berlin, ISBN: 978-36-42213-80-9
[90] Yuriyama, M., and Kushida, T., "Sensor-Cloud Infrastructure – Physical Sensor Management with Virtualized Sensors on Cloud Computing", NBiS 2010: 1–8.

[91] Mobile Edge Computing Will Be Critical For Internet-of-Things and Distributed Computing, Online: http://blogs.forrester.com/dan_bieler/16-06-07-mobile_edge_computing_will_be_critical_for_internet_of_things_and_distributed_computing

[92] Stamatakis, G., Elias, T., and Apostolos, T. (2018). "Energy Efficient Policies for Data Transmission in Disruption Tolerant Heterogeneous IoT Networks." *Sensors* 18.9: 2891.

[93] Stamatakis, G., Elias, Z. T., and Apostolos, T. (2015). "Periodic collection of spectrum occupancy data by energy constrained cognitive IoT devices." *Wireless Communications and Mobile Computing Conference (IWCMC), 2015 International*. IEEE.

[94] Stamatakis, G., Elias, Z. T., and Apostolos, T. (2015). "A Two-Stage Spectrum Assignment Scheme for Power and QoS Constrained Cognitive CSMA/CA Networks." *Globecom Workshops (GC Wkshps), 2015 IEEE*. IEEE.

[95] Test Considerations for 5G New Radio, White Paper, Keysight Technologies, April 2018, online at: http://literature.cdn.keysight.com/litweb/pdf/5992-2921EN.pdf

[96] ETSI ISG Multi-access Edge Computing, online at: https://portal.etsi.org/MEC

[97] Serrano, M., and Soldatos, J. (2015). "IoT is More Than Just Connecting Devices: The OpenIoT Stack Explained" IEEE Internet of Things Newsletter, September, 8th 2015.

[98] Nagavalli, Y. (2018). Huawei Software, September 2018, Prepare Now for the 5G Monetization Opportunity, online at: http://telecoms.com/intelligence/prepare-now-for-the-5g-monetization-opportunity/

[99] Li, R. (2018). "Towards a New Internet for the Year 2030 and Beyond," Third Annual ITU IMT-2020/5G Workshop and Demo Day, Geneva, Switzerland, July 18, 2018, online at: https://www.itu.int/en/ITU-T/Workshops-and-Seminars/201807/Documents/3_Richard%20Li.pdf

[100] 5G Applications Market Potential & Readiness Matrix, Huawei Wireless X Labs and ABI Research, 2018, online at: https://www-file.huawei.com/-/media/CORPORATE/PDF/x-lab/5G-Applications-Market-Potential_Readiness-Matrix.pdf?la=en&source=corp_comm

4

End-to-end Security and Privacy by Design for AHA-IoT Applications and Services

Mario Diaz Nava[1], Armand Castillejo[1], Sylvie Wuidart[1], Mathieu Gallissot[2], Nikolaos Kaklanis[3], Konstantinos Votis[3], Dimitrios Tzovaras[3], Anastasia Theodouli[3], Konstantinos Moschou[3], Aqeel Kazmi[5], Philippe Dallemagne[4], Corinne Kassapoglou-Faist[4], Sergio Guillen[7], Giuseppe Fico[8], Yorick Brunet[4], Thomas Loubier[2], Stephane Bergeon[2], Martin Serrano[5], Felipe Roca[6], Alejandro Medrano[8] and Byron Ortiz Sanchez[9]

[1]STMicroelectronics, France
[2]Univ. Grenoble Alpes, CEA-LETI Minatec Campus 38000 Grenoble, France
[3]Information Technologies Institute, Centre for Research and Technology Hellas, Greece
[4]CSEM Centre Suisse d'Electronique et de Microtechnique SA, Switzerland
[5]Insight Centre for Data Analytics, NUI Galway, Ireland
[6]HOP UBIQUITOUS SL, Spain
[7]MYSPHERA, Spain
[8]Life Supporting Technologies-Universidad Politécnica de Madrid, Spain
[9]Televes SA, Spain

Abstract

The chapter aims at describing the cybersecurity and privacy methodologies and solutions that the architecture defined in the ACTIVAGE Large-Scale Pilot, and the corresponding implementation in nine *Deployment sites* should follow to secure the IoT system and protect the personal data from potential malicious cyber-attacks and threats. It further presents common definitions, methods and repeatable processes to analyse and address all potential threats in terms of cybersecurity and privacy that might occur during the exploitation phase of the project.

4.1 Introduction

The Internet and mobile revolution have transformed our world. The Internet of Things (IoT) has significantly emerged over the last few years, aiming to change our lives by forming a massive ecosystem where interconnected devices and services collect, exchange and process data in order to adapt dynamically to a context to offer a variety of services. By 2020, market analysts expect between 20 and 50 billion connected devices in the world. With all the benefits originating from the use of IoT technology, also come a range of ever-increasing challenges and security threats including data manipulation, data theft, and cyber-attacks. For instance, the ransomware landscape has dramatically shifted in 2017 and organizations bore the brunt of the damage caused by new, self-propagating threats such as WannaCry and Petya. Most recently, a report from Symantec ISTR [1] revealed that there were 470 thousand ransomware infections in 2016 and 319 thousand in the first-half of 2017.

The threats and risks related to the Internet of Things devices, systems and services are of manifold and they evolve rapidly. With a great impact on citizens' safety, security and privacy, the threat landscape concerning the Internet of Things is extremely wide and evolves rapidly. Hence, it is important to understand what needs to be secured to develop sophisticated security measures to protect the IoT infrastructure. Information (or data) lies at the heart of an IoT system, feeding into a continuous cycle of sensing, decision-making, and actions. The billions of "things" can be the target of intrusions and interferences that might dramatically jeopardize personal privacy. Since IoT is seen as a key enabler for creating new services and improving overall quality of life, consumers need to have trust and confidence about their data being secured and protected, therefore, making the cybersecurity of IoT systems an essential part.

Currently, there are no official guidelines available for trust of IoT devices, in addition, there is no regulatory compliance defined for minimum-security requirements. Despite the existence of many security guidelines in general, the literature lacks primary guidelines to help adopt security measures and standards for the IoT systems.

The European Union (EU) is working on several fronts to promote cyber resilience across the EU. It published several proposals in a 'cybersecurity package' in September 2017 [2]. Furthermore, the EU set up the Large-Scale Pilots to deploy IoT systems in five main areas [3]. The main goals of these LSPs is to solve key practical issues such as interoperability, security and privacy, business models, validation of IoT powered applications and services at large-scale, etc. In this context, this chapter reports the initial outcomes obtained from security and privacy performed in the ACTIVAGE project[1] [4]. These activities contribute into mainly two areas:

- Technological – a secure large-scale deployment of connected objects.
- Societal – related to the project context, which is to create a smart environment for the ageing well of elderly people allowing the collection of sensitive personal data.

As in ACTIVAGE, the experimentations will involve around 7,000 users across 9 *Deployment Sites* (DSs)[2], the consortium has a great concern when it comes to the security and privacy related challenges and an opportunity to resolve these issues with the help of large-scale validation and testing. Platforms using public communication infrastructure will interconnect many IoT devices, which are inherently weakly secured. Several services will process confidential data by requiring control over the propagation of access control in the spirit of the General Data Protection Regulation (GDPR) [5]. GDPR is a primary law regulating how companies/organizations protect EU citizens' personal data.

This chapter gives an overview of the end-to-end security and privacy impact analysis performed in order to provide actionable recommendations. The outcomes are in the shape of guidelines and framework related to the cybersecurity and privacy aspects. The security risk analysis is

[1]ACTIVAGE project is a key factor in the IoT for the "Active and Healthy Ageing" (AHA) domain producing evidence of the IoT value on fostering the deployment of AHA solutions in Europe, through the integration of advanced IoT technologies across the value chain, demonstrating multiple AHA-IoT applications at large-scale in a usage context, in real operational conditions. IoT for the AHA domain is a strategic element for the creation of dynamic ecosystems to answer and prevent the challenges faced by health and social care systems. Differently from other sectors, "AHA-IoT" services are provided to persons taken individually and it takes place across all domains, as persons live in houses, neighbourhoods, cities, rural areas, mountains and valleys, access to transport systems, drive cars, go to shopping centres, airports, theatres, etc. Persons are the most extraordinary producers of individual's data: production and consumption of personal data across domains has become the front-line of concern, data privacy, security, authentication, access consent, ownership, storage management. In summary, ACTIVAGE is an LSP that brings together the IoT and AHA communities to demonstrate the value of the first with respect to successful implementation of AHA solutions in terms of QoL for Citizens, Sustainability of Health and Social Care Systems and Economical and industrial Growth in Europe.

[2]A Deployment site is a city or a region in the European Union in where a full large-scale pilot is set.

conducted at each layer of an IoT system and its deployment procedure. The objective is twofold: to bring an awareness of the security risks to the stakeholders involved in each deployment site and the provision of solutions/recommendations – concerning the technologies and services to be deployed for security and privacy of the IoT infrastructure.

The chapter aims at describing the cybersecurity and privacy methodologies and solutions that the ACTIVAGE architecture and the corresponding deployment sites should follow in order to secure the IoT system and data from potential malicious cyber-attacks and threats. It further presents common definitions, methods and repeatable processes to analyse and address all potential threats in terms of cybersecurity and privacy that might occur during the exploitation phase of the project. The whole process takes into account:

- Typical cybersecurity and privacy risks due to the IoT context.
- DSs particularities in terms of cybersecurity needs (e.g. data relevance).
- Relevance and effectiveness of cybersecurity and privacy mechanisms already foreseen by the DSs security managers.

In this work, an IoT system is divided into four layers (domains): device, gateway, cloud and application. The security and privacy analysis is performed throughout the entire system starting from the device domain to the application domain. It also considers the overall system life cycle, i.e. the analysis process is applied not only for the operation phase but also at configuration, installation, maintenance and removal phases.

The rest of the chapter is organized as follows. Section 4.2 presents the global objectives and requirements for cybersecurity and privacy in the context of AHA-IoT ecosystem. Section 4.3 presents the main recommendations on Cybersecurity and Privacy in IoT. Sections 4.4 and 4.5 present the methodologies undertaken for security and privacy and the recommendations in this context. Section 4.6 illustrates, through example use cases, some security and privacy solutions harnessed from the top-down approaches and their associated recommendations. Finally, Section 4.7 concludes this chapter.

4.2 Global Objectives and Requirements

4.2.1 Security

In an information system, the key objectives and requirements are defined to prevent unauthorized access, use, disclosure, modification, or removal of important data or information. CIA (Confidentiality, Integrity and

Availability) triad is a common and globally accepted model that is used to secure important information. The main cybersecurity objectives [6, 7] are:

- *Confidentiality*: no improper disclosure of information.
- *Integrity*: no improper modification of information (alteration, deletion or creation).
- *Availability*: no improper impairment of functionality.

In order to reach above objectives, the typical cybersecurity properties or requirements are listed as follows:

- *Authorization*: the rules on who is allowed to read, modify or delete which information.
- *User and entity authenticity*: the assurance that the other party is the intended communication peer, no "man-in-the-middle" scenario.
- *Integrity (data and service authenticity)*: the data is not altered during transmission (accidentally or intentionally).
- *Confidentiality*: the exchanged data cannot be overheard or made available to a third party.
- *Timeliness and validity of the data*: for example, protection against message replay.
- *Non-repudiation of the transaction*: the assurance that a transaction is auditable.

In addition, system integrity requirements include a system protection against physical and logical attacks, a secure software update mechanism and the monitoring and reaction capability to system malfunction. The mechanisms to achieve these requirements are the following:

- *Access Control*: selective restriction of access to data or services.
- *Entity authentication*: for example, a cryptography-based "handshake" scheme.
- *Message cryptographic protection*: encryption and data authentication.
- *Temporization of data*: use of nonces, timestamps, counters against replay attacks.
- *Code signing*: use of cryptographic hash to validate authenticity and integrity of the code.
- *Cryptographic key establishment*: a scheme to allow key exchange between two parties.
- *OS and hardware security*: protection mechanisms such as root of trust, secure boot, etc.

Additional requirements on the cybersecurity solutions are scalability and usability, which focus on the identification and access control methodology combined with usability of human interfaces. Furthermore, the system management deals with the management of the keys, the configuration, installation, replacement of devices, and the monitoring and malfunction detection.

If security and privacy are already big challenges on IT systems, these challenges become much more important on the IoT systems considering that the attack surface has significantly been enlarged as well as the amount of data generated and handled [8]. Furthermore, the impact becomes more important considering that IoT devices have not enough processing capabilities, in contrast to IT systems, and they have a limited autonomy because they work most of the cases on batteries. They use generally different wireless connectivity solutions not compliant with existing security standards. Last but not the least, the nature of the applications, for instance AHA, requires a high level of security to keep end-to-end data integrity, confidentiality and service availability. The AHA users are very concerned by these aspects.

Secure IoT systems with high level of personal data protection are mandatory to keep the users' trust. These aspects are essential to deploy massively the IoT technologies in the coming years.

4.2.2 Privacy

Concerning the objectives and recommendations for the privacy this work uses the General Data Protection Regulation (GDPR) (EU 2016/679)[3] as the basis. In addition, the Data Protection Impact Assessment (DPIA) process is used to put in place such regulation. The article 1 of GDPR defines the following objectives:

- This Regulation lays down relating to the protection of natural persons with regards to the processing of personal data and rules relating to the free movement of personal data.
- This regulation protects fundamental rights and freedoms of natural persons and in particular their right to the protection of personal data.
- The free movement of personal data within the Union shall be neither restricted nor prohibited for reasons connected with the protection of natural persons with regard to the processing of personal data.

[3]Regulation (EU) 2016/679 of the European Parliament and of the Council of 27 April 2016.

The GDPR defines also the following requirements:

The protection of the rights and freedoms of natural persons with regard to the processing of personal data requires that appropriate legal, technical and organizational measures be taken to ensure that the requirements of this Regulation are met.

In order to be able to demonstrate compliance with this regulation, the data controller should adopt internal policies and implement measures that meet in particular the principles of data protection by design and data protection by default. Such measures could consist, inter alia, of:

- Minimizing the processing of personal data.
- Pseudonymising personal data as soon as possible.
- Transparency regarding the functions and processing of personal data.
- Enabling the data subject to monitor the data processing.
- Enabling the controller to create and improve security features.

4.3 Recommendations on Cybersecurity and Privacy in IoT

4.3.1 Security

Security is a complex and critical concern for any manager of interconnected digital assets. Many private companies [9], public bodies [10] and standardization/harmonization institutes (e.g. RFC 2196 Site Security Handbook) have published recommendations aiming at improving the quality and consistency of the security levels across interconnected systems. Such recommendations target system managers, organization officers, service providers, infrastructure owners, product manufacturers, developers, end users and indirectly also attackers. In fact, as promoted by security experts, every security measure, mechanism and algorithm must rely on publicly available specifications. Recommendations are elaborated and publicized proactively [15, 11] and reactively [11]. Interestingly some of them are associated to supporting tools [10].

All these sets of recommendations present diverse facets of similar rules and recommendations. It is not possible to include the whole list in this chapter. However, recommendations insist on the fact that security is a continuous process with integrated improvement procedure, based on the continuous evaluation of the in-place security. Therefore, external inspection such as auditing is a must. Self-auditing and internal expertise are strongly required, but by far not enough. External companies offer services to analyse the implemented security, including security standards, such as ISO/IEC 27001 and 27002, the NIST Cybersecurity Framework.

4.3.2 Privacy

When developing, designing and using applications, services and products that aim to process personal data to fulfil their task, the developers/producers of such products, services and applications are recommended to take into account the right to data protection. It is important to make sure that controllers and processors are capable enough to fulfil data protection obligations. Furthermore, the principles of data protection by design and by default should be also taken into consideration in the context of public tenders.

A report by ENISA (the European Union Agency for Network and Information Society) elaborates on what needs to be done to achieve privacy and data protection by default [13]. It specifies that encryption and decryption operations must be carried out locally, not by a remote service, because both keys and data must remain in the power of the data owner if greater privacy needs to be achieved. The report specifies that outsourced data storage on remote clouds is practical and relatively safe, as long as only the data owner, not the cloud service, holds the decryption keys.

In literature, there are additional principles and guidelines available that can be used to achieve privacy and data protection by default, also known as *privacy by design*. Privacy by design [14] is a concept, developed in the 90's, to address the ever-growing and systemic effects of Information and Communication Technologies (ICTs), and of large-scale networked data systems. The objectives of privacy by design – ensuring privacy and gaining personal control over one's information, and, for organizations, gaining a sustainable competitive advantage – may be accomplished by practicing the following 7 foundational principles:

1. Proactive not reactive; preventative not remedial.
2. Privacy as the default setting.
3. Privacy embedded into design.
4. Full functionality – positive-sum, not zero-sum.
5. End-to-end security – full lifecycle protection.
6. Visibility and transparency – keep it open.
7. Respect for user privacy – keep it user-centric.

4.4 Security Approach

4.4.1 Methodology

To achieve the objectives defined above, a number of activities are performed to lay down the security and privacy policies in the context of ACTIVAGE project. For the purpose of security, activities include:

- Perform a reference risk analysis in the ACTIVAGE IoT environment in order to identify the general ACTIVAGE security requirements, which depend on the criticality of applications or services.
- Countermeasures to mitigate risks are identified at this stage.
- Create and elaborate the ACTIVAGE security questionnaire.
- Analyse questionnaires' responses and perform assessments for the DS' security requirements.
- Define the security cartography and recommendations for each deployment site.

The elaboration of the security questionnaire considered the following aspects:

- Collect relevant information allowing the identification of missing mechanisms to ensure full end-to-end cybersecurity and privacy for each of the DSs.
- Make it easy for the DSs security managers to reply. The DS security manager is in charge of the security and privacy aspects related to this DS.
- Make the DSs security managers aware of cybersecurity and privacy issues that have not yet been identified and support the other stakeholders to realize the high importance of these aspects that are critical considering the nature of the project, which includes data confidentiality, higher vulnerability by connecting "smart objects" to the system, etc.

Security analysis is performed based on the following assumptions:

- All IoT devices and elements constituting the DS meet safety requirements according to the existing norms and regulations in conformance to their original purpose. This falls into the responsibility of the device manufacturer or SW provider/service provider and of the DS manager. e.g., an electrical heater used to ensure the comfort of elderly people must respect basic norms for electrical heaters.
- DS security managers know the basic norms and regulations rules with which the devices, SW and services used must comply. The managers

should be able to provide the corresponding evidence and they should highlight any unconformity. Questionnaires take into account that the answers are given considering the country rules where the DS is deployed.

- Each of the service providers who plans to use the ACTIVAGE technology needs to upgrade and adapt the DS elements/settings/ components to the norms and the regulations in force at the time and in relation to the location of the commercial exploitation.

The general risk analysis adopted the ACTIVAGE Reference Architecture shown in Figure 4.1 and was carried along the following typical steps:

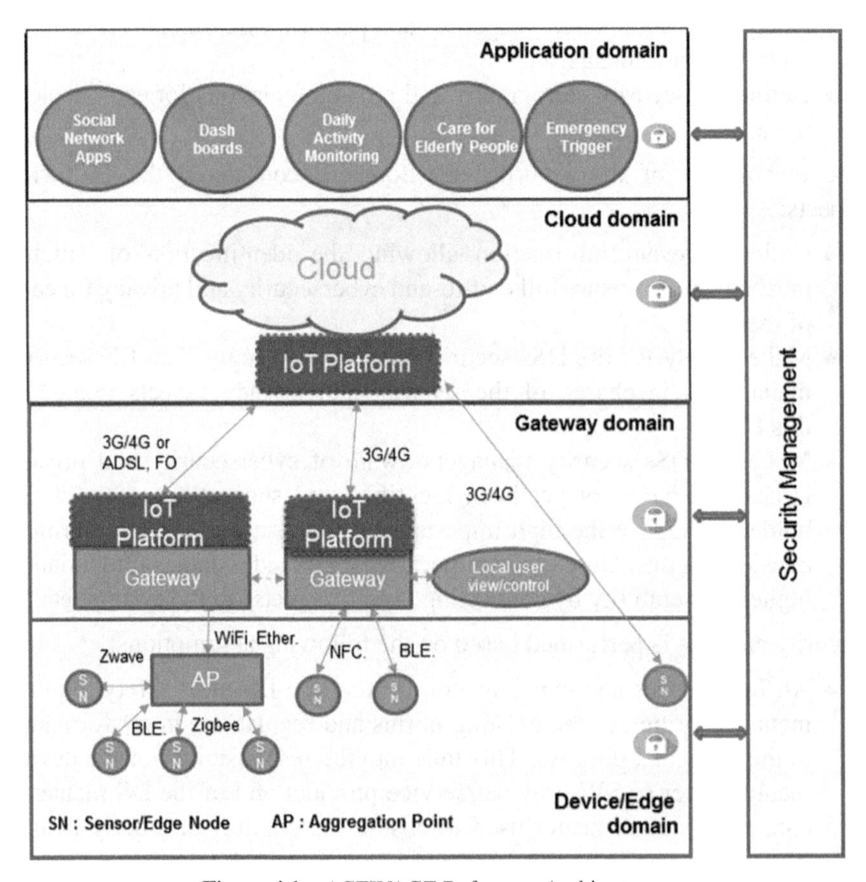

Figure 4.1 ACTIVAGE Reference Architecture.

- Identification and description of all the assets to be protected in the IoT system.
- Identification of all threats and vulnerabilities for each asset.
- Quantification of security risk caused by the threats and vulnerabilities, using a metric.
- Risk management: the decision on which risks to counter and which ones are acceptable.

The risk analysis leads to the definition of the appropriate security measures.

4.4.1.1 Assets identification and description

An Assets list was established as a guideline to be carefully analysed, completed (if needed) and used for each DS. It includes all data in the system, services, pieces of hardware, software, communication links and may be extended to intellectual property, brand reputation, buildings etc. The most important items in this list are given hereafter.

Data assets include application and management data. The typesets and formats should be defined in the data model.

Application data describe the elements or resources of the IoT system. They include, for example:

- Data describing all entities producing or consuming data (Identifiers and attributes of individuals, stakeholders, sensors).
- Data that are monitored and analysed by the IoT system in order to ensure the expected service (raw measurements, processed data elements).
- Decisions of the system that influence the subject's environment (guidance or prescriptions for individuals, environmental instructions for smart sensors, configuration instructions for devices).

Management data relate to system operation. They include, for example:

- Procedure, action plan descriptions (definition of all the planned actions in case of occurrence of an extreme event).
- Data storage organization definition (for example, a Grading Table, Detail Description predefines categories for data storage, such as Medical information, Medical report, Wellness information, Service, etc.).
- Access Rights Table, defining the access rights for each stakeholder profile.
- Transaction registers, logging the History of all operated transactions (communication channel, data, data user, time, etc.).

- Cryptographic material that may include log-in credentials, cryptographic secrets for authentication and encryption, root-of-trust information (e.g. trusted PKI public key), public key certificates, etc.

Assets also include hardware and software elements.

Communication channels. The connection between the devices and the IoT-Gateway is generally wireless (BLE, Z-wave; Zigbee, etc.). The connection between the IoT-Gateway and the Cloud can be an Internet connection. However, and many times, the IoT-Gateway is connected via Wi-Fi/Ethernet to a second Gateway that performs the Internet connection via 2G/3G/4G or a wired connection (XDSL, Cable, OF). On the application end, the connection between the user and the Cloud can be wired or wireless. The wired connection can be through the chain Lap and Desk Tops, LAN, Gateway. The Gateway allows connecting the user with the Internet network and this one to the Cloud. The wireless connection (2G/3G/4G) is done by having a direct connection between the Smart Phones and Tables directly to Internet having access to Web applications.

Component hardware. For example, typical hardware assets to consider at the low domains (Device and Gateway) are data storage units, processing units, power management blocks, sensing and actuating blocks as well as all device interfaces (e.g. I/O, JTAG ports, etc.) and device casing. Maturity and configuration must be assessed.

Component software and configuration information. Software must be analysed at all levels: OS, firmware, application embedded software, high-level application container. Boot mechanisms and system configuration at all IoT levels also need particular protection and are included in the assets list.

Trust associations (end-to-end security). Establishing an end-to-end security association, between the data source and their final destination, provides a higher and often necessary level of data protection. The data are not made available at any of the intermediate hops, since they are encrypted at their source and only the final data user is able to decrypt them.

4.4.1.2 Security risk analysis tools: Product or service compliance class, STRIDE, DREAD

The IoT Security Compliance Framework [15] and other guideline documents issued by the IoT Security Foundation are used to enhance best security practices during development and installation of an IoT product (or system or service). The Framework includes the definition of Compliance Classes for products and a series of criteria in order to validate their security depending on the targeted class. Applicability of the requirements on a product depends on its compliance class, which is expressed as a number between 0 to 4, increasing with security level. To define compliance classes, three levels of risk impact, BASIC, MEDIUM and HIGH, are defined for each of the three security objectives, namely confidentiality, integrity and availability. For instance, MEDIUM confidentiality corresponds to "*Devices process sensitive information (including Personally Identifiable Information – PII); limited impact if compromised*" and is required from class 2.

The risk analysis methodology followed in ACTIVAGE to identify the threats is based on the STRIDE Methodology, see Table 4.1. This Threat classification model was developed by Microsoft [18, 19], and helps answering the question "what can go wrong in the system?"

The risk mitigation technologies (Cybersecurity measures or Cybersecurity controls) against a STRIDE threat to apply on the system element under consideration depend on the element type, perspective (developer, administrator) and assessed risk level (DREAD rate). Recommendations by foundations or standard bodies give guidelines in this task, providing lists of Cybersecurity requirements depending on risk level (or compliance class) as well as best-practice tips [16, 17].

In the case study described here below, we identify the Cybersecurity controls to apply to each system element and gives an indication of:

- The compliance classes for which the control must be applied.
- The applicability level, which is defined as **mandatory** (the requirement shall be met, as it is vital to secure the product category) or as **advisory** (the requirement should be met unless there are sound reasons such as economic viability or hardware complexity, in which case the reasons for deviating from the requirement must be documented).

Table 4.1 STRIDE

Threat	Concerned Security Property
Spoofing	Authentication
Tampering	Integrity
Repudiation	Non-repudiation
Information disclosure	Confidentiality
Denial of service	Availability
Elevation of privilege	Authorization

4.4.1.3 ACTIVAGE as example of Risk Analysis

The Threat analysis is performed on **Device, Gateway, Cloud and Application** domains following the proposed IoT reference architecture. As an example, the STRIDE analysis applied to an IoT reference Device is detailed below.

Proposed Assets description of an IoT reference Device, see Figure 4.2:

- HW description, configuration integrity for IoT devices:
  - Connectivity (description and maturity): Communication Channel CC1
  - Processing (description and maturity): P1
  - Data Storage (description and maturity): DS1
    * Individual Subject id, Devices Id,
    * Raw Data (Individual Subject, Environmental, Devices and Services).
    * Processed Data (Individual Subject, Environmental, Devices and Services).
    * Instructions (Users, Environmental, Devices and Services).
    * Data grading table in (DS1) & Access right table in (DS1).
- In Device Data Flow (DF), the following analysis must be performed on:
  - Connectivity/Communication channels: BLE, Wi-Fi, LoRa, NB-IoT
    * Nature of Data: Individual Subject, Devices, Raw & Processed Data, Instructions (Users, Environmental, Devices & Services).

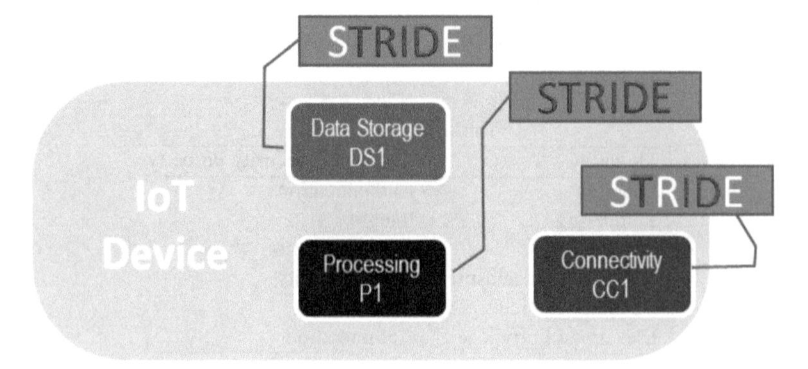

Figure 4.2 IoT device assets and STRIDE representation.

In this example, the threats concerning the related asset are identified in red bold characters in Figure 4.2:

- In DS1: Tampering, Repudiation, Information disclosure and Denial of service.
- In P1: Spoofing, Tampering, Repudiation, Information disclosure, Denial of service and Elevation of Privilege.
- In CC1: Tampering, Information disclosure and Denial of service.

Subsequently, an evaluation of the vulnerability of the IoT Device is performed. The question to be answered is: "What will be the impact of the attacks on the assets?" All the threats for every element are rated using DREAD method ranked from 1 to 3 point, where the DREAD rate refers to all the risks as defined in Table 4.2.

Table 4.2 DREAD ranking definition

Risk	Risk Property	Description/point
Damage potential	How great can be the damage?	1pt (low): Leaking trivial information 2pts (medium): Leaking sensitive information 3pts (high): Can subvert the security system
Reproducibility	How easy to reproduce?	1pt (low): Very difficult to reproduce, even with knowledge of the security hole 2pts (medium): Can be **reproduced**, but only with a timing window and a particular situation 3pts (high): Can be reproduces every time and doesn't require any particular situation
Exploitability	How easy to realize this threat?	1pt (low): Requires an extremely skilled person and in-depth knowledge every time to exploit 2pts (medium): A skilled programmer could make the attack, then repeat the steps 3pts (high): A novice programmer could make the attack in a short time
Affected users	How many users are affected?	1pt (low): Very small % of users, obscure feature; affects anonymous users 2pts (medium): Some users, non-default configuration 3pts (high): All users, default configuration, key customer
Discoverability	How easy to find this vulnerability?	1pt (low): The bug is obscure, and it's unlikely that users will work out damage potential 2pts (medium): located in a seldom-used part, and only a few users should come across it 3pts (high): The vulnerability is located in the most commonly feature and is very noticeable

See below a DREAD ranking based of on the proposed case study.

Table 4.3 DREAD ranking evaluation and analysis

Threat Applicable	DREAD Rate Evaluation	Analysis
Spoofing	$2, 3, 2, 2, 1 \rightarrow 2$	Weak Password
Tampering	$3, 2, 1, 2, 1 \rightarrow 1.8$	
Repudiation	$1, 2, 2, 2, 1 \rightarrow 1.6$	
Information disclosure	$3, 2, 1, 2, 1 \rightarrow 1.8$	
Denial of Service	$3, 3, 3, 1, 1 \rightarrow 2.2$	Physical port accessible
Elevation of Privilege	$3, 2, 2, 1, 1 \rightarrow 1.8$	

The result of the assessment can be compared to the minimum requirement of compliance class. As soon as the weaknesses are identified, the strategy to address the risk must be explicitly detailed. Basic risk strategies are mitigation, acceptation or transfer to a third party.

Table 4.4 Basic strategy analysis

Threat Applicable	Risk	Strategy	DREAD Rate
Spoofing	Mitigate	Secure boot process	$2, 2, 2, 2, 1 \rightarrow 1.8$
Tampering	Accepted		$3, 2, 1, 2, 1 \rightarrow 1.8$
Repudiation	Accepted		$1, 2, 2, 2, 1 \rightarrow 1.6$
Information disclosure	Accepted		$3, 2, 1, 2, 1 \rightarrow 1.8$
Denial of Service	Mitigate	All non-used ports are physically inaccessible	$3, 2, 1, 1, 1 \rightarrow 1.6$
Elevation of Privilege	Accepted		$3, 2, 2, 1, 1 \rightarrow 1.8$

4.5 Privacy Approach

4.5.1 Introduction

Nowadays, Privacy in Europe has gained a lot of visibility through the advent of the new General Data Protection Regulation (GDPR) entered in force on May 25th, 2018 in the European Union. Until recently, companies making business out of personal or other types of data systematically pushed privacy back. Entities promoting the privacy preservation and enforcement processes propose different approaches. In this chapter, the authors propose to develop a general methodology on Privacy to define a privacy impact analysis for a given IoT System and provide recommendations and guidelines in order to minimize the Privacy threats. The complete methodology is described hereafter. It is under deployment in the Deployment sites of the ACTIVAGE project.

Moreover, and in complement of the Security methodology described in Section 4.4, the authors made an analysis of the GDPR to identify the Privacy modules/services/articles that should be implemented in any IoT system of the ACTIVAGE project. This analysis allowed identifying some use cases that are well suited to be implemented using a Blockchain based technology, as described in the Section 4.6.3.

4.5.2 Methodology to Perform Privacy Analysis and Recommendations

Figure 4.3 shows the Privacy methodology proposed in order to perform risk privacy analysis on an IoT system. This is the methodology we have used in ACTIVAGE for this purpose. The expected outcomes are the identification of the countermeasures/recommendations for this IoT system to minimize the risks of privacy threats: data theft, data misuse or any other malicious usage. This methodology is addressed to any non-professional data protection manager to facilitate, him/her, the implementation of the GDPR regulation.

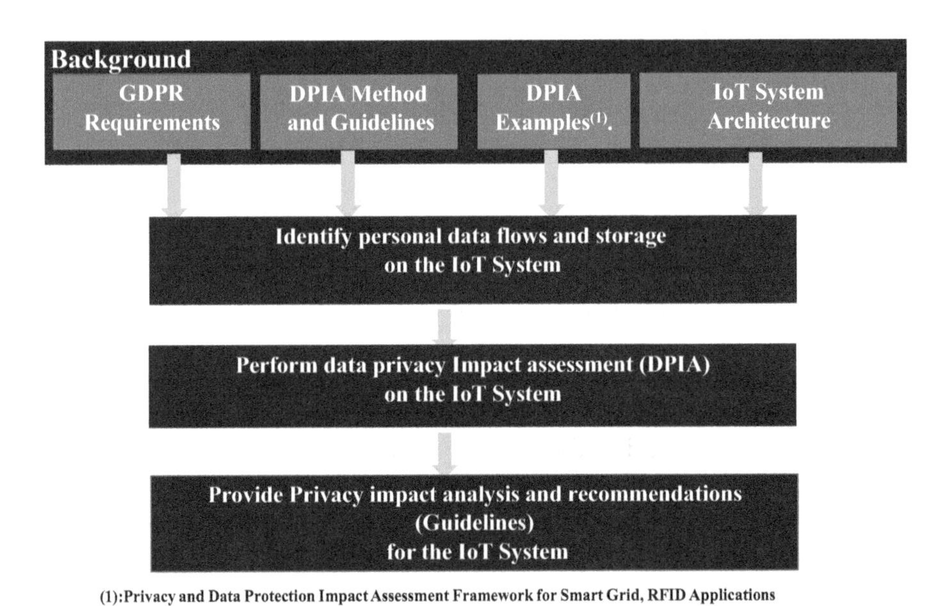

Figure 4.3 Privacy methodology.

This methodology consists in the execution of the following four main steps:

- **Background** – A good acknowledge of the following elements is required: What is the GDPR?

  What is a DPIA and how should be performed? What are the IoT System architecture and topology where the Data will be generated, stored, processed and exploited (and by whom) to identify security rights? In order to get the answers to these questions, the following documents are available [5, 20–24].
- **Identify personal data flow and storage** – For any IoT system, it is required to know its complete and detailed architecture and topology as discussed in Section 4.4. This information allows "easily" the identification of assets, data flows, data storage, process units, users, etc. and their location.
- **Perform Data Impact Performance Assessment** – (DPIA) This step is key in the methodology. The importance of this step and the way to develop it are described with more details in the next paragraph.
- **Provide Privacy Impact Analysis and Recommendations** – This step provides the DPIA analysis results of the IoT system under study and the recommendations proposed to deploy the system with good Privacy properties.

4.5.3 Data Protection Impact Assessment (DPIA)[4]

GDPR introduces the concept of a Data Protection Impact Assessment (DPIA)[5] [20] and strongly recommend carrying out one for each system concerned. This paragraph addresses the following questions: what is a DPIA?, when a DPIA is mandatory and how to carry it?, and what are the main elements containing a DPIA?

4.5.3.1 What is a DPIA?

"A DPIA is a process designed to describe the processing, assess the necessity and proportionality of a processing and to help managing the risks to the rights and freedoms of natural persons resulting from the processing of personal data. DPIAs are important tools for accountability, as they help controllers not only to comply with requirements of the GDPR, but also to

[4]This information contained in this paragraph was extracted from [20].

[5]The term "Privacy Impact Assessment (PIA) is often used in other contexts to refer to the same concept", for more information see [21–23].

demonstrate that appropriate measures have been taken to ensure compliance with the Regulation. In other words, *a DPIA is a process for building and demonstrating compliance*".

Under the GDPR, non-compliance with DPIA requirements can lead to fines imposed by the competent supervisory authority. Failure to carry out a DPIA[6] can each result in an administrative fine of up to 10M€, or in the case of an undertaking, up to 2% of the total worldwide annual turnover of the preceding financial year, whichever is higher.

4.5.3.2 When is a DPIA mandatory?

Where a processing is "likely to result in a high risk to the rights and freedoms of natural persons". Table 4.5 gives some examples where a DPIA is required.

Table 4.5 Examples where DPIA is required

Examples of Processing	Possible Relevant Criteria	DPIA Required?
A hospital processing its patients' genetic and health data (hospital information system).	• Sensitive data • Data concerning vulnerable data subjects	Yes
The use of a camera system to monitor driving behavior on highways. The controller envisages using an intelligent video analysis system to single out cars and automatically recognize license plates.	• Systematic monitoring • Innovative use or applying technological or organizational solutions	Yes
A company monitoring its employees' activities, including the monitoring of the employees' work station, internet activity, etc.	• Systematic monitoring • Data concerning vulnerable data subjects	Yes
An online magazine using a mailing list to send a generic daily digest to its subscribers.	—	Not necessarily

4.5.3.3 When should the DPIA be carried out?

"*prior to the processing*". This is consistent with data protection by design and by default principles. The DPIA should be started as early as practical

[6]For instance, when the processing is subject to a DPIA, or carrying out a DPIA in an incorrect way, or failing to consult the competent supervisory authority where required.

in the design of the processing operation even if some of the processing operations are still unknown. As the DPIA is updated throughout the lifecycle project. It will ensure that data protection and privacy are considered and promote the creation of solutions that promote compliance.

4.5.3.4 What is the DPIA minimum content?

The GDPR does not formally define the concept of a DPIA as such, but it sets out its minimum features as follows:

- Its minimal content is specified as follows:
    a) A systematic description of the envisaged processing operations and the purposes of the processing, including, where applicable, the legitimate interest pursued by the controller.
    b) An assessment of the necessity and proportionality of the processing operations in relation to the purposes.
    c) An assessment of the risks to the rights and freedoms of data subjects.
    d) The measures envisaged to address the risks, including safeguards, security measures and mechanisms to ensure the protection of personal data and to demonstrate compliance with this Regulation taking into account the rights and legitimate interests of data subjects and other persons concerned.
- Its meaning and role are clarified: "In order to enhance compliance with this Regulation where processing operations are likely to result in a high risk to the rights and freedoms of natural persons, the controller should be responsible for the carrying-out of a data protection impact assessment to evaluate, in particular, the origin, nature, particularity and severity of that risk".

Figure 4.4 illustrates the generic iterative process for carrying out a DPIA. It should be underlined that the process depicted here is iterative: in practice, it is likely that each of the stages is revisited multiple times before the DPIA can be completed. Furthermore, this process should be regularly performed to evaluate the IoT system evolution over the time.

Practical recommendations (necessary but not sufficient) when carrying out a DPIA

The basic recommendation is to collect only required personal data to minimize the risk of non-compliance. It excludes the "just in case" approach in

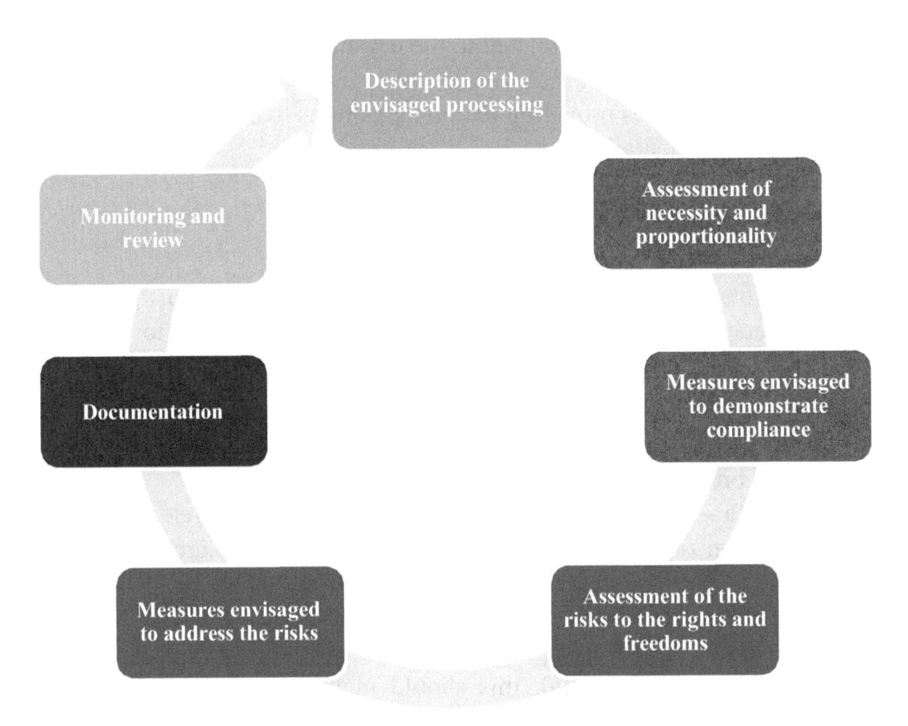

Figure 4.4 Process for carrying out a DPIA.

which unjustified data is collected for future uses, even when they may be justified.

It requires a complete audit of the data already in possession of the various stakeholders (processors, etc.), and the data must be kept based on the principle of usefulness for the subject and necessity for the service.

4.5.4 GDPR Analysis for Implementation

To cope with GDPR in an IT system and more particularly on IoT based system (as such foreseen in ACTIVAGE where security and privacy are of high importance according to AHA applications supported), a first analysis was performed on the set of articles constituting the GDPR. They were analysed and classified as follows:

- Legal: Articles related with legal issues.
- Technical: Articles requiring a technical implementation.

- Accountability: Articles related to the organization/company Governance.
- Principles: Articles providing recommendations to be considering in the GDPR implementation.

Table 4.6 gives the details of this analysis. It is composed of three columns indicating (from the left to the right): the type of article (Legal, Technical, etc.), the type of service and the article description concerned by the GDPR.

On top of this first analysis, Varonis[7] recommends focusing on the following technical aspects during the implementation phase to meet the GDPR [25]:

- **Data classification** – Know where personal data is stored on the IT/IoT system. This is critical for both protecting the data as well as following through on requests to correct and erase personal data.
- **Metadata** – With GDPR requirements for limiting data retention, basic information on when and why the data was collected are required, as well as its purpose. Personal data residing in IT/IoT systems should be periodically reviewed to see whether it needs to be saved for the future.
- **Governance** – GDPR highlights the need to get back to basics. For enterprise (or AHA data), this should include understanding who is accessing personal data in the AHA file system, who should be authorized to access, and limiting file permission based on users' actual roles – i.e., role-based access controls.
- **Monitoring** – The breach notification requirement places a new burden on data controllers. Under the GDPR, the IT/IoT security mantra should be "always be monitoring". Data protection controllers need to spot unusual access patterns against files containing personal data, and promptly report an exposure to the local data authority. Failure to do so can lead to enormous fines, particularly for multinationals with large global revenues.

The analysis performed in this section contributed to identify several Privacy uses cases to be implemented using the innovative and pervasive Blockchain as a potential technology to provide robust and efficient IoT solutions on security and privacy. The following section describes these developments.

---

[7]Varonis is a pioneer in data security and analytics, specializing in software for data security, governance, compliance, classification, and analytics.

Table 4.6 GDPR Analysis in view of its implementation

Type of Article	Provided Function or Service	GDPR Article
Legal/Principle		*Article 5 – Basic principles related to data Security*
Legal/Technical	Establish access controls and protected regulated data.	*Article 6 – Lawfulness of processing Subject's consent*
Legal/technical	Establish access controls and protected regulated data.	*Article 7 – Conditions for consumer Consent*
Legal/technical	Establish access controls and protected regulated data.	*Article 13 and 14 – Information and access to personal data*
Technical	Automatically discover and classify GDPR affected data	*Article 15 – Right of access by the data subject.* Enable to provide the data subject remote access to his or her personal data *Article 16 – Right to rectification* Be able to rectify specific data. *Article 17 – Right to erasure ('right to be forgotten').* Be able to discover and target specific data and automate removal *Article 18 – Right to restriction of processing* *Article 20 – Portability rights* Develop interoperable formats that enable data portability.
Technical	Audit and Traces control, protection against cyber-attacks and internal threats	*Article 30 – Records of processing activities.* Implement technical and organizational measures to properly process personal data
Technical	Establish access controls and protected regulated data.	*Article 25 – Data protection by design and by default.* Embrace accountability and privacy by design as a business culture • Collect only the required data • Give access only to the right people • Availability to prove and demonstrate
Legal/technical	Management of incidents and notifications	*Article 33 – Notification of a personal data breach to the supervisory authority.* Prevent and alert on data breach activity; have an incidence response plan in place

(*Continued*)

Table 4.6 Continued

Type of Article	Provided Function or Service	GDPR Article
	Security Review	*Article 32 – Security of processing (Ensure confidentiality, integrity and availability).* Ensure least privilege access; implement accountability via data owners; provide reports that policies and processes are in place and successful. *Article 34 – Communication of a personal data breach to the data subject.* *Article 35 – Data protection impact assessment (DPIA/Risk analysis).* Quantify regularly data protection risk profiles.
Accountability	Governance	*Article 37 – Designation of the data protection officer.* *Article 38 – Position of the data protection officer.* *Article 39 – Tasks of the data protection officer.*

4.6 Security and Privacy Implementation

4.6.1 Introduction

This section presents two use cases selected to illustrate the interest and the importance to follow a top-down approach for security and privacy. During the end-to-end security risk analysis and DPIA performed on the IoT systems of the ACTIVAGE project, this approach allowed the identification of the recommendations and solutions to put in place to improve the security of some IoT system components as well as the services/functions to cope with GDPR privacy requirements. It is clear that the Privacy services must run on top of a Secure IoT system.

The first use case presents the countermeasures implemented to secure the data storage of the Raspberry PI Gateway used in some Deployment sites of ACTIVAGE. The second use case presents several scenarios where the Blockchain technology can be used to provide efficient solutions on security and privacy for the ACTIVAGE's Deployment sites.

4.6.2 Securing a Gateway

The Gateway in an IoT device to Cloud architecture is a key element as it marks the frontier between the public and private domains. In this position in the architecture, the Gateway is indeed both an entry path from inside to outside and reverse.

In a worst-case scenario, somebody gaining access to a Gateway gains access to other Gateways, by reproducing the attack at a massive scale. In the ACTIVAGE context, Gateways are often deployed in homes, and thus it is not possible to master the physical access to the hardware. Moreover, the Gateway, in a residential place, might be stolen more easily than a server in a data centre might be.

The Raspberry PI is a popular platform for its low cost, stability and good support. In experimental projects such as ACTIVAGE, it is the platform of choice to be used as a Gateway. The analysis done from ACTIVAGE questionnaires on IoT devices used in the 9 Deployment sites has shown that at least 4 out 9 deployment sites are considering using such hardware platforms as reference for their experiments.

However, the Risk analysis performed on the Raspberry PI has identified potential weaknesses regarding security. A major weakness concerns the SD Card mass storage. Due to its removable nature, this mass storage can be easily accessed from a third-party system by simply removing the SD Card and plugging it to a computer.

In this way, the content would be cleared and read/write operation unauthenticable making it easy for a hacker to read out and even replace sensitive information such as user's password, SSH private keys or other credentials that could enable privileged access to the entire system. Table 4.7 illustrates the impact assessment of the different stride attributes for the mass storage of the gateway in the ACTIVAGE context (deployment in residential homes). The initial DREAD rates on the third column shows potential impacts. The last column shows new rates while mitigating the risks with a secure element. A first counter-measure for this weakness would be to encrypt the entire SD Card, thus, it requires storing the encryption key in a safe place, which is readable by the processor and the firmware while booting the OS located on the storage. A common solution for such a safe storage is to use a Trusted Platform Module (TPM). TPMs are standardized electronic components which have security related functions such as random number, hash and key generators, encryption and decryption hardware engines and offers facilities to store in secure manner keys or sensitive data such as Platform Configuration Registers. These components are used for example for secured boot in UEFI bios. Table 4.7 shows on the two last right columns the new DREAD rate while using such a component, with highest risks reduce to a safer impact level.

Table 4.7 DREAD impact assessment

Threat Class	STRIDE Security Property	DREAD Rate	Mitigation Choices	Mitigation Technology	New DREAD Rate
T	Integrity(I)	2,2,2,1,2 => 1.8	Data storage shall be temper-resistant File system shall be adapted to the technology (read/write cycles) Data shall be backed up	Use a secure element to store security information in order to: Encrypt application's partition Manage strong authentication at network level and application level	2,2,2,1,2 => 1.8
R	Confidentiality(C)	2,2,2,1,2 => 1.8	Read and write operation shall require authentication		2,1,1,1,2 => 1.4
I	Confidentiality(C)	3,2,3,2,3 => 2.6	Data storage shall be encrypted		1,1,1,1,2 => 1.2
D	Availability(A)	2,2,2,1,2 => 1.8	Removable storage devices shall be proscribed Data storage resources shall be monitored to avoid being saturated		2,2,2,1,2 => 1.8
E	Authorization(I)	3,2,2,2,2 => 2.2	Write and Read permission shall be tuned in the file system		1,1,1,1,2 => 1.2

Figure 4.5 Raspberry PI model 3 with TPM dedicated hat in white.

Figure 4.5 shows a prototype hat for a Raspberry PI embedding a TPM manufactured and assembled in this form by STMicroelectronics. Customization of the Linux Kernel for enabling TPM support was also made with the appropriate device tree modification. The TPM is provisioned with security credentials bound to the ACTIVAGE Public key Infrastructure (PKI), ensuring security credential lifecycle up to the revocation of gateways that are suspected to be compromised. This PKI delivers certificates that can be used for the OS and application layer, with state-of-art cryptography scheme.

At the application level, ongoing work is focused on using the TPM secure function whenever possible. A first step consists in the partial encryption of the SD Card. Indeed, while the kernel is located on a clear partition for booting up, the application section is located into a LUKS partition which key is located onto the TPM. It prevents somebody reading the SD card from another platform. Future work will be to encrypt the entire SD card with the decryption within the boot loader.

Other work consists in emulating a PKCS11 interface from the TPM. PKCS11 is a standardized public key cryptography standard specifically related to tokens. The use of this standard enables trustful communication for the establishment of TLS or SSL tunnels, which can be used for the traffic between the Gateway and the Cloud. Use of such tunnels enables encrypted and authenticated communications and prevents the Gateway from being detectable on public network as no IP ports need to be opened for incoming connections. Other use of this interface is under investigation for future work regarding IoT device provisioning.

4.6.3 Blockchain in Smart Homes

Recently, the Blockchain technology has been applied for the healthcare industry [26] but also in IoT-based Smart Homes [27], reducing the time required to access patient information, enhancing interoperability and improving data quality, while reducing maintenance costs. A Blockchain is a continuously growing list of immutable records, called blocks, which are linked and secured using cryptography. Thus, the adoption of Blockchain is a very promising technology towards enhancing the security, privacy and trust.

As described in the previous sections, ACTIVAGE gives special focus on GDPR compliance. Blockchain can act as a very useful tool towards achieving GDPR compliance [28], mainly by serving as a trusted decentralized repository for identification purposes. However, it has to be ensured that: a) no personal data are stored on the Blockchain, b) cryptographic data deletion should be used to give to the end-user the "right to be forgotten". Blockchain can also enhance security as it can enable IoT devices to connect securely and reliably avoiding the threats of device spoofing and impersonation. Every IoT device can be registered in the Blockchain and will have an ID that will uniquely identify this device in the universal namespace.

In the context of ACTIVAGE project, a trusted management solution, based on Blockchain technologies, has been proposed considering the results of other H2020 project implementations such as GHOST/H2020, myAirCoach/H2020. ACTIVAGE will find in this technology a convenient solution to cope with:

- Privacy regulation based on GDPR.
- The integrated healthcare and AHA implications for data and devices protection.
- An adequate trusted mechanism for IoT-based devices, users and systems within the smart Home environment.

The concept of distributed ledger technologies can be introduced within ACTIVAGE to support different use case scenarios such as:

- Requesting/giving/updating permissions for accessing personal data of the involved user.
- Device registration.
- Timely firmware updates.

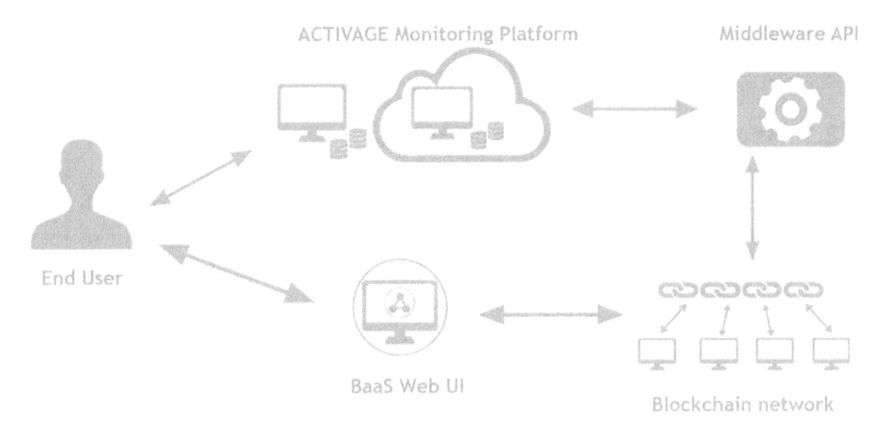

Figure 4.6 ACTIVAGE monitoring platform – BaaS platform architectural overview.

- User authentication & authorization.
- The secure data transfer between endpoints, users and healthcare network components.

Towards the formulation of a secure and trusted environment using information traceability mechanisms and the spreading of the data in AHA information systems, a related ACTIVAGE Blockchain framework has been introduced (see Figure 4.6) consisting of the following main components:

- **BaaS Web UI** (The Blockchain-as-a-Service (BaaS) Web UI) – It is a web front-end for accessing the functionalities provided by the Blockchain network that has been implemented within myAirCoach H2020 project.
- **Middleware API** – The Middleware API enables the communication between the ACTIVAGE Monitoring Platform and the Blockchain network. For this purpose, RESTful web services are used over the HTTPS protocol.
- **Blockchain network** – This is the network of Blockchain nodes where information regarding the various transactions are being stored.
- **ACTIVAGE decentralised Monitoring Platform** – This is a decentralised platform where raw data gathered from the sensors installed in the smart homes of the elderly users are stored and further analysed towards identifying patterns related to user activity (e.g. habits, sleeping times, etc.) and further identifying abnormal events that may be related to emergencies. Through the ACTIVAGE Monitoring Platform and all the other Blockchain components, a trusted environment is offered to the formal/informal carers/end-users as well as to the elderly users/patients.

In the following paragraphs, several use case scenarios for using Blockchain technology within ACTIVAGE are described.

4.6.3.1 Register in BaaS/give consent

In this scenario, the user accesses the registration form in the BaaS Web UI by clicking on the relevant link. After filling the registration form with their data and accepting the Terms of Service, a verification email is sent to their email address. By clicking on the hyperlink, included in the corresponding email, the user is redirected to the BaaS Web UI and their email is verified. After the email verification process, the user can Login the BaaS Web UI. The transaction related to user registration is logged in the Blockchain.

4.6.3.2 Register in the ACTIVAGE monitoring platform through BaaS

A user is able to register to the ACTIVAGE Monitoring Platform from the BaaS Web UI. Thus, the user first logs in to BaaS with his/her account, goes to "Platforms > Not Registered Platforms", chooses ACTIVAGE from the list and clicks on the "Register" button. Then, user is redirected to the ACTIVAGE Monitoring Platform and fills in the Registration Form. Similarly, to the previous scenario, an email verification process is followed for the completion of user registration in the ACTIVAGE Monitoring Platform. The next time that the user logs in the BaaS Web UI, ACTIVAGE is among his/her "Registered Platforms". Again, the transaction related to user registration is logged in the Blockchain.

4.6.3.3 Register in the ACTIVAGE monitoring platform with BaaS

In this scenario, user fills in the Registration Form in the ACTIVAGE Monitoring Platform (option: Register via BaaS). The ACTIVAGE Monitoring Platform sends the valid credentials of the user to the Middleware API through a RESTful Web Service. Then, the Middleware API sends the registration request to the user via email and redirects them to the BaaS Web UI Registration Form. The user registers using the BaaS Registration Form and this transaction of the newly Registered User is logged in the Blockchain network.

4.6.3.4 Registration of new devices and software updates

In ACTIVAGE, Blockchain can be applied not only for the secured registration and authorization of users, but also for the envisaged IoT-based devices that are being installed in the smart Home environment supporting also the timely update of firmware, patches, etc., in order to be performed only by authorized users.

4.6.3.5 Login/Logout

When the user logs in/out to/from the ACTIVAGE Monitoring Platform, a corresponding request for user login/logout is automatically sent to the middleware API over a RESTful Web Service. The Middleware API updates the list with online Users that are kept within the Blockchain by adding/removing the User to/from the list. Thus, all login/logout processes are logged in the Blockchain network.

4.6.3.6 Request/Give/Update permissions for accessing personal data

In this scenario, depicted in Figure 4.7, a caregiver asks for permission to access the personal data of an elderly person through the ACTIVAGE monitoring platform. This request is sent to the Middleware API, which logs it in the Blockchain by also sending a request for permission approval to

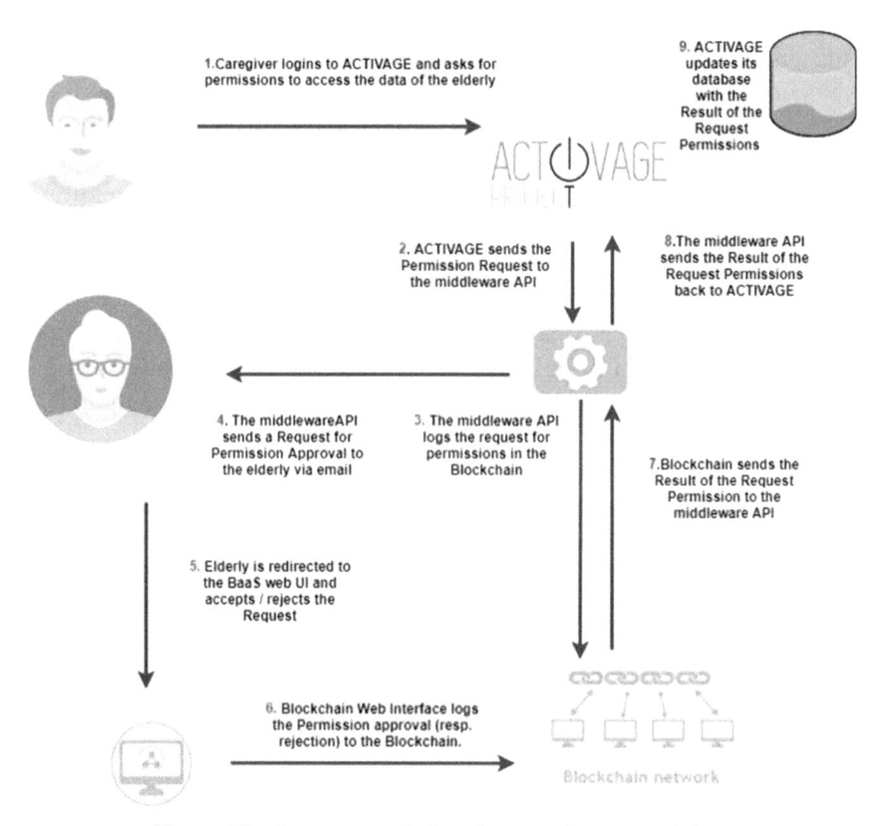

Figure 4.7 Request permissions for accessing personal data.

the elderly via email. By using the hyperlinks included in the email, the elderly is directed to the BaaS Web UI where he/she can accept or reject the request and the corresponding approval/rejection is also logged in the Blockchain network. Through the Middleware API, the result is sent back to the ACTIVAGE Monitoring Platform and based on the decision of the elderly the caregiver is able or unable to access the personal data of the elderly.

These scenarios give a good overview of the possibilities offered using Blockchain technology in AHA applications and more particularly its implementation and validation through the ACTIVAGE project in order to ensure security and privacy in its deployment sites.

4.7 Conclusions

In this chapter, two complementary methodologies were presented one for security and the other for privacy in order to address the challenges presented in the previous paragraphs. They were developed to help the IoT System developers of ACTIVAGE to secure their systems and implement correctly personal data protection to cope with the GDPR requirements. These methodologies follow a twofold approach a top down and an end-to-end. These approaches concern from one side the security risk analysis to identify in advance potential threats and find the countermeasures to mitigate/avoid them. From the other side, a privacy approach to put in place the GDPR following a DPIA analysis to identify the system characteristics and evaluate the risks related to the personal data and its protection. This work, developed in the frame of the ACTIVAGE project, can be also reused for any other IoT system considering the high constrains on security and privacy required by AHA applications.

Finally, the solutions presented give a good overview of the possibilities offered by the use of the Secure element component to secure IoT devices (Gateways and Sensor nodes) and the Blockchain technology in AHA applications. Both technologies will take an important place in the implementation and validation of the security and privacy requirements of the ACTIVAGE's Deployment sites to provide secure IoT systems with a high level of personal data protection and thus to increase the users' trust.

Future work will put in place and validate these methodologies and the potentials solutions to secure the 9 Deployment sites of ACTIVAGE project as well as the protection of the personal data of each of the seven thousands of patients "elderly people" participating in the project.

Acknowledgement

This research project has received funding from the European Union's Horizon 2020 research and innovation programme ACTIVAGE under grant agreement N° 732679.

The activities concerning the Secure Gateway has received funding from the French National Research Agency in the framework of the "Investissements d'avenir" program (ANR-10-AIRT-05)".

References

[1] Internet Security Threat Report ISTR Ramsonware 2017 An ISTR Special Report July 2017.

[2] Proposal for a Regulation OF THE EUROPEAN PARLIAMENT AND OF THE COUNCIL on ENISA, the "EU Cybersecurity Agency", and repealing Regulation (EU) 526/2013, and on Information and Communication Technology cybersecurity certification "Cybersecurity act"), online at: https://eur-lex.europa.eu/legal-content/EN/TXT/?qid=1505290611859&uri=COM:2017:477:FIN

[3] Internet of Things European Large-Scale Pilots Programme, online at: https://european-iot-pilots.eu

[4] ACTIVAGE Large-Scale Pilot project, online at: https://www.activageproject.eu/

[5] General Data Protection Regulation (GDPR) REGULATION (EU) 2016/679 OF THE EUROPEAN PARLIAMENT AND OF THE COUNCIL of 27 April 2016 on the protection of natural persons with regard to the processing of personal data and on the free movement of such data, and repealing Directive 95/46/EC Official journal of the European Union, online at: https://gdpr-info.eu/

[6] NIST FIPS 199, Standards for Security Categorization of Federal Information and Information Systems (February 2004), online at: https://nvlpubs.nist.gov/nistpubs/FIPS/NIST.FIPS.199.pdf

[7] NIST SP 800-53A, NIST Special Publication 800-53A. Revision 4. Assessing Security and Privacy. Controls in Federal Information. Systems and Organizations (December 2014), online at: https://nvlpubs.nist.gov/nistpubs/SpecialPublications/NIST.SP.800-53Ar4.pdf

[8] Mahmoud Elkhodr, Seyed Shahrestani and Hon Cheung, The Internet of Things: New Interoperability, Management and Security Challenges

International Journal of Network Security & Its Applications (IJNSA) Vol. 8, No. 2, March 2016.
[9] Managing security recommendations in Microsoft Azure Security Center, online at: https://docs.microsoft.com/en-us/azure/security-center/security center-recommendations
[10] Baseline Security Recommendations for IoT - Interactive tool, European Union Agency for Network and Information Security, online at: https://www.enisa.europa.eu/topics/iot-and-smart-infrastructures/iot/baseline-security-recommendations-for-iot-interactive-tool
[11] CERN Computer Security Recommendations, in CERN Computer Security, online at: https://security.web.cern.ch/security/recommenda tions/en/index.shtml
[12] Security Recommendations to Prevent Cyber Intrusions, US-CERT, Official website of the Department of Homeland Security, on line at, https://www.us-cert.gov/ncas/alerts/TA11-200A
[13] Privacy and Data protection by Design ENISA, Retrieved 2017 04-04, online at: https://www.enisa.europa.eu
[14] Ann Cavoukien, Privacy by Design, The 7 foundational Principles Information & Privacy Commissioner Ontario, Canada Originally Published: August 2009, Revised: January 2011.
[15] IoT Security Compliance Framework, Release 1.0 2016, IoT Security Foundation
[16] IoT Trust Framework, v2.0 – Released Jan 5, 2017, OTA (Online Trust Framework)
[17] Top 10 IoT security issue categories, OWASP proposal.
[18] Stride THREAT Modelling developed by Microsoft: https://msdn.microsoft.com/enus/library/ee823878%28v=cs.20%29.aspx, https://fr.slideshare.net/marcomorana/application-threat-modeling-presentation, http://www.cs.berkeley.edu/~daw/teaching/cs261f12/hws/IntroductiontoThreatModeling.pdf, http://resist.isti.cnr.it/freeslides/security/williams/RiskBasedSecurityTesting.pdf, https://users.encs.concordia.ca/~clark/courses/1601-6150/scribe/L04c.pdf
[19] Ronen, A. Shamir, A. O. Weingarten and C. O'Flynn, "IoT Goes Nuclear: Creating a ZigBee Chain Reaction," 2017 IEEE Symposium on Security and Privacy (SP), San Jose, CA, 2017, pp. 195–212. doi: 10.1109/SP.2017.14
[20] Guidelines on Data Protection Impact Assessment (DPIA) and determining whether processing is "likely to result in a high risk" for the purposes of Regulation 2016/679; Adopted on 4 April 2017.

[21] Recommendations for a privacy impact assessment framework for the European Union, Deliverable D, online at: http://www.piafproject.eu/ref/PIAF D3 final.pdf
[22] RFID PIA Tool: GS1 EPC/RFID Privacy Impact Assessment Tool, online at: http://www.gs1.org/pia
[23] Data Protection Impact Assessment Template for Smart Grid and Smart Metering systems, Smart Grid Task Force 2012–14, March 18th, 2014.
[24] Mario Diaz Nava and al., "Report on IoT Devices" Deliverable 3.6 of ACTIVAGE/H2020/LSP, March 2018.
[25] GDPR A practical Guide, Varonis.
[26] Blockchain: A healthcare Industry view, online at: https://www.capgemini.com/wp-content/uploads/2017/07/blockchain-a healthcare industry view 2017 web.pdf
[27] Ali Dorri, Salil S. Kanhere, and Raja Jurdak, Blockchain in Internet of Things: Challenges and Solutions: A healthcare Industry view, online at: https://arxiv.org/ftp/arxiv/papers/1608/1608.05187.pdf
[28] Grant Thornton, GDPR & blockchain – Blockchain solution to General Data Protection Regulation, online at: https://www.grantthornton.global/globalassets/_spain_/links-ciegos/otros/gdpr--blockchain.pdf

5

Use Cases, Applications and Implementation Aspects for IoT Interoperability

Regel Gonzalez-Usach[1], Carlos E. Palau[1], Matilde Julian[1], Andrea Belsa[1], Miguel A. Llorente[2], Miguel Montesinos[2], Maria Ganzha[3], Katarzyna Wasielewska[3] and Pilar Sala[4]

[1]Universitat Politècnica de València, Spain
[2]PRODEVELOP, Spain
[3]Systems Research Institute Polish Academy of Sciences, Poland
[4]MySphera, Spain

Abstract

Interoperability in IoT is currently a very complex and difficult challenge in IoT. Lack of interoperability drastically constrains potential benefits from the interconnection of smart objects and hampers the incipient evolution of IoT (Ambient Intelligent Environments, natural transparent human-oriented interfaces, integration with machine learning mechanisms, blockchain security and Artificial intelligence). INTER-IoT solution for interoperability enables platform-to-platform interoperability, across any IoT layer and any application domains. In this chapter, INTER-IoT solution for platforms' integration is applied to relevant use cases in the domains of e-Health, AHA, AAL, Transport and Logistics. Furthermore, innovative aspects and elements of the INTER-IoT are explained, and the benefits of its implementation.

5.1 Introduction

This chapter is about the enablement of IoT interoperability though a novel framework provided by the INTER-IoT project [1, 2]. In particular, it is focused on the uses cases and applications of the INTER-IoT framework, and on the innovative aspects of its implementation.

Interoperability is one of the major challenges in IoT and has a vital importance in the exploitation of all the potential benefits that can be achieved through this new technology paradigm. Without interoperability, possibilities and benefits from the use of IoT are significantly constrained [3].

INTER-IoT project provides an open cross-layer framework with its own associated methodology and integration tools to enable interoperability among heterogeneous Internet of Things (IoT) platforms [2]. INTER-IoT will enable the quick and effective development of smart IoT applications and services, on top of heterogeneous IoT platforms interconnected, and independently from the domain (thus between one or several application domains).

This chapter explains relevant uses cases of the INTER-IoT framework, focused on different application domains (Transportation & Logistics, e-Health, Active & Healthy Ageing, and others). In addition, it provides a state-of-the-art of the current situation of interoperability in IoT, and an overview of new approaches employed in the INTER-IoT implementation.

5.2 Current Interoperability State of the Art

Regarding the implementation of IoT, insufficient interoperability among platforms tends to provoke major issues both at the technical and business levels [3, 4]. Typical problems are the impossibility of integrating non-interoperable IoT devices into non-homogeneous IoT platforms as well as the inability of developing applications and services over several platforms and different domains. Other important setbacks are the paucity of IoT technology penetration, avoidance of customers and companies in employing IoT technology, cost increases in general, impossibility of reusing of technical solutions and low user satisfaction.

Furthermore, lack of interoperability slows and even impedes the incipient evolution of IoT. Ambient Intelligent Environments require seamless interoperability among elements and interfaces. Also, interoperability is essential for the creation of natural transparent human-oriented interfaces of Smart Systems, and it has vital importance for the IoT integration with Artificial Intelligence and the inclusion of new mechanisms such as blockchain security.

In recent years, many solutions have been implemented at different levels, from the device layer to complete IoT platforms, due to a great interest of both business and research institutions in investigating and developing

IoT technology. However, there is no reference standard for IoT and the development of one is not expected in the foreseeable future [3]. Hence, IoT deployments present high heterogeneity at all layers (device, networking, middleware, application service, data/semantics), which restricts interoperability among their elements and among them. Though many projects have dealt with the development of IoT architectures in diversified application domains, not many projects have addressed interoperability and integration issues among platforms (a clear exception are Butler and iCore [29]). Furthermore, no proposals up to the date of the INTER-IoT project approval have been put forward to deliver a general, fully reusable and systematic approach to solve multiple interoperability problems existing in the IoT platforms technology.

The main goal of the INTER-IoT project is to offer a solution for the lack of interoperability in the Internet of Things by providing an open framework that facilitates "voluntary interoperability" among heterogeneous IoT platforms, at any level an IoT deployment (device, network, middleware, application or data & semantics), and across any IoT application domain [2, 6]. Therefore, INTER-IoT guarantees a transparent and effective integration of heterogeneous IoT technology [2, 5].

By using the proposed approach, IoT platform heterogeneity can be turned from a crucial problem to a great advantage, as there will be no need to wait for a unique standard for an interoperable IoT. Instead, interoperable IoT, even on a very large scale, can be created through a bottom-up approach.

The majority of current existing sensor networks and IoT deployments work as standalone entities, and represent isolated islands of information, unable to communicate, interoperate and share information with other IoT systems and platforms due to the use of different standards and to their high inner heterogeneity. In the infrequent cases in which there is an integration effort of IoT elements, it is generally performed at the device or data layer, seeking only the collection of data from smart devices. However, there are many other levels of an IoT deployment, in which it is very beneficial to have interoperability, and many other relevant objectives. Differently from current interoperability approaches, INTER-IoT uses a layer-oriented approach to exploit in depth functionalities of each different layer (device, networking, middleware, application services, data & semantics) [1].

Among the different types and levels of interoperability, a main challenge is inter-platform interoperability, which is addressed on the INTER-IoT project.

5.3 Inter-IoT New Approaches for Implementation

5.3.1 Multilayer Approach

Differently from current interoperability solutions, that typically follow a global approach, INTER-IoT uses a layer-oriented approach for performing a complete exploitation of each different layer functions [2]. Despite of the research and development challenge that the design of a layer-oriented approach represents, in comparison to a global approach, it can potentially provide a very tight and superior bidirectional integration between different IoT platforms. Therefore, a multilayer-oriented approach can potentially offer improved performance, adaptability, flexibility, modularity, reliability, privacy, trust and security.

This layer-oriented approach is composed by several interoperability solutions, addressed specifically to each level or layer of an IoT deployment: Device-to-Device (D2D), Networking-to-Networking (N2N), Middleware-to-Middleware (MW2MW), Application & Services-to-Application & Services (AS2AS), Data & Semantics-to-Data & Semantics (DS2DS).

Each interoperability infrastructure layer has a strong coupling with adjacent layers and provides an interface. Interfaces are controlled by a meta-level framework to provide unrestricted interoperability. Every interoperability mechanism can be accessed through an API. The interoperability infrastructure layers can communicate and interoperate through the interfaces. This cross-layering allows to achieve a deeper and more complete integration. Next, the different layers and associated tools are detailed:

Device layer (D2D): Currently applications and platforms are tightly coupled, preventing their interaction with other applications and platforms, sensors and actuators communicate only within one system, certain platforms do not implement some important services (i.e. discovery), or do so in an incompatible way. Roaming elements can be missing or inaccessible. IoT Device software is never platform independent as companies create proprietary software. These facts present enormous difficulties for the achievement of interoperability. At the device level, D2D solution will allow the transparent inclusion of new IoT devices and the device-to-device interoperation with other smart objects (legacy). D2D interoperability will allow boosting the growth of IoT ecosystems. As a potential solution, INTER-IoT proposes a D2D gateway that allows any type of data forwarding, making the device layer flexible by decoupling the gateway into two independent parts: a physical part that only handles network access and communication protocols, and a virtual part that handles all other gateway operations and services. When connection is lost, the virtual part remains functional and will answer the API and Middleware requests. The gateway will follow a modular approach to allow the addition of optional service blocks, to adapt to the specific case.

Network layer (N2N): Currently the immense amount of traffic flows generated by smart devices is extremely hard to handle. The scalability of the IoT systems is difficult. Also creating the interconnections between gateways and platforms is a complex task. N2N solution aims to provide transparent roaming (support for smart devices mobility) and their associated mobility information. It will also allow offloading and roaming, what implies the interconnection of gateways and platforms through the network. INTER-IoT solution at network level uses paradigms such as SDN and NFV, and achieves interoperability through the creation of a virtual network, with the support of the N2N API. The N2N solution will allow the design and implementation of fully interconnected ecosystems.

Middleware layer (MW2MW): At the middleware level, INTER-IoT solution will enable seamless resource discovery and management of IoT smart objects in heterogeneous IoT platforms. Interoperability at the middleware layer is achieved through the establishment of an abstraction layer and the attachment of IoT platforms to it. Different modules included at this level will provide services to manage the virtual representation of the objects, creating the abstraction layer to access all their features and information. Among the offered services, there are component-based interoperability solutions within the middleware based on communication using mediators, bridges and brokers. Brokers are accessible through a general API. Interoperability at this layer will allow a global exploitation of smart objects in large-scale multi-platform IoT systems [7].

Application & Services layer (AS2AS): INTER-IoT allows the use of various services among different IoT platforms. Our approach enables discovery, catalogue and composition of services from different platforms. AS2AS will also provide an API as an integration toolbox to facilitate the development of new applications that integrate existing heterogeneous IoT services.

Semantics & Data layer (DS2DS): INTER-IoT solution for the DS2DS layer will allow a common meaning of data and information among different IoT systems and heterogeneous data sources, thus providing semantic interoperability. It is based on semantic translation of IoT platforms' ontologies to/from a common IPSM modular ontology. The Inter Platform Semantic

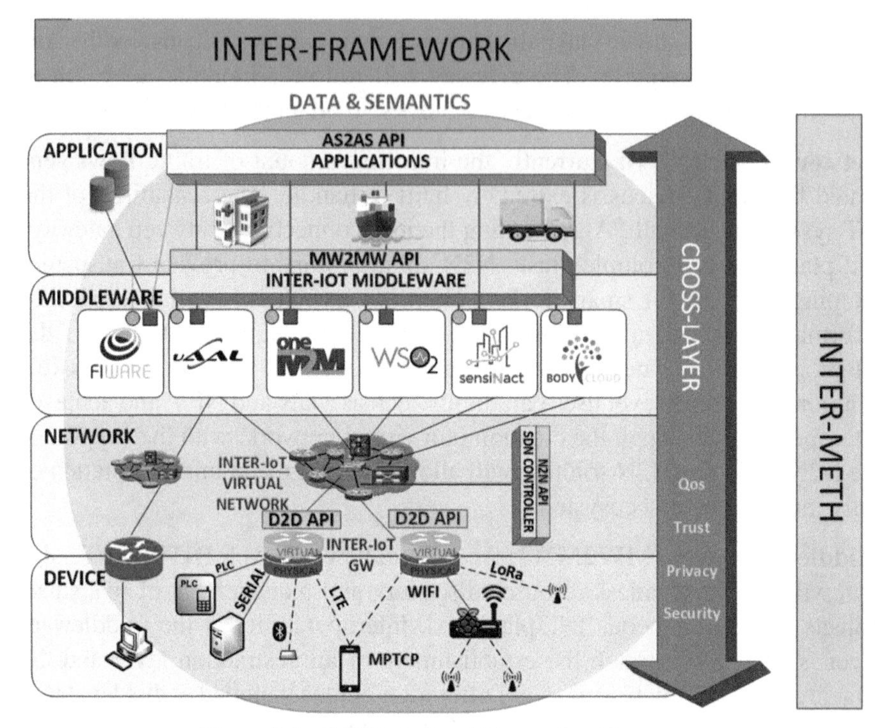

Figure 5.1 INTER-IoT multi-layered architecture.

Mediator (IPSM) component will be responsible for performing ontology-to-ontology translations of the information using ontology alignments. It will be necessary to define explicit OWL-demarcated semantics for each IoT artifact that would like to interoperate, communicate and collaborate [8, 9].

Cross-Layer guarantees non-functional aspects that are required across all layers, such as privacy, security, quality of service (QoS) and trust.

5.3.2 Virtualization of each INTER-IoT Layer Interoperability Solution

In order of providing the option of a quick set-up of each of the layers of the INTER-IoT framework, it is given the option of running a virtualized instance of each of them for rapidly implementing the INTER-IoT interoperability solution. This virtualization is performed by means of Docker [30] engine. Through the creation of Docker containers, the software layer components are separated from each other and from the underlying hardware and operating system. Despite of the virtualization, APIs to access to specific layers functionalities are secured, and protected through the use of security tokens and certificates, and by the assignment of specific permissions to each user or type of user.

Each layer solution can be deployed and implemented standalone, as far as they are independent from other layers' solutions. Thus, there is no need for a complete implementation to achieve interoperability on a specific layer. Though, the combined use of adjacent layers' solutions multiplies benefits, as enables some functionalities among them related to multiple layers.

5.3.3 Universal Semantic Translation

INTER-IoT offers a novel solution to provide automatic semantic translation among any pair of platforms [2]. DS2DS solution performs an ontology-to-ontology translation between two platforms, and thus it is able to provide universal semantic interoperability. The INTER-IoT approach for achieving semantic interoperability among heterogeneous IoT platforms is based on:

- The definition of explicit, OWL-demarcated, semantics for each IoT platform or artifact that is to interoperate, communicate and collaborate.
- An infrastructure that translates messages/data/communication from its native format to the common format used across the INTER-IoT infrastructure: an IoT Platform Semantic Mediator (IPSM) component that will be responsible for translating incoming information, representing semantics of artifact X to semantics of artifact Y. The IPSM will use ontological alignments to perform ontology-to-ontology translations.
- The existence of a common modular ontology of INTER-IoT, called GOIoTP [26].

The IoT Platform Semantic Mediator (IPSM) is a software component that performs semantic translation of data. In the context of the INTER-IoT, it is used to translate semantics of messages exchanged by IoT artifacts (platforms, gateways, applications, etc.) within the INTER-IoT software. It is composed of the IPSM Core and auxiliary components, i.e. Semantic Annotators, and exposes a REST API (for configuration). An additional Communication Infrastructure is required to enable communication between the IPSM and all other "artifacts" that are to use its semantic translation services.

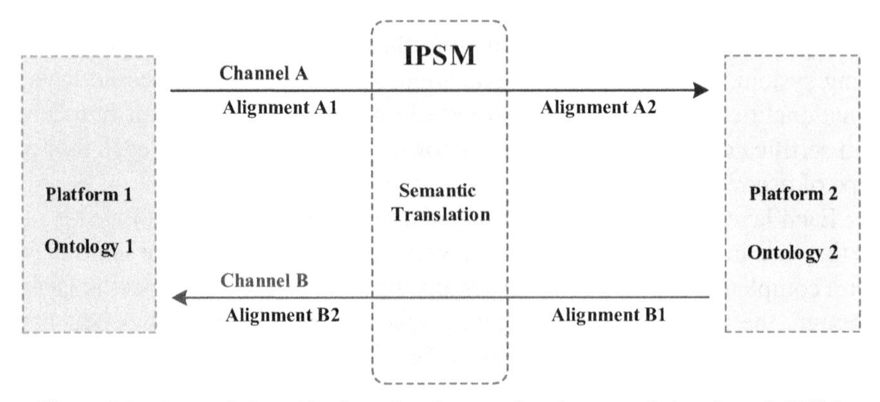

Figure 5.2 Semantic Inter-Platform Ontology-to-Ontology translation through IPSM.

The Semantic Annotators are located between the "outside world" and the IPSM. Their role is to produce RDF triples from data that they receive, e.g. from Bridges (component of MW2MW layer and the INTER-IoT middleware), and forward them to the IPSM Core through the Communication Channels, instantiated within the Communication Infrastructure. The IPSM Core performs the semantic translation of the RDF data, by applying pre-stored alignments (representing relationships between input and output ontologies). An instance of the IPSM can concurrently "service" multiple "conversations" taking place in separate Communication Channels. To achieve this goal, it can communicate with multiple instances of Semantic Annotators at the same time. Furthermore, each Alignment Applicator services a single Communication Channel and applies a separate alignment within the context of such channel.

Communication Channels work in publish-subscribe mode, which allows a single channel to serve both, one-to-one and one-to-many communication.

5.3.4 Methodology and Tools for Guiding the Implementation

A novel aspect of INTER-IoT is that provides a methodology to guide and ease the INTER-IoT framework implementation.

The INTER-METH methodology eases and offers guidance on the implementation of INTER-IoT in order to integrate different heterogeneous IoT platforms [10]. This makes it possible to achieve interoperability among the aforementioned IoT platforms and thus it enables to deploy fully functional IoT applications on top of them. There are currently no methodological approaches that might enable platform integration in a systematic and comprehensive way. It is a well-known truism that the utilization of an engineering methodology is of foremost importance at any domain (e.g. civil engineering, software engineering), maximizes, and ensures the effectiveness of the processes and actions to be performed. In sharp contrast with that, trying to manually apply complex techniques and methods in order to achieve platform integration would of necessity result in an unacceptably high rate of errors and bugs, which may instead be precluded via systematization and automation. The structure of the INTER-METH process can be seen in Figure 5.3. It is iterative in nature and comprises six successive stages: Analysis, Design, Implementation, Deployment, Testing and Maintenance. In principle, the output of each stage is the input of the following one. But in practice and depending on the particular circumstances being dealt with, is it possible to loop only specific steps of the process or else sets of successive ones, facilitating the adaptation to new components, and providing flexibility to this technique.

Additionally, INTER-IoT provides a set of tools, named INTER-CASE, that guide the implementation of the INTER-IoT framework, explaining the methodology for each specific implementation case. This set of programs offer step-to-step assessment and guidance in this process.

5.3.5 Middleware for the Interconnection of Platforms

This interoperability middleware has syntactic translators (bridges) that are able to convert the specific data format employed by an IoT platform to the INTER-IoT data format (JSON-LD), and vice versa. Thus, INTER-IoT middleware can provide syntactic interoperability among different IoT platforms. Platforms are therefore able to send or receive flows of information

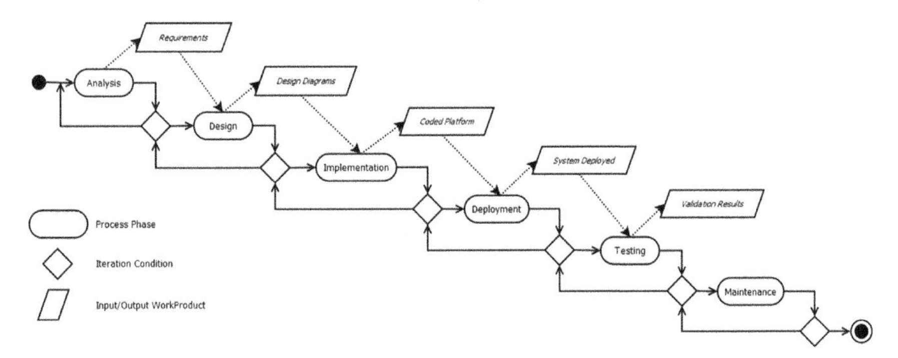

Figure 5.3 Process schema of INTER-METH.

in a data format understandable for them. INTER-IoT MW2MW represents a solution for interconnecting platforms at middleware level, and to enable interoperability among them [7].

In regard to the middleware structure (Figure 5.4), south from the Communication and Control block, the bridges manage the communication with the underlying platforms by translating requests and answers from and into messages for the queue. Different bridges might need to use HTTP, REST, sockets or other technologies to talk to the platforms, but these will be translated northwards into messages.They also pass the message content to the

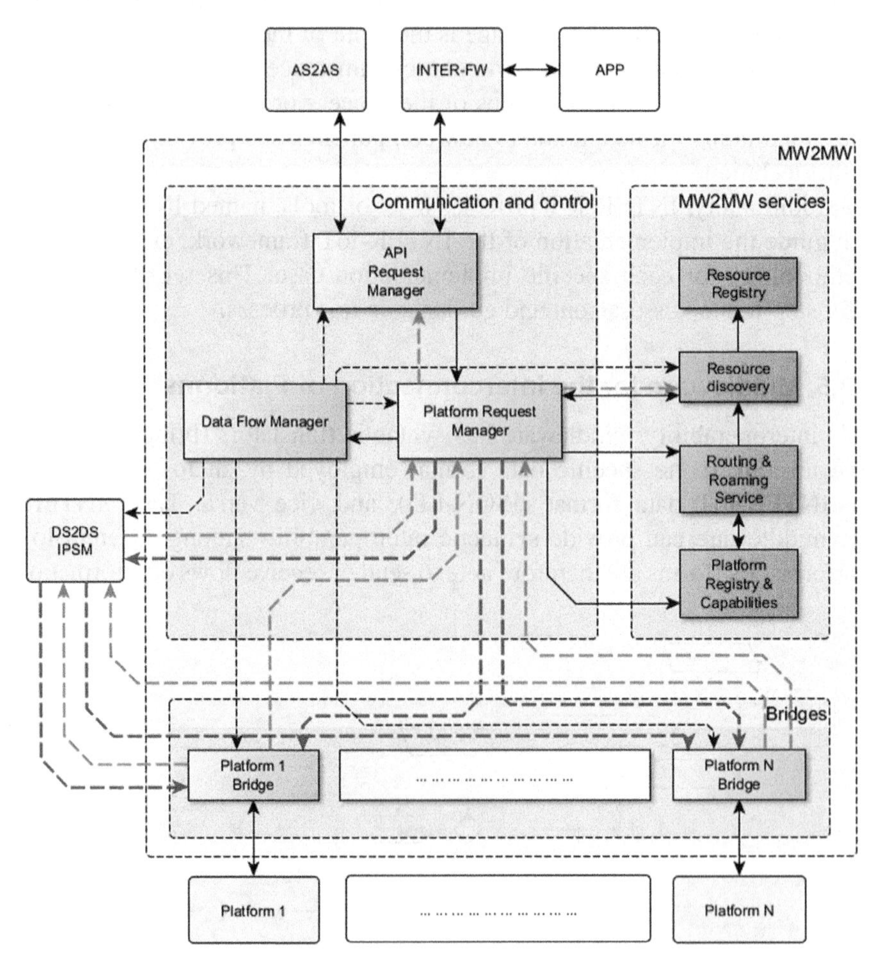

Figure 5.4 MW2MW structure.

IPSM, which is a service external to MW2MW that will allow for ontological and format translation between the platforms and a common language.

In the services group of components, the most important are the Platform Registry and Capabilities, that contains the information of all connected Platforms including their type and service capabilities, the Resource Discovery that creates requests to obtain the necessary information from the platforms, and the Resource Registry, that contains a list of resources (e.g. devices) and their properties that can be quickly consulted. In the second phase, the Routing and Roaming Service will be expected to allow the communication with a particular device independently of the platform it is currently connected to, while Authentication and Accountability (not shown) would provide services for the security and monitoring of all the actions.

5.3.6 Virtual Gateway

INTER-IoT provides a smart gateway that has the particularity that is partially virtual. This gateway provides IoT interoperability at the device level and has a modular design. Modularity in protocols and access networks is optimal. Any access network, protocol or middleware module can be inserted into the structure as long as its interface matches with the controller.

The device is build up in a way that once the system structure is functional a split-up can be realized. Part of the device gateway can be placed in the virtual world to allow device activity to higher level at all time. The device dispatcher will take care of connecting or simulating the actual platform. When connection is lost, the virtual part remains functional and will answer to requests of API and INTER-IoT middleware.

At the lowest level there are sensors and actuators. These are connected to the different input modules. These modules take care of connectivity with wireless smart objects.

This smart software gateway provides interoperability among very different network technologies and protocols. In addition to Wi-Fi and Bluetooth, INTER-IoT gateway supports network protocols and technologies specifically designed for IoT, such as CoAP, MQTT, LoRa and IQRF, as well as advanced techniques for offloading. Moreover, it is able to support the recent network protocol Multipath TCP [11], which is thought to be the successor of TCP in the Future Internet [12, 13], and it is massively used in smartphones due to its capability of bandwidth aggregation from different networks [13, 14] (e.g. such as 3G and Wi-Fi networks).

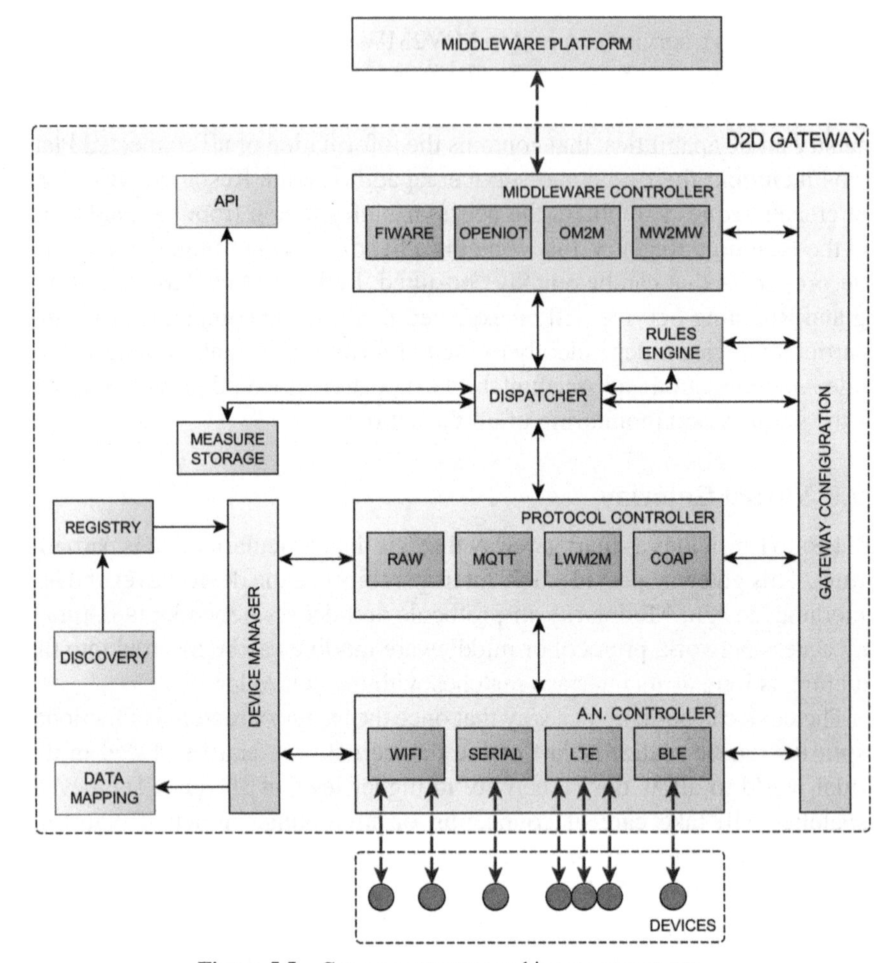

Figure 5.5 Gateway structure and inner components.

5.4 Inter-Iot Use Cases and Applications

INTER-IoT has two pilots INTER-LogP, as an interoperable solution in the seaport scenario, for port management, and INTER-Health, associated with the domain of e-Health.

5.4.1 e-Health (INTER-Health)

The INTER-IoT approach is case-driven, and it is implemented and tested in realistic large-scale pilots. One of its pilots, INTER-Health, is focused on the use case of INTER-IoT on e-Health [10], and it is tested on an Italian National Health Centre for m-health, involving 200 patients equipped with body sensor networks, wearable sensors and mobile smart devices for health monitoring.

This use case is based on the integration of two e-Health IoT platforms, and its goal is the development of an e-Health system through the integration of several IoT platforms and medical sensors. This system aims to monitor people's lifestyle in a decentralized and mobile manner for the prevention of health issues such obesity, caused by unappropriated diet and lack of physical activity [15]. These monitoring processes are meant to be decentralized from the healthcare centre to the monitored subjects' homes and supported in mobility by using on-body physical activity monitors. It is worth noting that, the strategic importance of such complete use case, is largely motivated by the fact that unhealthy lifestyles such as improper and hypercaloric diet and insufficient physical activity, are at the base of main chronic diseases [16, 17]. During the use case experimentation, the effectiveness of the novel system,

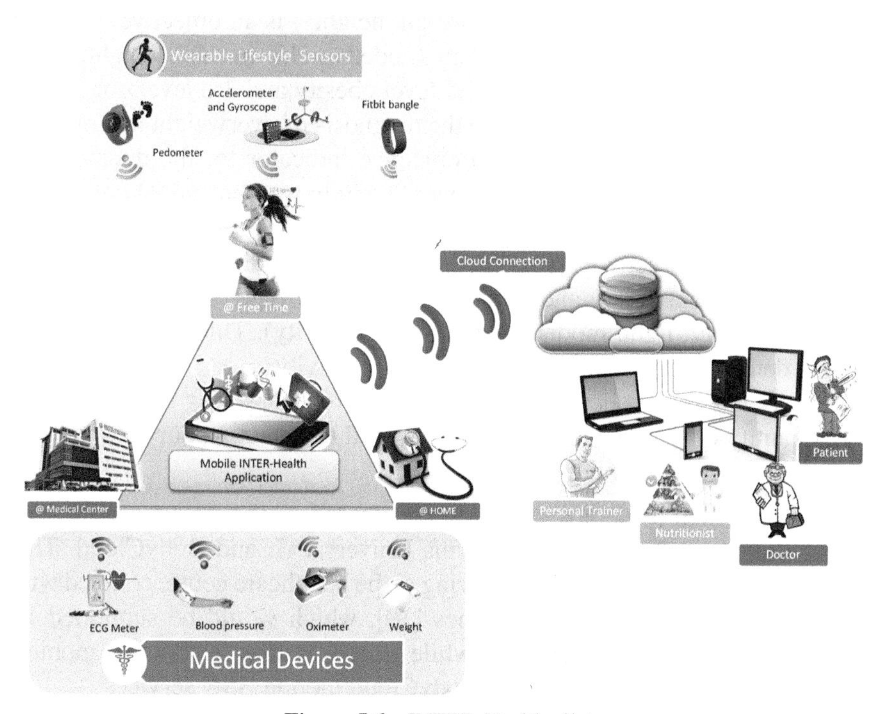

Figure 5.6 INTER-Health pilot.

in terms of lifestyle improvement indices, will be evaluated with respect to the current "manual" monitoring performed by conventional Healthcare Centres.

5.4.1.1 Lifestyle monitor: Medical perspective

There are a variety of indicators to measure and observe for preventing and/or detecting obesity, following the medical protocol given by the World Health Organisation (WHO) [16, 17]. Through these indicators, it is possible to determine the health status in terms of appropriate or inappropriate weight (levels vary from underweight, normal weight, overweight to obesity). These measurements can be collected in health centres by a healthcare worker (dietist or doctor). These include objective measurements (body mass index, blood pressure, weight, height, waist circumference) and subjective indices (eating habits and the practise of physical activity) [15]. For these reasons, the goal of this use case is to monitor a person's lifestyle in a decentralized manner with mobile sensors in order to prevent health issues. Specifically, the use case requires the monitoring of the following health indicators:

- The Body Mass Index (BMI) (weight/height^2) is an objective indicator of the health state of the patient (underweight, normal weight, overweight, first level obesity, second level obesity and 3rd level obesity) of subjects, also allowing to make the diagnosis of overweight and obesity.
- The waist circumference is an objective indicator for the diagnosis of overweight and obesity; values over 80 cm in women and 94 cm in men are considered pathological.
- The physical activity practice is a subjective indicator that detect the amount (hours/daily and hours/week) and the type of physical activity (no activity; light, moderate and intense activity). This measure is used to detect a poor lifestyle with physical inactivity.
- The eating habits is a subjective indicator for measuring the quality and quantity of the diet. This measure is used to detect a poor lifestyle due to improper diet and high-calories.

This use case will be deployed over the integrated system composed through the joint of the IoT e-Health platforms UniversAAL and BodyCloud. This will enable the computerized monitoring at the healthcare centre coupled with the monitoring at the patients' homes [19], which would be supported by the UniversAAL remote services, while BodyCloud will allow to monitor subjects' physical activity through BodyCloud mobile BSN services.

5.4.1.2 Platforms to integrate

BodyCloud

BodyCloud [18] is an IoT platform specifically addressed to the creation and management of Body Sensor Networks (BSNs). It has a Software-as-a-Service architecture, and it is capable of creating a smart gateway on smart phone devices that are able to receive and monitor health rates from medical wearable sensors. BodyCloud supports the management and storage of body sensor data streams and the offline and online analysis, of the stored data using software services hosted in the Cloud in order to enable large-scale data sharing and collaborations among users and applications in the Cloud and deliver Cloud services via sensor-rich mobile devices. BodyCloud endeavours to support a variety of specialized processing tasks and multi-domain applications and offers decision support services to take further actions based on the analysed BSN data.

The BodyCloud approach is based on four main components:

- The Body-side refers to an Android-based element for the monitoring of assisted living by means of smart wearable medical sensors, and the collection and upload of data to the Cloud through a smart phone that acts as a mobile gateway.
- The Cloud-side is a Software-as-a-Service element that provides Cloud services, such as storage.
- The Viewer-side refers to the Web browser-enabled component for the visualization of data.
- The Analyst side facilitates the analysis of data and the creation of BodyCloud applications.

UniversAAL

The IoT platform UniversAAL[1] (Universal Ambient Assisted Living) is specifically designed for the domain of Ambient Assisted Living and medical environments. UniversAAL is a platform that enables the creation of assistive systems by connecting different, heterogeneous technical devices to a single, unified network. UniversAAL also delivers the means to control this distributed system. Well-defined semantics is an important concern in medical and AAL environments, to lead to no ambiguity in measurements, units and terms employed [13]. In this regard, UniversAAL utilizes semantics

[1] http://www.universaal.info

in a very strict and well-defined manner, unlike many other IoT platforms, and employs W3C SSN ontology for IoT and smart devices.

Complementary Platforms

The aforementioned IoT platforms (i.e. BodyCloud and UniversAAL) have several high-level characteristics in common and differing aims and technology. Both are e-Health platforms that employ Bluetooth technology to interact with sensors. Moreover, both platforms employ Cloud data storage, cloud big data analysis and data visualization. Though, the two platforms have different specific objectives and are not interoperable from a technological point of view.

Their specific objectives are complementary: UniversAAL is focused on non-mobile remote monitoring based on non-wearable measurement devices, whereas BodyCloud provides monitoring of subjects in mobility through wearable devices organized as body sensor networks (BSN). Thus, their integration would produce a full-fledged m-Health platform atop of which multitudes of m-Health services could be developed and furnished.

5.4.1.3 INTER-IoT integration of health platforms

The integration of UniversAAL and BodyCloud is achieved through their interconnection through INTER-IoT, as can be seen on Figure 5.7. This integration is done across three layers (device, application and semantics). The middleware layer of the resulting integrated IoT system is based entirely on UniversAAL thus no interconnection is required across different platforms.

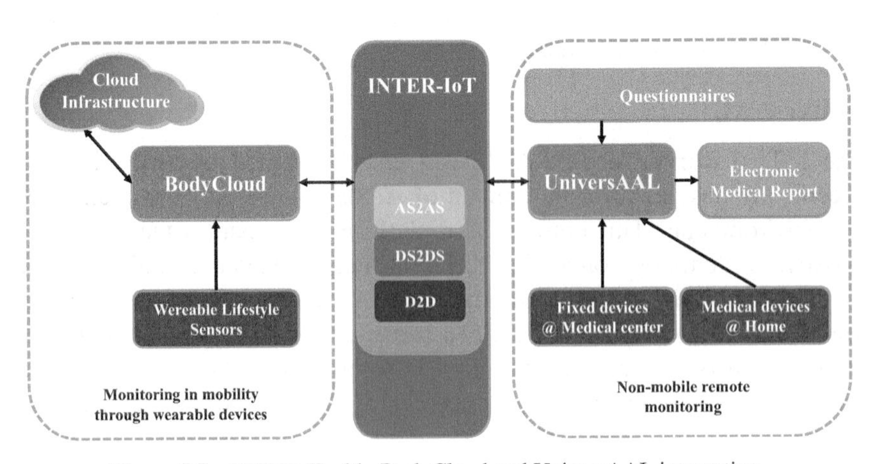

Figure 5.7 INTER-Health: BodyCloud and UniversAAL integration.

Therefore, INTER-IoT provides integration and transparent interconnection at the following levels:

- at device layer (D2D), enabling the new IoT system to communicate with the wireless medical devices supported by BodyCloud and with the fixed e-Health devices from the health centre or from the patients' houses.
- at the application and services layer (AS2AS), the applications for handling patients' reports the reports from UniversAAL are integrated in the overall systems in such a way that reports can be complemented with additional data from BodyCloud measurement applications.
- at the data and semantics layer (DS2DS), enabling semantic and syntactic interoperability among all platforms and systems.

The integration scheme of the aforementioned IoT platforms by means of INTER-IoT can be seen on Figure 5.10.

5.4.1.4 INTER-Health technical functionalities

The integrated IoT system has the following main functionalities:

- collection of objective (weight, height, body mass index, blood pressure or waist circumference) and subjective (questionnaires concerning the eating habits and the practice of physical activity) measures during the visits at the healthcare centre (based on UniversAAL);
- telemonitoring at the healthcare centre of subjective (questionnaires) and objective (weight, blood pressure, etc...) measures sent by the patients at home (based on UniversAAL platform);
- telemonitoring at the healthcare centre of the physical activities performed by patient at home with wearable devices (based on BodyCloud platform) report and visualization of all the measurements collected for analysis and interaction on treatments.

5.4.1.5 INTER-Health pilot

The main goal of the INTER-Health pilot is demonstrating how to foster a healthy lifestyle and how to prevent chronic diseases by monitoring subjects' physical characteristics, nutritional behaviour and activity [15, 19, 20].

The pilot consists of 200 test subjects: 100 subjects following traditional monitoring without IoT devices and 100 subjects with devices. The latter use the INTER-IoT solution. They attended a nutritional counseling session a medical and nutritional centre where their initial physical characteristics are measured, using IoT Devices on the premises (BMI, waist circumference, weight, blood pressure...). Each subject received a management program. Then at home, while they follow the program, they measure their characteristics using their phone and IoT devices.

The subjects will visit the medical and nutritional centre each 6 month for check-ups. The healthcare professional in charge of monitoring each user will have access to the history of all the measurements through a dedicated web application.

The assisted living environment created for INTER-Health enables the remote measurement of different physiological parameters by means of medical IoT devices such as weigh scale, blood pressure monitor and physical activity rate monitor [21, 22]. The aforementioned sensors interact with an IoT platform (BodyCloud or UniversAAL), and provide measurements through the connection with a smart gateway employing Bluetooth communication [23]. This smart gateway receives the measures from the devices and sends them to the platform via 2G/3G/4G/Wi-Fi/ADSL connectivity. The platform BodyCloud creates a smart gateway on a mobile phone, thus enables a smartphone to become an IoT gateway for the medical sensors. Doctors have access to a web medical application that allows them to follow up the

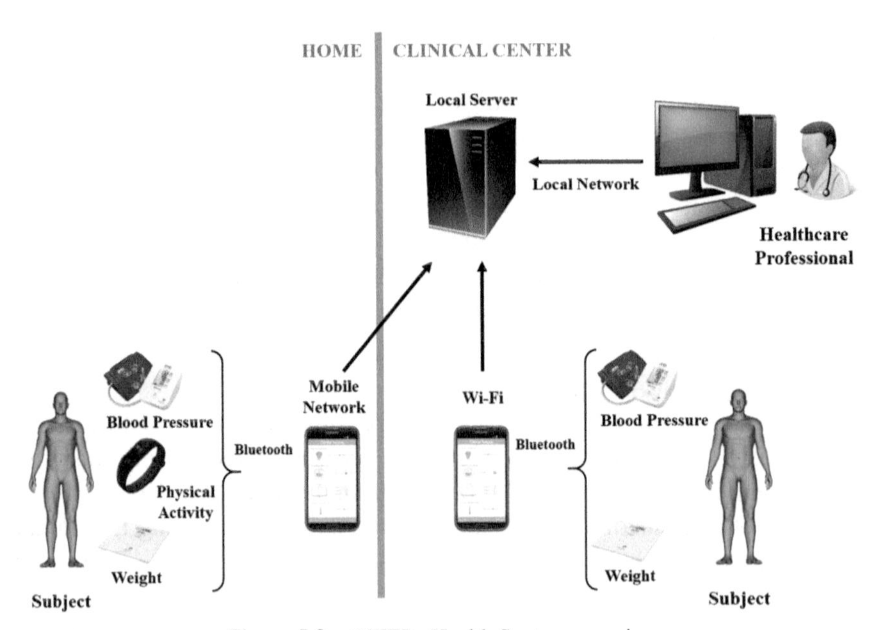

Figure 5.8 INTER- Health System overview.

monitoring of patients remotely at any moment, as well as to contact them via ITC communication tools (SMS, e-mail, telephone, and teleconference) and give medical assessment.

5.4.1.6 Benefits

INTER-IoT integration guarantees an effective and efficient interoperability between heterogeneous IoT platforms such as the two described e-Health IoT Platforms (i.e., UniversAAL and BodyCloud). The proposed interoperable approach will enable the development of new cross-platform services. Thus, the main benefit of the proposed approach consists of the availability of a more powerful IoT healthcare platform for lifestyle monitoring to implement new applications and services that the individual platforms could not support. Finally, the aforementioned monitoring process can be decentralized from the healthcare centre to the monitored subjects' homes, and supported in mobility by using on-body physical activity monitors connected to the novel fully integrated IoT environment. This approach, would reduce both the transfer costs of patients at medical centres and the waiting times, also obtaining constantly updated results to make the necessary adjustments in a faster and precise manner.

From a final user perspective, INTER-Health use case significantly benefits from INTER-IoT solutions:

- For outpatients subjects: improving the survey quality of their health status; improve the definition of risk behaviour; provide information on diets and physical activity more relevant with the health status and with the risks of the subject compared to the traditional methods; increasing the sensitivity of the screening of subjects who need intervention from the local doctor or of the hospitals (second and third level obesity, diabetes, etc); reduce the time spent in face-to-face contact with the nutritional outpatient and the number of travels.
- For public health services: increase efficiency with the same resources used; increase effectiveness through standardization of objective and subjective measurements; turning subjective ones, such as activity practice, into objective ones (by exploiting IoT wearable systems); enlarge the number and type of subjects that appeal to nutritional outpatient.
- For local doctors: lighten the taking charge of healthy subjects by the local doctor for guaranteeing greater availability toward pathological subjects; overall, the local doctor becomes a vehicle from a lower general incidence of healthcare costs on the income of citizens, improve

the care and diagnostics efficiency making directly available on the computer system of the local doctor, the data present on the platform used from the nutritional ambulatory.

5.4.2 Smart Transport & Logistics (INTER-LogP)

INTER-IoT has a pilot, called INTER-LogP, focused on a use case of Smart Transport & Logistics. INTER-IoT offers an interoperable solution in the seaport scenario for port management [7, 24]. The main objective of this pilot is to provide a service to control port access, monitor traffic and assist the operations at the port. Several systems will be able to identify trucks and drivers using different devices. This information can be shared under certain predefined rules through interoperability between the platforms involved; it can be used to monitor trucks inside the port by the Port Authority platform (due to security and safety purposes), and to manage more efficiently resources in the terminal. Moreover, this information is employed to avoid queues in the access gates to the port and the terminal.

The use IoT platforms in ports can potentially enable traffic and container monitoring, geolocation of cargo and vehicles, management of storage and cargo processes and improvement of services. These benefits can be multiplied through appropriate sharing of valuable information and cooperation among the different IoT platforms in port environments, creating synergies. This use case addresses the need of IoT platforms interoperation within port actors: such as container terminals, transport companies (road and maritime transportation), the port authority, and customers.

This pilot has been deployed in the port of Valencia, the most important port of the Mediterranean. The pilot is mainly composed by an Access Control System, and a Health Emergency System, which are possible fruit of the interoperability among platforms provided by INTER-IoT.

The platforms integrated through INTER-IoT belong to the main actors of the port: IoT platforms of the Port Authority, of one of the Port Container Terminal (NOATUM Valencia), and Intelligent Transportation Systems of several Road Haulier Companies. This interconnection is set at middleware level, employing the INTER-IoT MW2MW solution [7].

Important platforms and systems involved are:

- In the Container Terminal :
    - SEAMS : IoT platform for controlling container terminal machinery

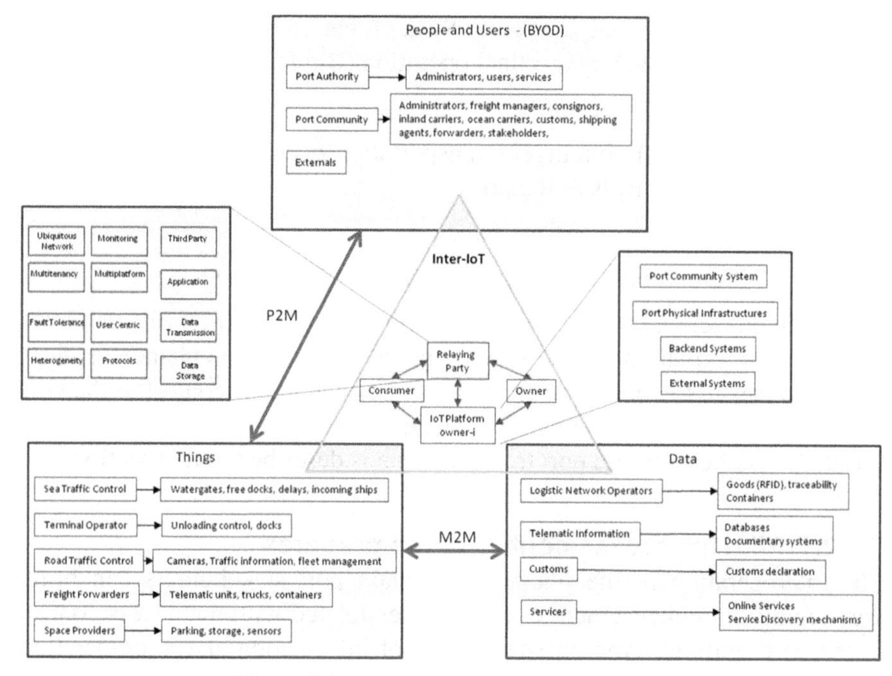

Figure 5.9 INTER-LogP use case approach.

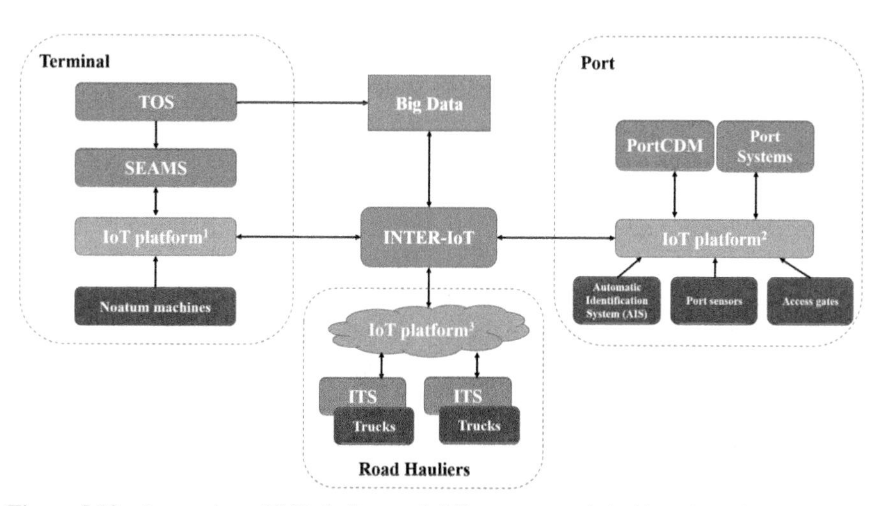

Figure 5.10 Integration of IoT platforms of different port stakeholders through INTER-IoT.

- TOS : Terminal Operating System that controls and handles data associated to any terminal operation (Big Data)

- From the Port of Valencia :
  - PORTCDM : Intelligent transportation system for the management of ships arrivals to the port
- From the Road Haulier Companies
  - Different Intelligent Transportation Systems (ITS)

This interconnection provides interoperability at middleware level, and flows of relevant information can be shared among platforms to enhance services and processes in the port. The enhancement of the access control to the port facilities is explained in the next subsection. Also, a new service combining e-Health emergencies and port transportation is described in this section.

5.4.2.1 Pilot for access control at the port area

The interoperation of the platforms of main port stakeholders can bring very significant enhancement to the services related with the access control to the port facilities. Appropriate sharing of information and interoperation among them lead to a more efficient access control in terms of time and cost efficiency, security, safety, minor waiting times and improved management. Platforms integrated through INTER-IoT are from diverse entities from the port environment: the port, a container terminal (NOATUM) and several road haulier transportation companies. Relevant information shared among platforms are the location of trucks inside the port, information regarding load and unload operations, and access controls.

The main benefits from these services are the collection and analysis of data regarding queues, congestion and temporary distribution of traffic, and to manage efficiently the resources. Relevant information obtained is the position of the trucks inside the port facilities, and its use it is important in the sake of safety and security. All these data can be shared between the port authority and the port terminals to improve operations.

5.4.2.2 Pilot for health accident at the port area

Starting from the access control pilot, trucks will be monitored once they enter in the port facilities. The Emergency Warning System (EWS) will be monitoring the data coming from the truck and the driver. In case it detects an accident or a medical problem, EWS will publish a notification to the port

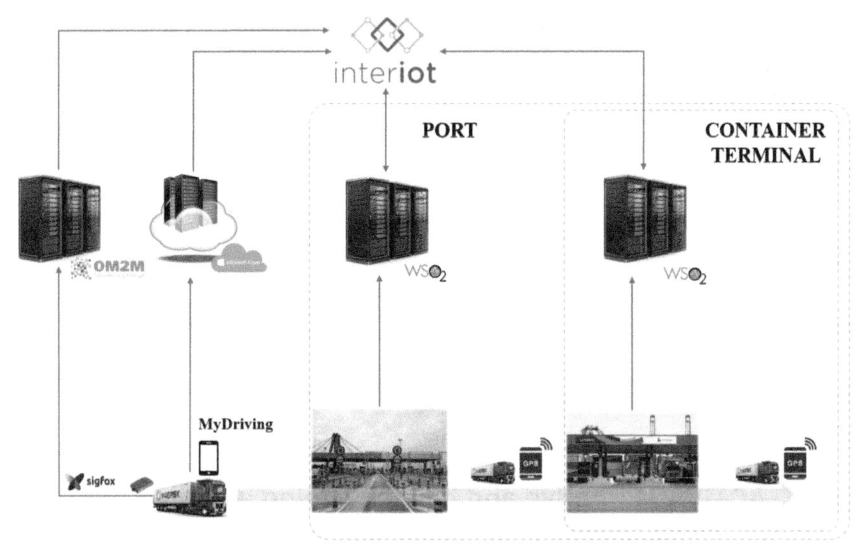

Figure 5.11 High-level view of the access control pilot.

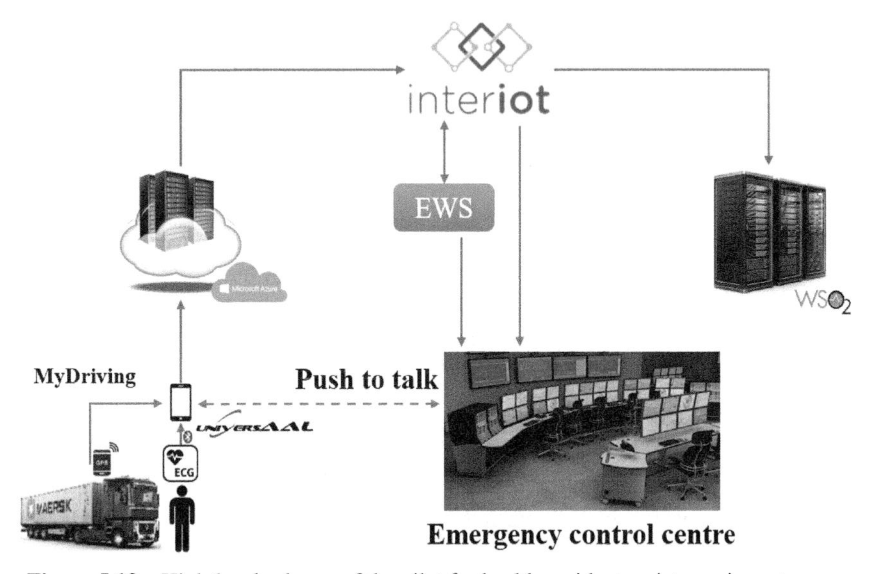

Figure 5.12 High-level scheme of the pilot for health accident assistance in port areas.

authority in a standard format (EDXL). Once the emergence control centre receives the notification, it will be possible to communicate with the driver through a push to talk protocol in the driver's mobile.

The main benefits we can get from this scenario are: apply in the port communications a standard format in accident reporting like EDXL, real-time identification of the location of the accident, direct communication with the closest control centre when an accident occurs and monitoring driver's health if it is necessary.

5.4.3 Active and Healhy Ageing (ACTIVAGE)

ACTIVAGE [27] is a H2020 LSP project that addresses the use of IoT technologies in the Active & Healthy Ageing (AHA) domain [25]. INTER-IoT framework is a core element in the ACTIVAGE system that enables interoperability at different levels and fulfills the ACTIVAGE's needs of interoperability.

5.4.3.1 ACTIVAGE: Active and healthy ageing initiative

ACTIVAGE project addresses the use of IoT technologies in the Active & Healthy Ageing (AHA) domain [25]. The objective of this project is to prolong and support the independent living of older adults in their cities and homes and responding to real needs of caregivers, service providers and public authorities. Hence, this project aims to improve the autonomy and quality of life of older adults and contribute to the sustainability of the health and care systems. The ACTIVAGE project has been designed as a Multi Centric Large-Scale Pilot consisting of nine interconnected Deployment Sites (DS) distributed over seven European countries. A DS can be defined as a cluster of stakeholders in the AHA value network, working together within a geographical space. Therefore, a DS includes users (elderly people, formal and informal caregivers), service providers, AHA services, health care/social care administration and technological infrastructures and technology providers. The DS make use of existing open and proprietary IoT platforms. Each DS utilizes a specific IoT platform, or two. The different IoT platforms employed in ACTIVAGE DS are FIWARE, SOFIA2, UniversAAL, SensiNact, OpenIoT, IoTivity and SENIORSOME.

The following DS have been defined in the ACTIVAGE project:

- DS1: Galicia (Spain) will make use of the SOFIA2 platform.
- DS2: Valencia (Spain) will provide services based on a combination of data from UniversAAL and Fiware platforms.
- DS3: Madrid (Spain) will deploy services based on UniversAAL.
- DS4: Region Emilia Romagna (Italy) will make use of the Fiware platform.

- DS5: Greece will offer services based on the IoTivity platform
- DS6: Isére (France) will provide services based on the SensiNact platform.
- DS7: Woqaz (Germany) will develop services based on the UniversAAL platform.
- DS8: Leeds (UK) will deploy services based on the IoTivity platform.
- DS9: Finland will make use of the proprietary IoT platform SENIOR-SOME.

Due to the lack of interoperability among IoT platforms, the definition of an interoperability framework is needed in order to create a European AHA ecosystem.

With this aim, ACTIVAGE will develop the ACTIVAGE IoT Ecosystem Suite (AIoTES), which is defined as a set of tools, techniques and methodology for interoperability between existing IoT platforms. The AIoTES Framework will provide interoperability among IoT platforms and ensure security and privacy. The different DS will connect to AIoTES and AHA applications will be deployed over this framework, thus allowing the integration of remote health-care services and wearable systems-based health-care

Figure 5.13 AHA Interoperable DS (Smart Home Clusters).

services in mobility, which will include remote medical measurements, local mobile physical detection and processing, and on-line and off-line analysis of lifestyle data.

ACTIVAGE aims to achieve interoperability at three stages:

- **Intra-deployment site interoperability,** which means that the services provided at each DS must be interoperable to each other. In order to achieve this, any the application should be able to access all the application data within the same DS. Moreover, the applications within a DS should support multiple IoT platforms and be able to be transferred between different platforms within s DS.
- **Inter-deployment site interoperability**: enables new services to be automatically incorporated into the ecosystem of the DS. This means that different DS should be able to exchange application data. Moreover, it should be possible to transfer an application that was designed for a DS to a different DS and new applications could be developed for multiple DS instead of being designed for a particular DS.
- **Interoperable external adopted solutions**: according to the needs of each specific DS, new solutions will be implemented within the DS. They will be interoperable according to the ACTIVAGE interoperability framework.

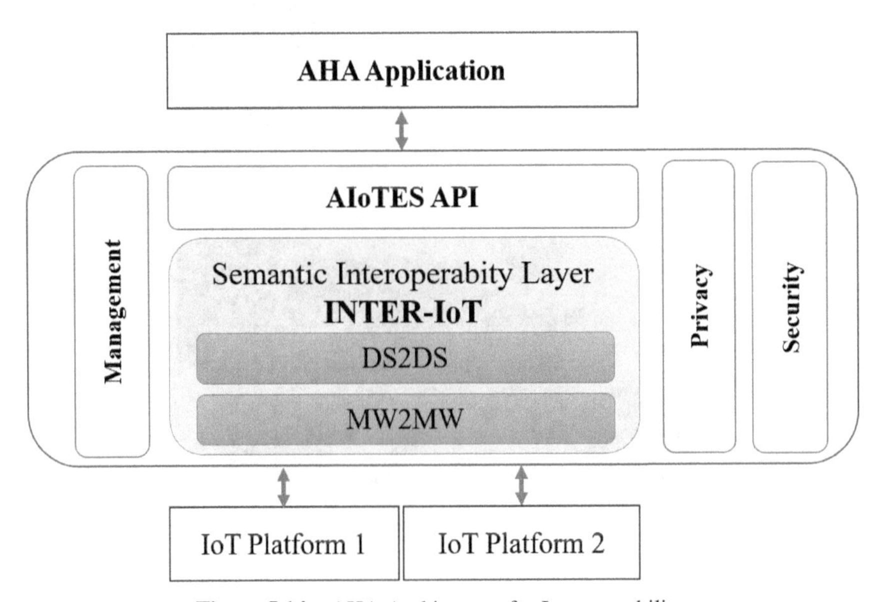

Figure 5.14 AHA Architecture for Interoperability.

These goals imply that the same data format must be used by the applications regardless of the IoT platforms being used in the DSs so as the applications can access any platform's data. Moreover, applications initially development for a DS can be extended for any other DS only by adapting them to AIoTES instead of making an adaptation for each individual platform. Overall, multiple AHA applications and IoT platforms may coexist in the same DS, thus contributing to fulfil the main goal of a DS. Therefore, it is required inter-platform and intra-platform interoperability among the different IoT platforms of the DS of this AHA intiative (FIWARE, UniversAAL, SOFIA2, OpenIoT, IoTivity and Seniorsome). Furthermore, this interoperability must be both syntactic and semantic, to allow the understandability of the information across platforms and a common interpretation of data shared among them. In this regard, INTER-IoT is the key component that makes it possible for the ACTIVAGE deployment to enable and ensure platform interoperability in DSs.

The required interoperability among IoT platforms is provided by the Semantic Interoperability Layer (SIL), which is a component of the AIoTES framework. INTER-IoT is the key component that provides inter-platform interoperability in this AHA deployment. Two components of INTER-IoT, namely, the MW2MW layer and the DS2DS layer (composed by the IPSM), have been incorporated in the SIL. The MW2MW layer connects to all the IoT platforms and provides a common abstraction layer to provide access to platform's features and information. An important function of the MW2MW is to convert the data to a common syntax, which is based on JSON-LD. Once the data is in the common format, the IPSM performs semantic translations. As a result, AIoTES provides semantic and syntactic interoperability among the different platforms, and enables information sharing and interoperation among them. This data shared will be understandable for the receiver platform, not only in terms of data format but also regarding the meaning of the received information.

In addition to these INTER-IoT components, the AIoTES framework includes an API, security and privacy protection components management functions. Security and privacy span across all the components of AIoTES in order to ensure the protection of sensitive data against unauthorized access. AIoTES management provides a set of tools that allow access to the information of a DS, such as platforms and devices, and mechanisms to facilitate the integration of the framework. Finally, AIoTES will provide a common API, which will allow a homogeneous access to the components of AIoTES in order to develop services and applications able to exchange data from different IoT platforms and produce new added value services. Hence, the AIoTES API will make possible the development of and ecosystem based on applications and services compatible with AIoTES.

5.4.3.2 Use Cases of ACTIVAGE

AAL and AHA systems can very significantly benefit from interoperability [31, 32]. The following Reference Use Cases respond to specific user needs (senior people and caregivers), to improve their quality of life and autonomy, in a AHA context, that require IoT interoperability:

Daily activity monitoring at home for informal caregivers support and for formal caregivers follow up in order to alert them about deviations of the elderly persons' habits, allowing early interventions while extending independency. Wireless sensors like presence, magnetic contact, power measurement, proximity, are deployed at the home of the elderly. A gateway transmits the information to a Cloud where calculation on activities, trends and risks is performed.

Integrated care for older adults under chronic conditions. This use case combines daily activity monitoring at home and the use of medical devices for health monitoring. The combination of IoT technologies with eHealth solutions in one single integrated IT system, and the integration of care protocols from entities traditionally working separately will promote the coordination among care providers, joint response to emergencies, better planning of resources and more effective interventions. This will lead to economic savings and a better quality of life for people with chronic disease.

Monitoring assisted persons outside home and controlling risky situations. This use case combines wearable devices or smartphones and the Smart City infrastructure in order to promote socialization and activity. The Smart City infrastructure tracks the wearable devices and request for help if certain rules are met in order to help persons at risk.

Emergency trigger. The system automatically reports an emergency when a critical situation is detected. Wireless or wired sensors and "panic" buttons distributed in the home environment in strategic situations linked to a gateway that forwards the emergency to a call-centre system. Other complex scenarios might involve the processing of data in the private or hybrid cloud and then the emergency is triggered. Compared to state-of-the-art systems at home, emergency works when the user requests for help, but also when the environment detects the emergency and the person cannot (unconsciousness, fall, gas).

Exercise promotion for fall prevention and physical activeness using wearable and ambient sensors.

Cognitive stimulation for mental decline prevention in order to extend the time elderly people live independently. This use case combines behavioural monitoring at home and outside, and interventions, such as the promotion of mental and physical exercises and gaming, making use of apps in tablets or smartphones and peripheral connected devices.

Prevention of social isolation by means of communication tools at home. This use case promotes social interaction and mobility though the use of video-based system and apps connected to the Smart City infrastructure, which provides data about events, and linking to other peers. In addition, continuity between home (home sensors) and outdoors scenarios (smart phone as a sensor) provides seamless information about users' social activity. Social engagement keeps depression and decline away.

Comfort and safety at home. This use case includes climate and light control, perimeter safety, energy control and home automation.

Support for transportation and mobility. This use case includes adapted route planning for elderly persons both in cities and between different cities. Routes can be computed making use of the Smart City data about traffic conditions and other mobility aspects and personalized according to goals such as exercise promotions or finding the easiest/fastest route.

5.4.4 Other Potential Use Cases

INTER-IoT can be employed in any domain or across domains where there is a need of IoT interoperability. Thus, its use is not limited to the aforementioned use cases and can be utilized in the most various IoT environments, allowing very different aims that are enabled or partially enabled through interoperability.

A clear example is the Smart Cities use case, which greatly benefits from the synergies and cooperation among different systems and platforms that provide different city services. In this case, there is an enormous need of interconnection that is limited by the typical interoperability problems in the IoT realm. The application of the INTER-IoT framework is able to solve the integration of heterogeneous platforms and systems within a Smart City, provide numerous benefits to the citizens and enable the creation of new useful services fruit of this interoperability.

5.5 Conclusions and Outlook

In this chapter, it has been described the current problem of lack of interoperability in the heterogeneous Internet of Things realm, and the usefulness of INTER-IoT for solving this important problem and enabling the integration and interoperation of heterogeneous IoT platforms at all layers and across multiple domains.

The effective application of INTER-IoT for solving the lack of interoperability among platforms has been explained and demonstrated in several use cases associated to different application domains. First, the usefulness of INTER-IoT has been analysed in a e-Health and AAL use case, in which the interoperability framework is implemented. In this regard, INTER-IoT enables the integration and interoperability of IoT platforms and provides a more powerful solution that the individual solution provided by each one of those platforms. These advantages are a consequence of the enablement of synergies and the sum of capabilities of all the integrated platforms.

Second, INTER-IoT has been a key integrator element in an interoperable solution for efficient port management. This use case is focused on the domain of Transportation and Logistics. INTER-IoT enables the interconnection of several platforms at middleware level, and the syntactic and semantic interoperability of any information shared among them, despite of the different data formats, standards, message structure and semantics. Because of this interconnection of platforms and sharing of relevant data among key entities, several management processes in the port can be very significantly improved. Also, the interoperability provided by INTER-IoT demonstrates that enables the existence of new services, fruit of the new information sharing and the possibilities of cooperation among platforms.

Third, INTER-IoT interoperability framework is employed in an AHA and AAL use case for enabling an assisted living environment in elder homes, to allow ancient people to live at home in a safe and autonomous way. INTER-IoT allows different IoT platforms to interoperate with the ACTIVAGE system around Europe, to enable this autonomous life of elderly people.

Finally, other potential use cases are mentioned, as far as INTER-IoT framework can be successfully employed in any domain and use case that has a need of IoT interoperability at any level (e.g. Smart Cities).

Regarding implementation aspects, INTER-IoT employs several innovative elements to provide enhanced functionality and has clear positive differentiators from other interoperability approaches. First, INTER-IoT has a layered approach to guarantee tight interoperability on each of the different layers (device, network, middleware, application, data and semantics), compared to a more global approach. Also, due to this multilayer approach, any of the INTER-IoT layer solutions can be employed in a standalone way, providing more flexibility and adaptation to specific IoT cases. Additionally, to guarantee a quick and easy implementation, INTER-IoT gives the option of running virtualized interoperability solutions for each layer through Docker. This virtualization enormously facilitates the deployment of the INTER-IoT solutions. Moreover, INTER-IoT has a huge concern on security, and layer solutions and APIs are securitized.

The INTER-IoT interoperability framework provides innovative elements, such as a universal semantic platform-to-platforms translator, a middleware that enables the interconnection and interoperation of any platform at middleware level, despite of the standards and formats employed, and a partially virtualized gateway. Furthermore, INTER-IoT implementation is guided and eased through a novel methodology (INTER-METH) specifically designed with this aim.

Interoperability in IoT, and more specifically among platforms, represents one of the most important challenges in IoT, and interoperability solutions such as INTER-IoT can potentially unlock immense benefits from the use of smart technology, and a huge integrator and enabler of services on top of IoT deployments. INTER-IoT can be used in the middle future to enable interoperability solutions among the most diverse use cases and domains in which IoT interoperability is required, solving modern society problems to let technology improve people's daily life, and propel European economy. Also, INTER-IoT facilitates a key element for the evolution of IoT; interoperability is essential for the creation of natural human interfaces in IoT systems, the existence of Ambient Intelligent Environments or the integration of IoT with Artifical Intelligence.

Acknowledgements

This work has received funding from the European Union's "Horizon 2020" research and innovation programme as part of the "Interoperability of Heterogeneous IoT Platforms" (INTER-IoT) grant agreement N°687283 and as a part of "Activating Innovative Iot Smart Living Environments For Ageing Well" (ACTIVAGE) grant agreement N°732679.

List of Notations and Abbreviations:

Notations	Abbreviations
API	Application Programming Interface
IoT	Internet of Things
BSN	Body Sensor Network
AAL	Ambient Assisted Living
SaaS	Software as a Service
ITS	Intelligent Transportation System
EWS	Emergency Warning System
AHA	Active and Healthy Ageing
LSP	Large Scale Pilot
DS	Deployment Site

References

[1] INTER-IoT European project, Research and Innovation action – Horizon 2020, online at: http://www.inter-iot-project.eu/

[2] Ganzha, M., Paprzycki, M., Pawłowski, W., Szmeja, P., and Wasielewska, K. (2017). Semantic interoperability in the Internet of Things: An overview from the INTER-IoT perspective. Journal of Network and Computer Applications, 81, 111–124.

[3] Manyika, J., Chui, M., Bisson, P., Woetzel, J., Dobbs, R., Bughin, J., and Aharon, D. (2015). Unlocking the Potential of the Internet of Things. McKinsey Global Institute.

[4] S. Kubler, K. Framling et al. "IoT Platforms Initiative", In Digitising the Industry Internet of Things Connecting the Physical, Digital and Virtual Worlds (Eds. Ovidiu Vermessan and Peter Freiss), Rivers Publishers Series in Communications, vol. 49, pp. 265–292, 2016.

[5] Savaglio, C., Fortino, G., and Zhou, M. (2016, December). Towards interoperable, cognitive and autonomic IoT systems: an agent-based approach. In Internet of Things (WF-IoT), 2016 IEEE 3rd World Forum on (pp. 58–63). IEEE.

[6] Soursos, S., Žarko, I. P., Zwickl, P., Gojmerac, I., Bianchi, G., and Carrozzo, G. (2016, June). Towards the cross-domain interoperability

of IoT platforms. In Networks and Communications (EuCNC), 2016 European Conference on (pp. 398–402). IEEE.

[7] Yacchirema, D., Gonzalez-Usach., R, Esteve, M., and Palau, C. (2018, April) IoT interoperability applied to the domain of port and logistics, Transport Research Arena Conference 2018.

[8] Ganzha, M., Paprzycki, M., Pawłowski, W., Szmeja, P., and Wasielewska, K. (2016, April). Semantic technologies for the IoT-an Inter-IoT perspective. In Internet-of-Things Design and Implementation (IoTDI), 2016 IEEE First International Conference on (pp. 271–276). IEEE.

[9] Ganzha, M., Paprzycki, M., Pawłowski, W., Szmeja, P., and Wasielewska, K. (2017, September). Alignment-based semantic translation of geospatial data. In Advances in Computing, Communication and Automation (ICACCA)(Fall), 2017 3rd International Conference on (pp. 1–8). IEEE.

[10] Pace, P., Aloi, G., Gravina, R., Fortino, G., Larini, G., and Gulino, M. (2016, December). Towards interoperability of IoT-based health care platforms: the INTER-health use case. In Proceedings of the 11th EAI International Conference on Body Area Networks (pp. 12–18). ICST (Institute for Computer Sciences, Social-Informatics and Telecommunications Engineering).

[11] Gonzalez-Usach, R., Pradilla, J., Esteve, M., and Palau, C. E. (2016, April). Hybrid delay-based congestion control for multipath tcp. In Electrotechnical Conference (MELECON), 2016 18th Mediterranean (pp. 1–6). IEEE.

[12] Gonzalez-Usach, R., and Kühlewind, M. (2012, August). Implementation and evaluation of coupled congestion control for multipath TCP. In Meeting of the European Network of Universities and Companies in Information and Communication Engineering (pp. 173–182). Springer, Berlin, Heidelberg.

[13] GONZALEZ-USACH, R. (2014). Design and Evaluation of a Delay-Based Algorithm for Multipath TCP (Doctoral dissertation).

[14] Gonzalez-Usach, R., Rene, O., (2018, April) Wi-Fi thermograph for remote cold chain monitoring with Multipath TCP support. In Transport Research Arena 2018.

[15] S.M.R. Islam, D. Kwak, M.H. Kabir, M. Hossain, K. Kwak “The Internet of Things for Health Care: A Comprehensive Survey”, IEEE Access, Vol. 3, pp. 678–708, 2015.

[16] World Health Organization, "Global status report on non-communicable diseases", 2010. http://www.who.int/nmh/publications/ncdreport2010/en/
[17] World Health Organization, "Obesity: Preventing and Managing the Global Epidemic", WHO Obesity Technical Report Series 894, 2000. http://www.who.int/nutrition/publications/obesity/WHO TRS 894/en/
[18] BodyCloud: A Cloud-assisted Software Platform for Pervasive and Continuous Monitoring of Assisted Livings using Wearable and Mobile Devices, http://bodycloud.dimes.unical.it
[19] C. Fernndez-Llatas, A. Martinez-Romero, A.M. Bianchi, J. Henriques, P. Carvalho, V. Traver "Challenges in personalized systems for Personal Health Care" IEEE-EMBS International Conference on Biomedical and Health Informatics (BHI), IEEE, pp. 356–359, 2016.
[20] S.C. Mukhopadhyay, "Wearable Sensors for Human Activity Monitoring: A Review", IEEE Sensors Journal, Vol. 15, Issue 3, pp. 1321–1330, 2015.
[21] R. Gravina, C. Ma, P. Pace, G. Aloi, W. Russo, W. Li, G. Fortino "Cloudbased Activity-aaService cyberphysical framework for human activity monitoring in mobility", Future Generation Computer Systems, 2016.
[22] G. Fortino, R. Gravina, W. Li, C. Ma "Using Cloud-assisted Body Area Networks to Track People Physical Activity in Mobility", International Conference on Body Area Networks (BodyNets 2015), pp. 85–91, 2015.
[23] Aloi, G., Caliciuri, G., Fortino, G., Gravina, R., Pace, P., Russo, W., and Savaglio, C. (2017). Enabling IoT interoperability through opportunistic smartphone-based mobile gateways. Journal of Network and Computer Applications, 81, 74–84.
[24] Gonzalez-Usach, R., Sarabia, D., Esteve, M., and Palau, C. (2018, April). Smart Interoperable Dynamic Lighting, Transport Research Arena Conference 2018.
[25] Bousquet, J., Kuh, D., Bewick, M., Standberg, T., Farrell, J., Pengelly, R., and Camuzat, T. (2015). Operational definition of active and healthy ageing (AHA): a conceptual framework. The journal of nutrition, health and aging, 19(9), 955–960.
[26] INTER-IoT Ontology, online at: http://docs.inter-iot.eu/ontology
[27] ACTIVAGE, online at: http://www.activageproject.eu/
[28] UNIVERSAAL, online at: http://www.universaal.info
[29] Open IoT Platforms: iCore-Butler demo, online at: http://open-platforms.eu/app_deployment/butler-icore-integrated-common-demo/

[30] Docker, online at: www.docker.com
[31] Gonzalez-Usach, R., Collado, V., Esteve, M., and Palau, C. E. (2017, May). AAL open source system for the monitoring and intelligent control of nursing homes. In Networking, Sensing and Control (ICNSC), 2017 IEEE 14th International Conference on (pp. 84–89). IEEE.
[32] Gonzalez-Usach, R., Yacchirema, D., Collado, V., and Palau, C. E. (2017, Nov). AmI open source system for the intelligent control for residences for the elderly. In InterIoT 2017 Conference.

6

Smart Data and the Industrial Internet of Things

Christian Beecks, Hassan Rasheed, Alexander Grass, Shreekantha Devasya, Marc Jentsch, José Ángel Carvajal Soto, Farshid Tavakolizadeh, Anja Linnemann and Markus Eisenhauer

Fraunhofer Institute for Applied Information Technology FIT,
Schloss Birlinghoven, Sankt Augustin, Germany
E-mail: Christian.Beecks@fit.fraunhofer.de

Abstract

Many modern production processes are nowadays equipped with cyber-physical systems in order to capture, manage, and process large amounts of sensor data including information about machines, processes, and products. The proliferation of cyber-physical systems (CPS) and the advancement of Internet of Things (IoT) technologies have led to an explosive digitization of the industrial sector. Driven by the high-tech strategy of the federal government in Germany, many manufacturers across all industry segments are accelerating the adoption of cyber-physical system and IoT technologies to gain actionable insight into their industrial production processes and finally improve their processes by means of data-driven methodology. In this work, we aim to give insights into our recent research regarding the domains of Smart Data and Industrial Internet of Things (IIoT). To this end, we are focusing on the EU projects MONSOON and COMPOSITION as examples for the Public-Private Partnership (PPP) initiatives Factories of the Future (FoF) and Sustainable Process Industry (SPIRE) and show how to approach data analytics via scalable and agile analytic platforms. Along these analytic platforms, we provide an overview of our recent Smart Data activities and exemplify data-driven analysis of industrial production processes from the process and manufacturing industries.

6.1 Introduction

Many modern production processes are nowadays equipped with cyber-physical systems in order to capture, manage, and process large amounts of sensor data. These sensor data include information about machines, processes, and products and are encountered in form of data streams. These data streams from the production site are then frequently integrated into cloud-based solutions by means of Internet of Things technologies in order to allow comprehensive data-driven investigations and process optimizations.

The proliferation of cyber-physical systems and the advancement of IoT technologies have led to an explosive digitization of the industrial sector. Driven by the high-tech strategy of the federal government in Germany, many manufacturers across all industry segments are accelerating the adoption of cyber-physical systems and Internet of Things technologies in order to gain actionable insight into industrial production processes and finally improve these processes by means of data-driven methodology.

The IoT is one of the key enabler for intelligent manufacturing and production. It facilitates the intelligent connectivity of smart embedded devices in factories and shop floors. Endowing the manufacturing and production site with technologies from the IoT, which is then also referred to as the IIoT, has become a technical prerequisite for a sustainable and competitive industrial production of the future.

Digitizing the industrial sector with cyber-physical systems, Internet of Things technologies, cloud computing services, and Smart Data analytics leads to the fourth industrial revolution, which is denoted as Industry 4.0. The importance of strengthen the European industry to become more sustainable and competitive is also taken into account by the European Commission. Within the EU Framework Programme for Research and Innovation the two Public-Private Partnership (PPP) initiatives Factories of the Future (FoF) and Sustainable Process Industry (SPIRE) aim to (i) help EU manufacturing enterprises to adapt to global competitive pressures by developing the necessary key enabling technologies across a broad range of sectors and (ii) support EU process industry in the development of novel technologies for improved resource and energy efficiency.

Turning industrial Big Data into structured and useable knowledge is one of the major data-centric challenges for enhancing production processes. Integrating data from heterogeneous systems and gaining insight into voluminous amounts of streaming sensor data with high variety and velocity requires scalable methods and techniques. Structuring knowledge in a way that it can be used to manage and improve industrial production processes is one of the objectives of Smart Data analytics. By improving Big Data to a higher degree of quality, Smart Data analytics aims to understand the following aspects:

- Purpose: What problem to solve with the data?
- People: Who is involved?
- Processes: What are the surrounding processes?
- Platform: Which IT infrastructure is necessary for realization?

The aforementioned aspects are also referred to as the 4Ps of Smart Data. They indicate the information to be gathered in addition to the sensor data from the production site in order to get a more complete understanding about the data and its surrounding entities. It is obvious that addressing the 4Ps within the Smart Data analytics process strongly relies on user-centered methods since many of the required information need to be discovered from non-documented data.

The Fraunhofer Institute for Applied Information Technology FIT has been conducting research and development on user-friendly smart solutions that blend seamlessly in business processes for about 30 years and has a strong experience in digitization, Industry 4.0 projects and IoT solutions. Having about 160 researchers with different scientific background, the Fraunhofer Institute for Applied Information Technology FIT is organized into five research departments:

- The User-Centered Computing department develops IT systems and technologies that focus on their users throughout their complete life cycle. Current work focuses on usability engineering, web compliance, and accessibility.
- The Cooperation Systems department develops and evaluates groupware and community systems for virtual teams and organizations. Our work on hardware and software of Mixed and Augmented Reality systems focuses on support for cooperative planning tasks.
- The Life Science Informatics department designs and implements complex biomedical information systems and creates novel software solutions for manufacturers and users in health care, biotechnology, drug research and social services. Focal areas are image-based navigation systems, information-intensive optical instruments, visual information analysis, multi-parametric molecular sensor technology and diagnostics as well as bio-analogue analysis of changing images.
- The Risk Management and Decision Support department offers decision and process support for application domains whose processes can be

characterized by their high level of complexity as well as their weak determination of process structures.

- The Fraunhofer Project Group Business & Information Systems Engineering, located in Augsburg and Bayreuth, has proven expertise at the interface of Financial Management, Information Management and Business & Information Systems Engineering. The ability to combine methodological know-how at the highest scientific level with a customer-focused and solution-oriented way of working, is our distinctive feature.

As part of User-Centered Computing department, the User-Centered Ubiquitous Computing group develops systems providing effective personal assistance that dynamically respond to user demands and at the same time adapt to new work practices. The group is focusing on the application domains Industry 4.0, Smart Cities and Energy Efficiency/Smart Grids and approach novel applied solutions via methods from the domains User-Centered Design, Internet of Things Platforms, and Smart Data.

In this chapter, we aim to give insights into our recent research into the domains of Smart Data and Industrial Internet of Things. To this end, we are focusing on the following EU projects:

- The MONSOON (MOdel based coNtrol framework for Site-wide OptimizatiON of data-intensive processes) project aims to establish a data-driven methodology to support the identification and exploitation potentials by applying multi-scale model based predictive controls in production processes. It offers an integrated real-time and dependable infrastructure easing in improving the efficient use and re-use of raw resources and energy across plant- and site-wide applications in heterogeneous and distributed production environments. EU funds it under SPIRE (Sustainable Process Industry through Resource and Energy Efficiency) research project that aims to develop an infrastructure in support of the process industries.
- The COMPOSITION (Ecosystem for COllaborative Manufacturing PrOceSses – Intra- and Interfactory Integration and AutomaTION) project has two main goals: The first goal is to integrate data along the value chain inside a factory into one integrated information management system (IIMS) combining physical world, simulation, planning and forecasting data to enhance re-configurability, scalability and optimisation of resources and processes inside the factory. The second goal is to create a (semi-)automatic ecosystem, which extends the local IIMS concept to

a holistic and collaborative system incorporating and inter-linking both the Supply and the Value Chains. The COMPOSITION project is funded under the Factories of the Future PPP.

In conjunction with both EU projects mentioned above, EXCELL is a twinning project addressing Big Data applications for cyber-physical systems in production and logistics Networks. The consortium of academics from Hungary, Great Britain, Belgium and Germany expands the scientific activities through central publications and active participation in scientific discourses. Priority Research Fields (PRFs) define the topic areas in which the partners work closely together to mutually train, support and empower each other with their knowledge and expertise. PRFs are for example cyber-physical systems and human system interaction, business-based Internet of Things and services, as well as data mining and data interoperability.

In the remainder of this chapter, we will first describe our research activities and results with respect to the EU project MONSOON, which is an example for the process industry, in Section 6.2. Afterwards, we will continue with the EU project COMPOSITION, which is an example for the manufacturing/discrete industry, in Section 6.3. We finally conclude this chapter in Section 6.4.

6.2 Process Industry

6.2.1 Introduction

The process industry is characterized by intense use of raw resources and energy, and thus represents a significant share of European industry in terms of energy, resources consumption and environment impact. In this area, even a small optimization can lead to high absolute savings, both economic and environmental. Predictive modelling techniques can be especially effective in optimization of production processes. However, the application of these techniques is not straightforward. Predictive models are built using the data obtained from production processes. In many cases, process industries must invest in the monitoring and data integration as well as in the development and maintenance of the underlying infrastructure for data analytics. Many other obstacles are also present, e.g., interoperability issues between software systems in production, difficulties in the physical monitoring of the production parameters, problems with the real-time handling of the data, or difficulties in defining relevant Key-Performance Indicators (KPIs) to support management. Therefore, the deployment of such predictive functions in production with reasonable costs requires consolidation of the available resources into shared cloud-based technologies. In the case of more flexible production environments, approaches that are even more significant are possible, such as the reinvention or redesign of the production processes. However, this is not applicable to major, capital-intensive process industries. In this case, the integration of innovations in the established production processes can be fundamental in their transformation from resource-consuming production into the "circular" model.

6.2.2 Reference Architecture

The high-level conceptual view of the reference architecture that is developed within the scope of the project MONSOON is depicted in Figure 6.1.

The platform is able to inter operate with the heterogeneous existing systems deployed in process industries at different layers of the SCADA pyramid (Control, Supervision, Management, Enterprise). It includes sensors or controllers (PLCs), SCADA (Supervision control and data acquisition), Management Information Systems (MES) and Enterprise Resource Planning (ERP). There are two main components of the architecture. The Real-time Plant Operations Platform deployed on-site and supports data collection, storage and interaction with the production systems respecting relevant constraints and satisfying data-intensive conditions. The Cross-Sectorial Data

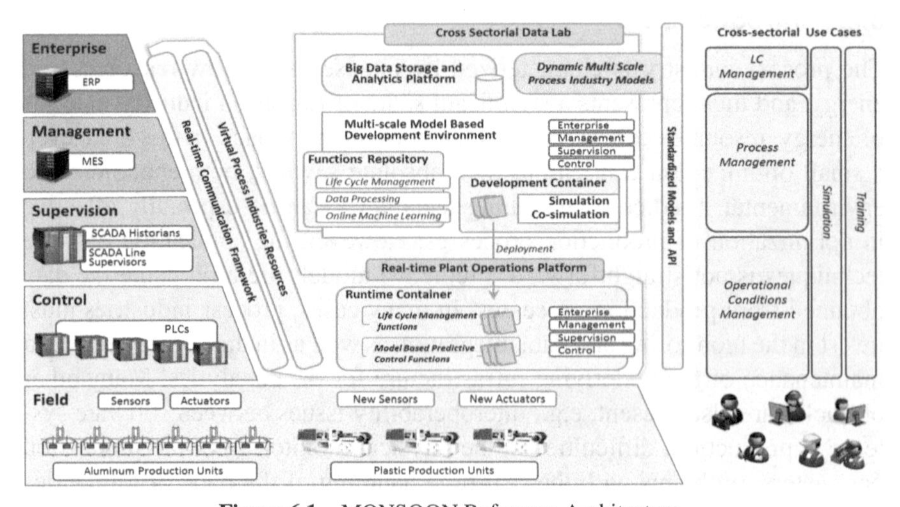

Figure 6.1 MONSOON Reference Architecture.

Lab supports the development of new dynamic model base multi-scale controls. All the relevant data from the production site are transferred to the Data Lab where it is stored and processed for optimization of production process. To validate and demonstrate the results, two real environments are used within the project: an aluminium plant in France and a plastic factory in Portugal. We have identified two main use cases for both domains.

For the aluminium sector, we focused on production of the anodes (positive electrodes) used in aluminium extraction by electrolysis. The first use case was targeted to predictive maintenance, where the main objective was to anticipate the breakdowns and/or highlight equipment/process deviations that affect the green anode final quality (e.g., anode density). The second use case dealt with the predictive anode quality control, where the goal was to identify bad anodes with a high level of confidence and scrap them to avoid sending them to the electrolysis area.

For the plastic domain, the use cases are from the area of production of coffee capsules, produced in large quantities. In this type of production, it is important to produce the correct diameter and height of the coffee capsules and to make sure that the holes at the bottom of the capsules are formed properly. Moreover it is also expected to predict the failures of molding machines and their stoppages based on the process parameters and sensor measurements during molding processes. While the data analysis process for the plastic domain is described in Section 6.2.6, we provide a short description of main components of the MONSOON platform along with their interfaces in the next sections.

6.2.3 Plant Operational Platform

The functional view of the architecture of the Plant Operations Platform is presented in Figure 6.2. It acts as an advanced semantic factory service bus and is in-charge of interacting with existing production systems deployed in a plant. The Plant Platform IT infrastructure and its associated Real-time Data Integration layer collect the operational raw data from the plant's systems necessary to the execution of the predictive functions. The acquired operational raw data and associated relevant information is also routed to the data lab where it is stored and used for analytics.

6.2.3.1 Real-time communication framework

It configures the dependable real-time communication infrastructure necessary to support operations of prediction functions. The *Monitoring Tools*

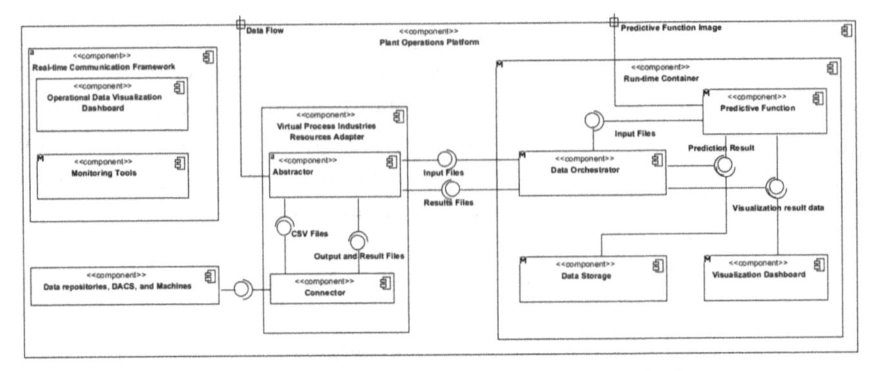

Figure 6.2 Functional View of Plant Operational Platform.

exploit and integrate existing solutions for real-time networking and QoS management and perform continuous (passive/active) monitoring of plant-wide process industry resources ensuring that communication-related malfunctions are properly detected. The *Operation Data Visualization Dashboard* provides a web user interface where operational managers can configure various real-time visualizations of operational data and monitor the deployed predictive functions. The visualized data can include operational data from the plant environment or predictions from the predictive functions executed in the Run-time Container.

6.2.3.2 Virtual process industries resources adapter

The main function of the Virtual Process Industries Resources Adapter (VPIRA) is data integration, mediation and routing. The *Connector* allows the integration of data from various SCADA, MES and ERP systems deployed on the plant site. It ensures that all heterogeneous process industry resources and systems are easily accessed and managed. The *Abstractor* is a distributed and scalable data flow engine aiming for routing integrated data to multiple destinations, e.g., run-time container or data lab. Routing of data from the source to target connectors can be dynamic depending on the type of data or actual content. The data flows can be re-configured in a flexible way, connecting multiple sources to the multiple targets, overcoming any data heterogeneity problems. Besides the flexible configuration interface, the Virtual Process Industries Resources Adapter provide a flexible programming interface to simply implement connectors or processors for new types of data sources and formats.

6.2.3.3 Run-time container

The Run-time Container executes the model based predictive functions and life-cycle management functions within the overall plant infrastructure. It ensures proper deployment and execution of predictive functions developed by means of the data lab, hence it manages all aspects of predictive functions life cycle. It is composed into four sub-components as described below:

- *Data Orchestrator*: coordinates the data flow between different components, such as transmit input data, store prediction result, and pass visualization result data to relevant components.
- *Predictive Function*: exports predictive function image from Function Repository and instantiate the execution of predictive function that perform real-time scoring of input operational data. It performs all operations required for the pre-processing of raw data into inputs for the specific predictive function and into process prediction output.
- *Data Storage*: stores the prediction results into a scalable database. The prediction results are also sent to the Operational Platform systems and the data lab for combining these real-time results with historical data analysis.
- *Visualization Dashboard*: displays prediction results and generates feedback instructions or alerts towards plant's systems to inform/warn the site operators to adjust the process regulation parameters.

6.2.4 Cross Sectorial Data Lab Platform

The Data Lab provides a collaborative environment where high amounts of data from multiple sites, and possibly from multitude of industry sectors, are collected, stored and processed in a scalable way. It enables multidisciplinary collaboration of experts allowing teams to jointly model, develop and evaluate distributed controls in rapid and cost-effective way. The Data Lab eases the definition of predictive control and life cycle management functions, allowing to work in a simulated environment or to exploit co-simulation by mixing stored data with data flowing in real-time from the real systems.

The Data Lab thus supports data science and automation experts interested to optimization and scheduling aspects by providing the suitable environment to mine, process, re-play production data. It allows modelling of the whole production process across the SCADA layers including the specification of the data dictionary of all inputs and outputs of the processing steps and their relations to the overall KPIs. The semantic models capture the site knowledge base for given application cases and used data analytics

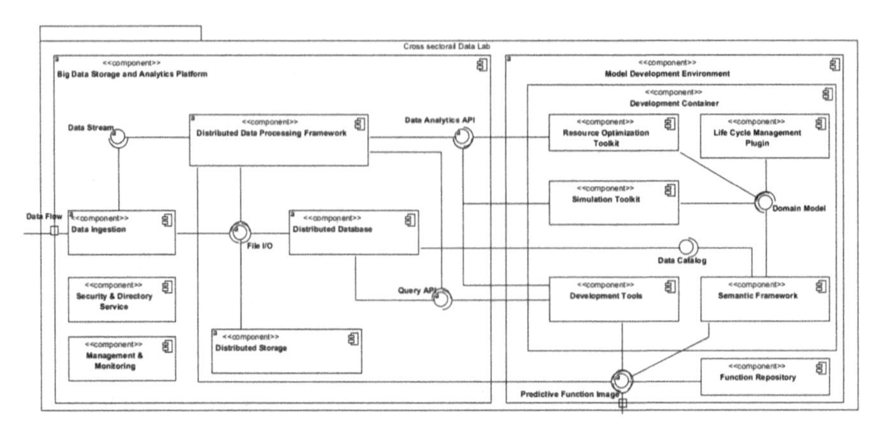

Figure 6.3 Functional View of Cross Sectorial Data Lab Platform.

methods allowing generalization of cases to existing good practices and transfer of the knowledge by adaptation of cases to new environment/site. The main outcome of the Data Lab is typically a single or multiple new predictive functions and life cycle management controls ready to be deployed in the Runtime Container of the Plant Operations Platform.

The components of the Cross Sectorial Data Lab are shown in Figure 6.3 and explained in the sections below.

6.2.4.1 Big data storage & analytics platform

The Big Data Storage and Analytics Platform provides resources and functionalities for storage as well as batch and real-time processing of the operational data from multiple site characterized as Big Data. The platform combines and orchestrates existing technologies from the Big Data and Analytic landscape and sets a distributed and scalable run-time infrastructure for the developed data analytics methods. It provides main integration interfaces between the site Operational Platform and the cloud Data Lab platform and the programming interfaces for the implementation of the data intensive analytics methods. The Big Data Storage and Analytics Platform consist of the following sub-components:

- *Distributed Storage*: provides a reliable, scalable file system with similar interfaces and semantics to access data as local file systems.
- *Distributed Database*: provides a structured view of the data stored in the platform using the standard SQL language, and supports standard RDBMS programming interfaces such as JDBC for Java or ODBC for Net platforms.

- *Distributed Data Processing Framework*: allows the execution of applications in multiple nodes in order to retrieve, classify or transform the arriving data. The framework provides Data Analytics APIs for processing large datasets via parallel and distributed computations.
- *Data Ingestion*: implements an interface for real-time communication between the Data Lab and Operation platforms. It also supports batch uploading of the historical data between the Data Lab and Operation platform.
- *Security & Directory Service*: provides user management and content authorization capabilities for the platform services.
- *Management & Monitoring*: provides the management, monitoring and provisioning of the platform services on the hosted environment.

6.2.4.2 Model development environment

The Model Development Environment provides tools and interfaces that cover the whole life cycle of planning, implementation, testing, validation and deployment of predictive functions and life-cycle management controls into the plant production supporting simulation/co-simulation features.

- *Development Tools*: provide the main collaborative and interactive interface for data engineers, data analysts and data scientists to execute and interact with the data processing workflows running on the Data Lab platform. Using the provided interface, data scientists can organize, execute and share data, and code and visualize results without referring to the internal details of the underlying Data Lab run-time infrastructure. The interface is integrated in form of analytical "notebooks" where different parts of the analysis are logically grouped and presented in one document. These notebooks consist of code editors for data processing scripts and SQL queries, and interactive tabular or graphical presentations of the processed data.
- *Semantic Modelling Framework*: provides a common communication language between domain experts, stakeholders and data scientists. A collaborative web interface is provided for the creation and sharing of semantic models in order to use the knowledge expressed in such models for the optimization of the production processes in the Simulation and Resource Optimization Framework.
- *Simulation Toolkit*: supports validation and deployment of predictive functions in order to optimize overall KPIs defined for the production process. The estimation of overall impacts can be used to test various "what if" scenarios, or for the automatic discrete optimization of the

production process by finding the optimal combination of predictive functions for various process phases.

- *Resource Optimization Toolkit:* optimizes the production process based on various indicators representing the performance of manufacturing process of the plant leveraging process data and knowledge extracted from analytics methods.
- *Life-Cycle Management Plugin*: serves as multi-disciplinary, transversal tool to evaluate environmental performance of a given production process for life-cycle environmental indicators, such as Global Warming Potential and Total Energy Requirement.

6.2.4.3 Function repository

The Function Repository provides a storage for predictive functions together with all settings required for the deployment of predictive functions, where they are available for production deployment or for the simulations and overall optimization of the production processes. The predictive functions are packaged as container images so that entire predictive function pipeline (including pre-processing and task specific evaluation) can be implemented within a virtualized container.

6.2.5 Deployment

The Data Lab Platform promises to combine and orchestrate existing technologies and open source frameworks from the Big Data landscape to establish a distributed and scalable run-time infrastructure for the data analytics methods. We present in Figure 6.4 the mapping of the platform components

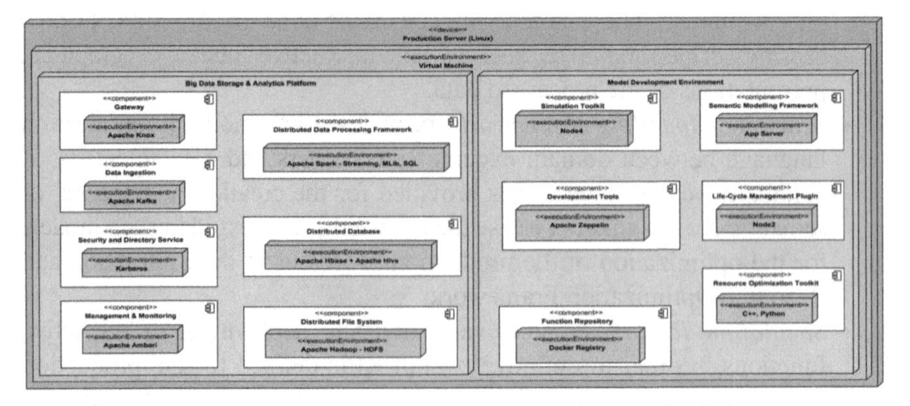

Figure 6.4 Components Mapping to Open-source Technologies.

to existing and emerging open source technologies selected and used during the initial deployment.

The initial deployment was performed with multiple virtual machines on an in-house physical infrastructure. It turned out that the overall deployment time and configuration management is the most critical aspect in realizing and operationalizing such a platform. It would be optimal to devise a uniform deployment strategy taking into account different deployment options for the platform such as on-premises, cloud/external provider or hybrid. It has also been learned that different demonstrative and use-case scenarios in both aluminium and plastic domains pose different infrastructure and data requirements. Hence, it is useful to define different deployment pipelines or modes for the platform where the right set of platform services are deployed and orchestrated accordingly instead of full stack deployment. Towards this goal, the Big Data Storage and Analytics Platform has been containerized to adapt a common deployment ground with the objective of easing the usage of common platform technologies and make integration with other services or applications easy. The containerization based on Docker framework is depicted in Figure 6.5.

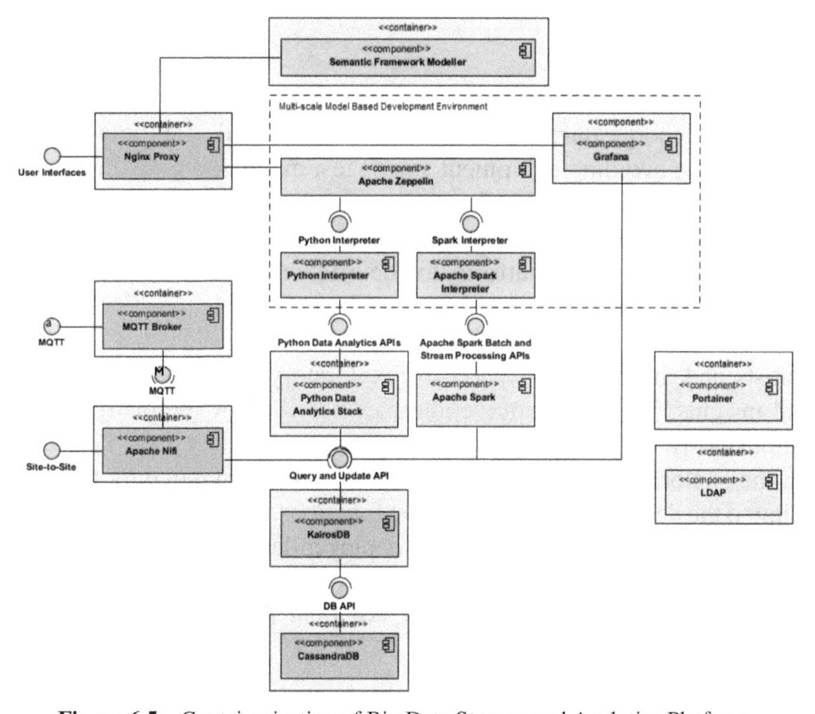

Figure 6.5 Containerization of Big Data Storage and Analytics Platform.

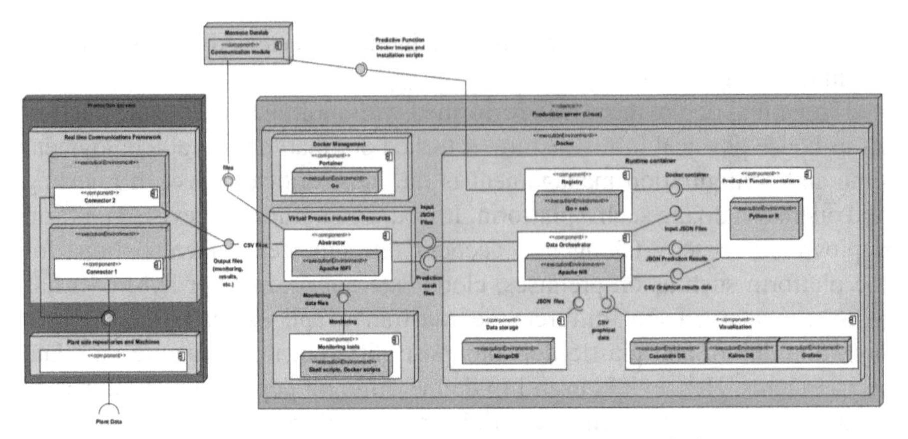

Figure 6.6 Deployment view of Plant Operational Platform.

The deployment of the Site Operational Platform with open source technologies mainly for Virtual Process Industries Resources Adaptor and Run-time Container is finally illustrated in Figure 6.6. It shows how predictive functions can be applied in factorial settings.

6.2.6 Data Analysis

Data analysis in process industries mainly aims to reduce the wastage of time, resource and energy during production processes. This can be achieved by several means: avoiding equipment stoppages, maintaining optimum configurations, early detection of a chain of events causing an anomaly etc. Data analysis is simplified by the components of the *Cross Sectorial Data Lab* which provides a single platform for data fetching, accessing and artefact development. The data collected from the plastic molding machines are stored in the *Big Data Storage & Analytic Platform.* These data are used by the data scientists and the process experts for exploratory analysis in order to gain initial insights. The collaborative interface provided by the platform is used simultaneously by the process expert and the data scientists. The findings from the exploratory analysis is used as the basis for modelling the process leading to the development of predictive functions. These functions are stored in the *Function Repository* which are deployed later in factory premises for real-time predictions. Although this process is generic enough to be applied in any kind of industrial environment, we shall limit our discussion to the plastic industry.

The objective of data analysis in the plastic industry is to anticipate the breakdowns and/or highlight equipment/process deviation that impacts the injection molding process and therefore to improve the quality of the produced coffee capsules. In general, there are two areas where waste parts can occur in plastic injection molding process: the molding tool and the molding process. During the long-term production of the coffee capsules, parameters of the injection molding process can slightly change due to various changes of the environment (temperature and humidity in the factory, deviations in the energy supply system, heating of oil temperature, deviations in the quality of the plastic granules, wearing of machine parts). The aim is to monitor technical parameters of the molding machine and raise an alarm if the deviation is increasing over the defined values. These long-term changes can also cause the stoppage of molding machines. Which in turn causes reduction of produced capsules. In addition, few of the initial cycles after restart are wasted during the calibration process producing defective capsules.

6.2.6.1 Data description

Two kinds of data have been collected in the first year from GLN site during the production of coffee capsules. The first data set is collected automatically from a Euromap63 interface recorded on molding machines and the second data set is collected during the experiments conducted by a process expert during their visit to the production site.

The first data set is unlabeled and contains sensor measurements of several coffee capsule production cycles. Each cycle lasts almost 7 seconds, except if it causes a breakdown. The data set has a total of 88 attributes representing temperatures, time taken for different stages, pressure, cylinder positions etc. All data were directly monitored by the injection molding machine and stored there. Of them, only 12 (heating belt temperatures, maximum cycle pressure, coolant temperatures, residual melt cushion, plastification time) are proposed as useful, and, particularly, their ranges/deviations over intervals instead of their values themselves are suggested to serve as explanatory variables.

The second data set [1] is manually labelled and comprises information about 250 production cycles of coffee capsules from the injection molding machine and their quality information. It contains 36 attributes reflecting the machine's internal sensor measurements for each cycle. These measurements include values about the internal states, e.g. temperature and pressure values, as well as timings about the different phases within each cycle. In addition, we also take into account quality information for each cycle, i.e., the number of non-defect coffee capsules which changes throughout individual production cycles. The quality of each capsule is inspected by the domain expert in different aspects. i.e. the capsules have permissible range of height and base diameter. Also each capsule should have uniform thickness and should not have holes. If any of these expectations are not met, the capsule is considered to be defective. If the number of produced high quality coffee capsules is larger than a predefined threshold, we label the corresponding cycle with high.quality, and otherwise we assign the label low.quality. The decision about the quality labels was made by domain experts.

Exploratory analysis is performed on the unlabeled data in order to discover hidden insights. On the other hand, basic machine learning algorithms are applied to the labelled data to classify the cycles based on their quality. In the upcoming subsections we discuss these two different approaches on these data sets.

6.2.6.2 Preliminary trend analysis of unlabeled data

The main aim of the preliminary analysis of is to get some initial overall insights that might be interesting for the process experts to be further analyzed. The first step was to understand the attributes and their correlations. This was followed by visual exploration of data with manual inspection followed by clustering the data to find significant relation between different cycles. Considering the huge amount of data generated by sensors, clustering usually takes lots of time. One strategy is to use the computation powers of the Data Lab clusters to perform these operations faster. If the algorithms for exploratory data analysis are deployed in the Data Lab, domain experts and data scientists can use the results simultaneously to get actionable insights.

One of the insight was repeating set of parameters in the data. This was found by using matrix profiles. A pattern obtained by applying matrix profiles is the decrease in plastification time and at the same time, increase in cycle time. Plastic domain expert cross checked these patterns and found out that this happens whenever there is an equipment stoppage due to lack of lubrication. Though the characteristics of these incidents are known, early prediction of the possible stoppage has not been found out with data analysis. The corresponding patterns are shown in Figure 6.7.

Preliminary trend analysis helps us to extract the knowledge hidden in voluminous unlabelled data sets. This process can be automated to get the best results in minimum time. In addition, in the MONSOON project we include many stakeholders such as process experts, machine supervisors and ground workers to actively contribute to the production process optimization.

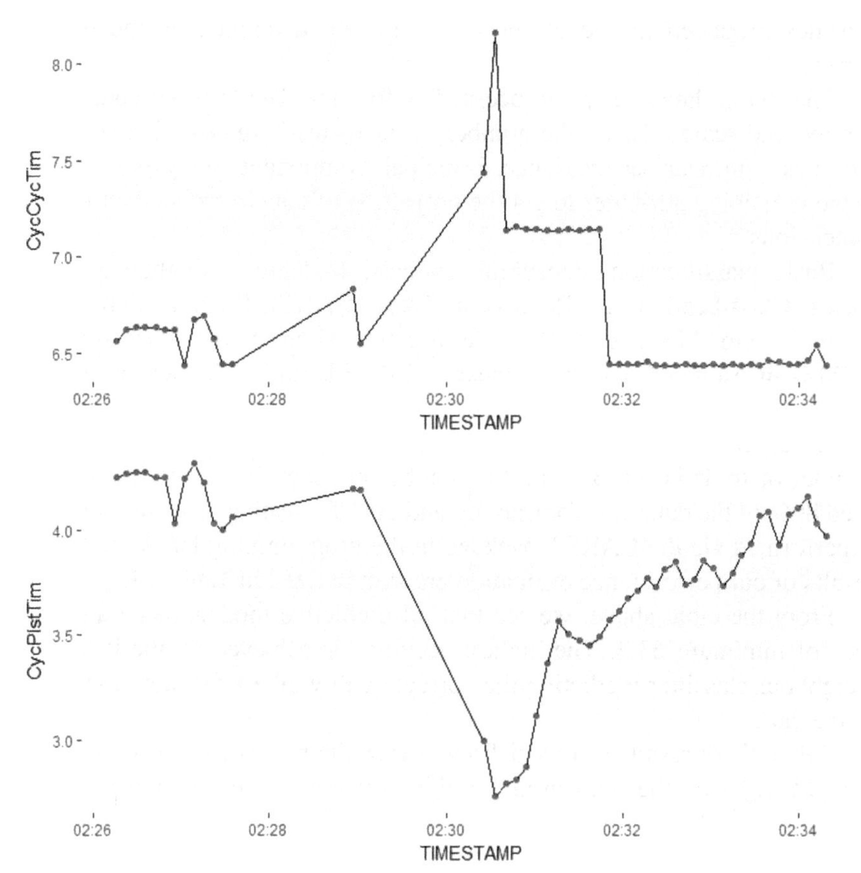

Figure 6.7 Increase in cycle time and decrease in plastification at the same time. The same pattern has repeated multiple times in the unlabelled set of plastic data. CycCycTim is cycle time and CycPlstTim is plastification time.

This is achieved with the help of a centralized Big Data analytics platform. On deploying the knowledge discovery algorithms in the Big Data analytics platform, the stakeholders can give live feedbacks. The data scientists further use these feedbacks for deriving conclusions. This is an ongoing work as part of the project.

6.2.6.3 Machine learning for labelled data

The goal of the machine learning process is to classify the injection molding cycles to high and low-quality cycles. As discussed earlier, the cycles are labelled as *high.quality* or *low.quality* based on the number of defective capsules produced in a cycle beyond a threshold defined by the process expert.

The initial dataset is pre-processed as follows. The labelled data is first centred and scaled. Later, the number of attributes is reduced by excluding the ones with near zero variance. Principal Component Analysis is applied to the remaining attributes to get the projection of data in reduced number of dimensions.

Basic classification algorithms, namely, k-Nearest Neighbour, Naïve Bayes, Classification and Regression Trees (CART), Random Forests and Support vector Machines (SVM) are investigated on the pre-processed data. SVM is investigated both with linear and RBF kernels. The performance of the models are measured in terms of balanced accuracy, precision, recall and F1 scores. K-fold cross validation is used to evaluate the performance. The number of folds is set to 5 and the number of repetitions is set to 100. We used 80% of the dataset is for training and 20% for testing. This investigation is performed via the CARET package in the programming language R. The results of our performance evaluation are summarized in Table 6.1.

From the table above, we see that all predictive models reach an accuracy of minimum 63%. The highest accuracy is achieved by the k-Nearest Neighbour classifier predicting the correct quality labels for more than 69% of the data.

Albeit these results were satisfying, these algorithms cannot be deployed straight away as the data used for this performance evaluation has been manually labelled by the experts. In the situations where the capsules are produced in millions per day, it is wiser to use the automatically labelled data for training the models and deploy them afterwards. One approach is to use the decision of the visual inspection systems in order to label the data. But this is not trivial since there is no one to one mapping between the optical inspection systems data and the actual cycle data. This is because multiple capsules belonging to different cycles and machines are passed to

Table 6.1 Classification results of different predictive models

	Balanced Accuracy	Precision	Recall	F1 score
k-NN	0.697	0.638	0.686	0.657
Naïve Bayes	0.643	0.604	0.563	0.578
CART	0.637	0.595	0.566	0.573
Random Forest	0.653	0.619	0.570	0.589
SVM (linear)	0.632	0.626	0.488	0.540
SVM (RBF)	0.663	0.643	0.563	0.594

the automatic visual inspection system at once making it harder to identify individual cycles belonging to a particular machine.

6.2.7 Summary

In this section, we have presented our recent research activities within the scope of the EU project MONSOON: As an example for the process industry, we have described the overall reference architecture facilitating cross-sectorial data analytics. As part of our ongoing work, we have also highlighted the analysis of sensor data arising from the plastic industry sector. In the following section, we will focus on the manufacturing industry.

6.3 Manufacturing/Discrete Industry

6.3.1 Introduction

As an example for the manufacturing industry, we focus on the EU project COMPOSITION. This project addresses the requirements of modern production processes, which stress the need of greater agility and flexibility leading to faster production cycles, increased productivity, less waste and more sustainable production. At the factory level, decisions need to be supported by detailed knowledge about the production process and its interplay with external entities. Unfortunately, historical and live data that generates this knowledge is becoming more and more distributed and few solutions are available that can easily tackle the implied challenges. Moreover, factories are becoming less isolated in the productive tissue of nations and several suppliers and third-party service providers need to be contacted and coordinated to implement decisions taken at the factory level.

In such a worldwide and dynamic environment, the ability of automatizing the preliminary coordination and negotiation activities involved in setting up supply chains for specific needs, in an open marketplace-like fashion, could greatly improve the ability of factories to quickly react to external challenges and driving forces.

6.3.2 Intra-factory Interoperability Layer Part of the COMPOSITION Architecture

In this chapter, we will address the COMPOSITION architecture in the data analytics context. The intra-factory interoperability layer has two main goals: the first one is to provide an infrastructure to combine distributed

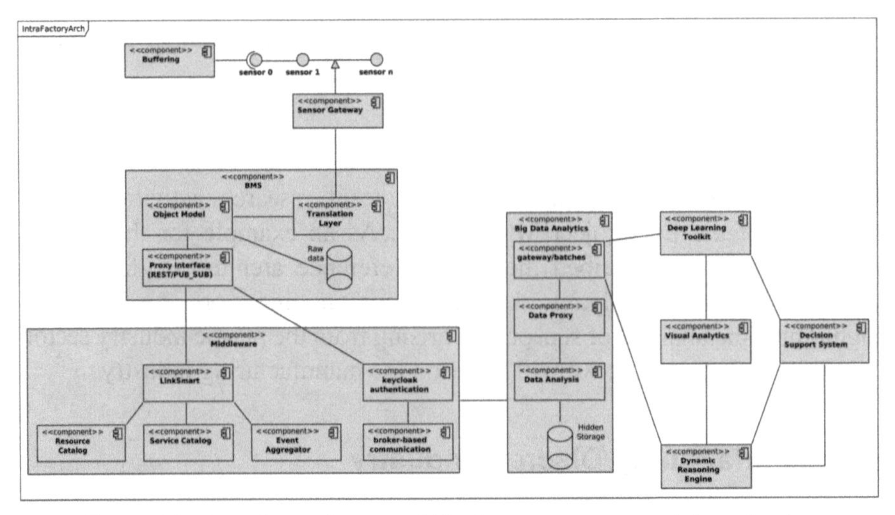

Figure 6.8 Intra-factory interoperability layer components and dependencies.

data in the integrated information management system and to do data analytics, the second one is to ensure the conformity between communications among interconnected components. Figure 6.8 shows the relevant part of the architecture.

The components of the architecture are introduced and described in the following:

- The *BMS* is provided by a project development stakeholder and is the translation layer providing shop floor connectivity from sensors to the COMPOSITION system. Raw data storage is added for offline debug purposes.
- The *Middleware* is the main recipient in which the interoperability of single components act.
- *LinkSmart* is a well-known middleware solution per se and is customized to satisfy the requirements of the COMPOSITION project. LinkSmart comprises the following components:
    - The *Service Catalog* works as service index and provides security information for service intercommunication.
    - The *Event Aggregator* parses messages to ensure homogeneity in data streams.
    - *Keycloak* is a virtual layer that ensures authorization and authentication. Like all security related measures, it is deployed by the Security Framework.

  - The *broker-based intra-factory communication system* manages all internal communication.
- The *Big Data Analytics* component provides Complex Event Processing (CEP) capabilities for the data provided by the intra-factory integration layer
- The *Hidden Storage* is an optional storage not accessible from the outside in which aggregated data are stored for debug purposes, i.e. re-bootstrapping already trained artificial neural networks belonging to the Deep Learning Toolkit and to the Dynamic Reasoning Engine.
- The *Visual Analytics* component is the reporting interface of the Decision Support System and Simulation and Forecasting Toolkit.
- The *Dynamic Reasoning Engine* is part of the Simulation and Forecasting Toolkit.
- The *Decision Support System* uses process models to guide the production process.

Having a fist overview of the components of the COMPOSITION project and their dependencies, we continue with describing our approach to smart data analysis in the following section.

6.3.3 The Complex-Event Machine Learning methodology

Manufacturing in assembly lines consist of a set of hundreds, thousands or millions of small discrete steps aligned in a production process. Automatized production processes or production lines thereby produce for each of those steps small bits of data in form of events. Although the events possess valuable information, this information loses its value over time. Additionally, the data in the events usually are meaningless if they are not contextualized, either by other events, sensor data or process context. To extract most value of the data, it must be processed as it is produced, to be more precise in real-time and on demand. Therefore, in case of Big Data Analyses we propose the usage of Complex-Event Processing for the data management coming from the production facilities. In this manner, the data is processed in the moment when it is produced, extracting the maximum value, reducing latency, providing reactivity, giving it context and avoiding the need of archiving unnecessary data.

The Complex-Event Processing service is provided by the LinkSmart® Learning Agent (LA). The LA is a Stream Mining service that provides the utilities to manage real-time data for several purposes. On the one hand, the LA provides a set of tools to collect, annotate, filter, aggregate, or cache the real-time data incoming from the production facilities. This set of tools facilitates the possibility to build applications on top of real-time data. On the other hand, the LA provides a set of APIs to manage the real-time data lifecycle for continuous learning. Moreover, the LA can process the live data to provide complex analysis creating real-time results for alerting or informing about important conditions in the factory, that may be not be seen at first glance. Finally, the LA allows the possibility to adapt to the productions needs during the production process.

The Complex-Event Machine Learning (CEML) [2] is a framework that combines Complex-Event Processing (CEP) [3] and Machine Learning (ML) [4] applied to the IoT. This means that the framework was developed to be deployed everywhere, from the edge of the network to the cloud. Furthermore, the framework can manage itself and works autonomously. The following section briefly describes the different aspects that CEML covers. The framework must automate the learning process and the deployment management. This process can be broken down in different phases: (1) the data must be collected from different sensors, either from the same device or in a local network. (2) The data must be pre-processed for attribute extraction. (3) The learning process takes place. (4) The learning must be evaluated. (5) When the evaluation shows that the model is ready, the deployment must take place. Finally, all these phases happen continuously and repetitively, while the environment constantly changes. Therefore, the model and the deployment must adapt as well.

6.3.3.1 Learning agents architecture

We utilize LinkSmart® LA following a modular architecture with loosely coupled modules responsible for different tasks. Figure 6.9 illustrates the architecture of the LA. The data and commands come via communication protocols implemented by Connectors (Figure 6.9 shows two example implementations, REST and MQTT). The connectors transfer the information to the Feeders, which process the data accordingly to the API logic. This logic depends on whether it is an insertion of new raw data, request of simple data processing (statement) or a machine learning request (CEML request). The data is inserted into the execution environment (in this case EsperEngine[1]), while the data processing requests are deployed in the same engine for the

[1]Esper is an open-source Java-based software product for Complex event processing (CEP) and Event stream processing (ESP) that analyzes series of events for deriving conclusions from them. See http://www.espertech.com/

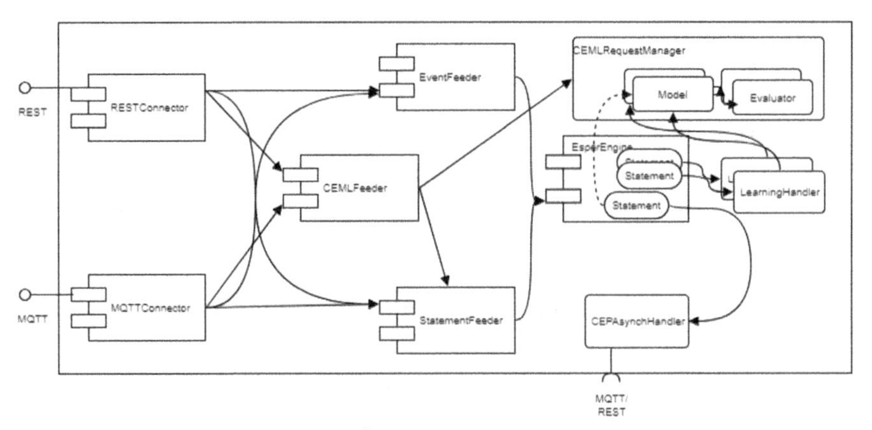

Figure 6.9 LinkSmart® Learning Service Architecture sketch.

processing of the raw data. The CEML request has a more complex behaviour. Each CEML request is managed by its own CEMLManager, which contains and coordinates the model(s), evaluator for each model, and several statements. Finally, all output of any process (Statement) in the execution pipeline (EsperEngine) is captured or managed by a Handler. If the process should be prepared and sent through a communication protocol, then it will be handled by a Complex-Event Handler: An Asynchronous Handler, if the protocol is asynchronous (e.g. MQTT); or Synchronous Handler, if the protocol is synchronous (e.g. HTTP).

6.3.3.2 Data propagation phase

Data in the IoT is produced in several places, protocols, formats, and devices. Although this article does not address the problem of data heterogeneity in detail, the learning agents require a mechanism to acquire and manage the heterogeneity of the data. The mechanism must be scalable and, at the same time, the protocol should handle the asynchronous nature of IoT. Finally, the protocol must provide tools to handle the pub/sub characteristics of the CEP engines. Therefore, we have chosen MQTT[2], a well-established Client Server publish/subscribe messaging transport protocol. The topic based message protocol provides a mechanism to manage the data heterogeneity by making a relation between topics and payloads. It allows deployments in several architectures, OS, and hardware platforms; basic constraints at the edge of the

[2]MQTT is a machine-to-machine (M2M)/"Internet of Things" connectivity protocol. Source http://mqtt.org/

network. The protocol is payload agnostic and as such allows for maximum flexibility to support several types of payloads.

6.3.3.3 Data pre-processing (munging) phase

Usually ML is tied to stored datasets, which incurs several drawbacks. Firstly, the learning can take place only with persistent data. Secondly, usually the models generated are based on historical data, not current data. Both constrains, in the IoT, have direct consequences. It is neither feasible nor profitable to store all data. In addition, embedded devices do not have much storage capacity, which makes it impossible to use ML algorithms on them. Furthermore, IoT deployments are commonly exposed to ever-changing environments.

Using historical data for off-line learning could cause outdated models to learn old patterns rather than current ones, producing drifted models. Although some IoT platforms like COMPOSITION support storage of historical data, it may be too time and space consuming to create large enough times series. Therefore, there is also a need for non-persistence manipulation tools. This is precisely what the CEP engine provides in the CEML framework. This means, the CEP engine decides which data and how the data is manipulated using predefined CEP statements deployed in the engine. Each statement can be seen as a topic, to which each learning model is subscribed. Any update of the subscribers provides a sample to be learnt in the learning phase.

6.3.3.4 Learning phase

There is no pre-selection of algorithms in the framework. They are selected by the restrictions imposed by the problem domain. For example, in extreme constrained devices, algorithms such as Algorithm Output Granularity (AOG) [5] may be the right choice. In other cases where the model changes quickly, one-shot algorithms may be the best fit. Artificial Neural Networks are good for complex problems but only with stable phenomena. This means that the algorithm selection should be made case-by-case. Our framework provides mechanisms for the management and deployment of the learning models, and the process of how the model is fed with samples. In general, the process is based on incremental learning [6] albeit with online and non-persistent data. The process can be summarized as follows: the samples, without the target provided in the last phase, are used to generate a prediction. The prediction will then be sent to the next phase. Thereafter, the sample is applied to update the model. Thus, all updates are used for the learning process.

6.3.3.5 Continuous validation phase

This section describes how the validation of the learning models is done inside the CEML. This phase does not influence the learning process nor validate the CEML framework itself.

ML model validation is a challenging topic in real-time environments and the evaluation for distributed environments or embedded devices is not addressed extensively in the literature, which is why we think it needs further research. There are two addressed strategies. Either we holdout an evaluation dataset by taking a control subset for given time-frame (time window), or we use *Predictive Sequential*, also known as *Prequential* [7], in which we assess each sequential prediction against the observation. The following section describes the continuous validation we applied for a **classification** problem, even though it can be applied for other cases as well.

Instead of accumulating a sample for validation, we analyse the predictions made before the learning takes place. All predictions are assessed each time an update arrives. The assessment is an entry for the confusion matrix [8], which is accumulated in an *accumulated confusion matrix*. The matrix contains the accumulation of all assessed predictions done before. In other words, the matrix does not describe the current validation state of the model, but instead the trajectory of it. Using this matrix, the accumulated validation metrics (e.g. Accuracy, Precision, Sensitivity, etc.) are being calculated. This methodology does have some drawbacks and advantages, explained more extensively in [9].

6.3.3.6 Deployment phase

The continuous validation opens the possibility for making an assessment of the status of the model each time a new update arrives, e.g. if it is accrued or not. Using this information, the CEML framework has the capability to decide if the model should or should not be deployed into the system at any time. If the model is behaving well, then it should be deployed, otherwise it should be removed from the deployment. The decision is made by user-provided thresholds w.r.t. evaluation metrics. If a threshold is reached, the CEML inserts the model into the CEP engine and starts processing the streams using the model. Otherwise, if the model do not reach the threshold, it is removed from the CEP engine.

6.3.3.7 Assessment

In [6] 13 issues for learning in the IoT where left open. The CEML framework addresses 10 out of the 13 challenges as follows:

- Handling the continuous flow of data streams: This is done by the stream statements inside the CEP engine using continuous streams for learning an evaluating.
- Unbounded memory requirements: The use of CEP engines in stream windows allows the intelligent usage of the memory as is needed, dropping it otherwise.
- Transferring data mining results over a wireless network with limited bandwidth: This is partially handled. MQTT is a reliable low-bandwidth lightweight protocol developed for satellite monitoring of pipelines. Nevertheless, this paper does not address the physical layer.
- Modelling changes of mining results over time: The CEML is a continuous automatic learning mechanism. The learning models will adjust as they learn.
- Interactive mining environment to satisfy user requirements: The IoT Learning agent provide an REST API. Thus, update the learning request is possible, as well as, obtaining live or on-demand updates.
- Integration between data-stream management systems and ubiquitous data-stream mining approaches: The CEML provides a REST API for managing each kind of request independently. Thus, the learning request can be managed as a whole, including the involved streams. Besides, the streams can be managed individually as single stream statement. Additionally, the MQTT API provides a multi-cast API so that in distributed multi-agent deployment, the agents can be managed as one, as groups, or as one entity.
- The relationship between the proposed techniques and the needs of real-world applications: Legal, ethical and technical reasons are part of the motivation. E.g. the storage constrains or the legal constraints in the health domain.
- Data pre-processing in the stream-mining process: This is handled in the pre-processing phase of the CEML.
- The technological issue of mining data streams: The implementation presented here shows that the system behaves in a real-time environment.
- The formalization of real-time accuracy evaluation: This is addressed by the Double-Tumble-Window Evaluation.

In addition to the Complex-Event Machine Learning approach based on the open-source IoT platform LinkSmart, we also describe another approach carried out in the scope of the project COMPOSITION in the next section.

6.3.4 Unsupervised Anomaly Detection in Production Lines

In addition to the previously introduced framework, which primarily allows for an exploitation of supervised machine learning algorithms, this chapter focuses on an alternative unsupervised approach that was also implemented in the scope of the project COMPOSITION. This method was used as a further extension to optimize the detection of machine errors in production lines at early stages.

In the last couple of years, the importance of cyber-physical systems in order to optimize industry processes, has led to a significant increase of sensorized production environments. Data collected in this context allows for new intelligent solutions to e.g. support decision processes or to enable predictive maintenance.

One problem related to the latter case is the detection of anomalies in the behaviour of machines without any kind of predefined ground truth. This fact is further complicated, if a reconfiguration of machine parameters is done on-the-fly, due to varying requirements of multiple items processed by the same production line. As a consequence, a change of adjustable parameters in most cases directly leads to divergent measurements, even though those observations should not be regarded as anomalies.

In the scope of the project COMPOSITION, the task of detecting anomalies for predictive maintenance within historical sensor data from a real reflow oven was investigated. While the oven is used for soldering surface mount electronic components to printed circuit boards based on continuously changing recipes, one related problem was the unsupervised recognition of potential misbehaviours of the oven resulting from erroneous components. The utilized data set comprises information about the heat and power consumption of individual fans. Apart from additional machine parameters like a predefined heat value for each section of the oven, it contains time-annotated sensor observations and process information recorded over a period of more than seven years.

As one solution for this problem, we will present our approach named Generic Anomaly Detection for Production Lines, short GADPL. The hereafter-presented description of GADPL is based on the stage-wise implementation of the algorithm. After an initial clustering of similar input parameters and a consecutive segmentation, we will discuss the representation of individual segments and the corresponding measurement of dissimilarity.

6.3.4.1 Configuration clustering

In many companies, as well as in the case of the project COMPOSITION, a single production line is often used to produce multiple items according to different requirements. Those requirements are in general defined by varying machine configurations consisting of one or more adjustable parameters, which are changed 'on-the-fly' during runtime. For a detection of deviations with respect to some default behaviour of a machine, this fact raises the problem of invalid comparisons between sensor measurements of dissimilar configurations. If a measurement or an interval of measurements is identified as an anomaly, it should only be considered as such, if this observation is related to the same configuration as observations representing the default behaviour. Therefore in advance to all subsequent steps, at first all sensor measurements have to be clustered according to their associated configuration. For the sake of simplicity, we are only discussing the process within a single cluster in the following subsections, although one has to keep in mind that each step is done in parallel for all clusters.

6.3.4.2 Segmentation

As a result of the configuration-based clustering, the data is already segmented coarsely. However, since this approach describes unsupervised anomaly detection, the idea of a further segmentation is to create some kind of ground truth, which reflects the default behaviour of a machine. In this section, we will see how the segmentation is utilized to implement this idea. In an initial step, a maximum segmentation length is defined, in order to specify the time horizon, after which an anomaly can be detected. Assuming a sampling rate of 5 mins per sensor, the maximum length of a segment would consequently be $(60 \times 24)/5 = 288$ to describe the behaviour on a daily basis. Although a decrease of the segment length implies a decrease of response time, it also increases the computational complexity and makes the detection more sensitive to invalid sensor measurements. In this context, it needs to be mentioned that in this stage segments are also spitted, if they are not continuous with respect to time as a result of missing values. Another fact that has to be considered is the transition time of configuration changes. While the input parameters associated with a configuration change directly, the observations might adapt more slowly and therefore blur the expressiveness of the new segment. To prevent this from happening, the transition part of all segments, which have been created due to configuration changes, is truncated. If segments become smaller than a predefined threshold, they can be ignored in the upcoming phases.

6.3.4.3 Feature extraction

Having a set of segments for each configuration, the next step is to determine the characteristics of all segments. While the literature presents multiple approaches to describe the behaviour of time series, we will focus on common statistical features extracted from each segment. Nonetheless, the choice of features is not fixed, which is why any feature suitable for the individual application scenario can be used. One example for rather complex features could be the result of a kernel fitting in the context of Gaussian processes, accepting a decrease in performance. Since the goal is to capture comparable characteristics of a segment, we compute different real-valued features and combine them in a vectorised representation. In the case of the project COMPOSITION, we used the mean to describe the average level, the variance as a measure of fluctuation and the lower and upper quartiles as a coarse distribution-binning of values. Due to the expressiveness of features being dependent from the actual data, one possible way to optimize the selection of features is the Principal Component Analysis. Simply using a large number of features to best possibly cover the variety of characteristics might have a negative influence on the measurement of dissimilarity. The reason for this is the partial consideration of irrelevant features within distance computations. Moreover, since thresholds could be regarded as a more intuitive solution compared to additionally extracted features, this replacement would lead to a significant decrease in the number of recognized anomalies. Apart from the sensitivity to outliers, the reason is a neglect of the inherent behaviour of a time series. As an example, consider the measurements of an acoustic sensor attached to a motor that recently is sending fluctuating measurements, yet within the predefined tolerance. Although the recorded values are still considered as valid, the fluctuation with respect to the volume could already indicate a nearly defect motor. Finally, one initially needs to evaluate appropriate thresholds for any parameter of each configuration.

6.3.4.4 Dissimilarity measurement

So far, we have discussed the exploitation of inherent information, extracted from segmented time series. The final step of GADPL is to measure the level of dissimilarity for all obtained representatives. Since no ground truth is available to define the default behaviour for a specific configuration, the algorithm uses an approximation based on the given data. One problem in this regard is the variability of a default behaviour, consisting of more than one pattern. Therefore, a naive approach as choosing the most occurring representative, would already fail for a time series consisting of two equally appearing patterns captured by different segments, where consequently half of the data would be detected as anomalous behaviour.

As one potential solution GADPL instead uses the mean over a specified size of nearest neighbours, depicting the most similar behaviour according to each segment. The idea is that even though there might multiple distinct characteristics in the data, at least a predefined number of elements represent the same behaviour compared to the processed item. Otherwise, this item will even have a high average dissimilarity with respect to the most similar observations and can therefore be classified as anomaly.

Here, for the vectorised feature representations, any suitable distance function is applicable. In the context of the project COMPOSITION we decided to use the Euclidean distance for a uniform distribution of weights, applied to normalized feature values. To further increase the performance of nearest neighbour queries, we exploited the R*-tree as a high-dimensional index structure. Given the dissimilarity for each individual representative together with a predefined anomaly threshold, GADPL finally emits potential candidates having an anomalous behaviour.

The application of GADPL is illustrated in Figure 6.10. The upper part shows the segmentation of time annotated power consumption data in percent. The lower part illustrates the result of the dissimilarity measurement, where the red rectangle indicates classified anomalies.

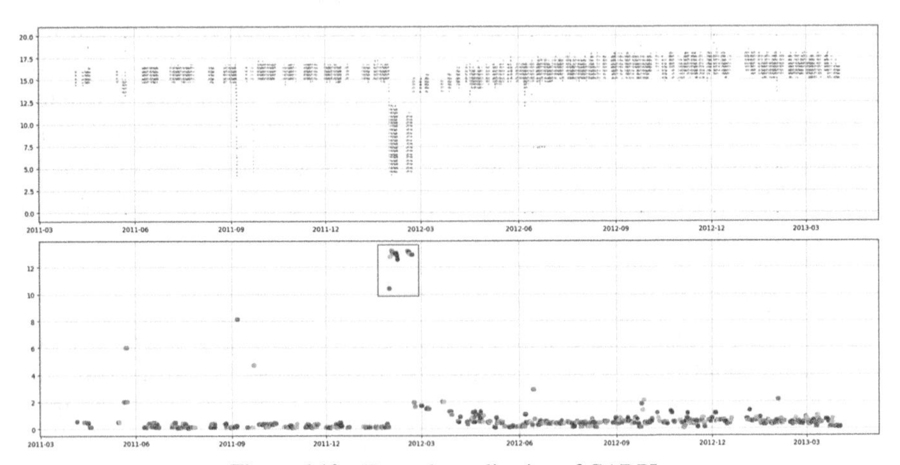

Figure 6.10 Example application of GADPL.

6.3.5 Summary

In this section, we have presented our recent research activities within the scope of the EU project COMPOSITION. As an example for the manufacturing industry, we have briefly described the COMPOSITION architecture along with one of its main components: the open-source IoT platform Link-Smart. As part of our ongoing work, we have also described our corresponding research activities regarding the analysis of sensor data from manufacturing industry.

6.4 Summary and Conclusions

In this work, we have given insights into our recent research activities with regard to the domains of Smart Data and Industrial Internet of Things. To this end, we have focused on the EU projects MONSOON and COMPOSITION as examples for the Public-Private Partnership (PPP) initiatives Factories of the Future (FoF) and Sustainable Process Industry (SPIRE). We have shown two different but conceptually similar architectures for scalable and agile data analytics. In addition, we have provided an overview of our recent Smart Data activities and have exemplified ongoing data-driven analysis of industrial production processes from the process and manufacturing industries.

We conclude that data-driven investigations, either applied in process industry or manufacturing industry, require a solid platform for handling data analytics at scale. The proposed architecture of the Cross Sectorial Data Lab in combination with the open-source IoT platform LinkSmart seem to be promising developments, which are applicable to any industrial sector.

Acknowledgements

The projects underlying this work have received funding from the European Union's Horizon 2020 research and innovation programme under grant agreement No 723650 (MONSOON) and grant agreement No 723145 (COMPOSITION). This work reflects only the authors' views and the commission is not responsible for any use that may be made of the information it contains.

References

[1] C. Beecks, S. Devasya and R. Schlutter, "Data Mining and Industrial Internet of Things: An Example for Sensor-enabled Production Process Optimization from the Plastic Industry," *International Conference on Industrial Internet of Things and Smart Manufacturing (accepted),* 2018.

[2] J. Á. Carvajal Soto, M. Jentsch, D. Preuveneers und E. Ilie-Zudor, "CEML: Mixing and Moving Complex Event Processing and Machine Learning to the Edge of the Network for IoT Applications," *Proceedings of the 6th International Conference on the Internet of Things,* pp. 103–110, 2016.

[3] D. Bonino, J. A. Carvajal Soto, M. T. Delgado Alizo, A. Alapetite, T. Gilbert, M. Axling, H. Udsen und M. Spirito, "Almanac: Internet of things for smart cities," *2015 3rd IEEE International Conference on Future Internet of Things and Cloud (FiCloud),* pp. 309–316, 2015.

[4] G. Cugola und A. Margara, "Processing flows of information: From data stream to complex event processing," *ACM Computing Surveys (CSUR),* p. 15, 2012.

[5] C. Andrieu, N. De Freitas, A. Doucet und M. I. Jordan, "An introduction to MCMC for machine learning," *Machine learning,* pp. 5–43, 2003.

[6] M. M. Gaber, S. Krishnaswamy und A. Zaslavsky, "Advanced Methods for Knowledge Discovery from Complex Data," *On-board Mining of Data Streams in Sensor Networks,* pp. 307–335, 2005.

[7] N. A. Syed, S. Huan, L. Kah und K. Sung, "Incremental learning with support vector machines," *Citeseer,* 1999.

[8] A. P. Dawid, "Present position and potential developments: Some personal views: Statistical theory: The prequential approach," *Journal of the Royal Statistical Society. Series A (General),* pp. 278–292, 1984.

[9] G. J. Vachtsevanos, I. M. Dar, K. E. Newman und E. Sahinci, "Inspection system and method for bond detection and validation of surface mount devices." USA Patent 5,963,662, 1999.

7

IoT European Security and Privacy Projects: Integration, Architectures and Interoperability

Enrico Ferrera[1], Claudio Pastrone[1], Paul-Emmanuel Brun[2], Remi De Besombes[2], Konstantinos Loupos[3], Gerasimos Kouloumpis[3], Patrick O' Sullivan[3], Alexandros Papageorgiou[3], Panayiotis Katsoulakos[3], Bill Karakostas[4], Antonis Mygiakis[5], Christina Stratigaki[5], Bora Caglayan[6], Basile Starynkevitch[7], Christos Skoufis[8], Stelios Christofi[8], Nicolas Ferry[9], Hui Song[9], Arnor Solberg[10], Peter Matthews[11], Antonio F. Skarmeta[12], José Santa[12], Michail J. Beliatis[13], Mirko A. Presser[13], Josiane X. Parreira[14], Juan A. Martínez[15], Payam Barnaghi[16], Shirin Enshaeifar[16], Thorben Iggena[17], Marten Fischer[17], Ralf Tönjes[17], Martin Strohbach[18], Alessandro Sforzin[19], Hien Truong[19], John Soldatos[20], Sofoklis Efremidis[20], Georgios Koutalieris[21], Panagiotis Gouvas[22], Juergen Neises[23], George Hatzivasilis[24], Ioannis Askoxylakis[24], Vivek Kulkarni[25], Arne Broering[25], Dariusz Dober[26], Kostas Ramantas[27], Christos Verikoukis[28], Joachim Posegga[29], Domenico Presenza[30], George Spanoudakis[31], Danilo Pau[32], Erol Gelenbe[33,34], Sławomir Nowak[34], Mateusz Nowak[34], Tadeusz Czachórski[34], Joanna Domańska[34], Anastasis Drosou[35], Dimitrios Tzovaras[35], Tommi Elo[36], Santeri Paavolainen[36], Dmitrij Lagutin[36], Helen C. Leligou[37], Panagiotis Trakadas[37] and George C. Polyzos[38]

[1]Istituto Superiore Mario Boella, Italy
[2]AIRBUS CyberSecurity, France
[3]INLECOM Systems Ltd, United Kingdom
[4]VLTN BVBA, Belgium
[5]CLMS Hellas, Greece

[6]IBM Ireland Ltd, Ireland
[7]Basile Starynkevitch, CEA, France
[8]EBOS Technologies Ltd, Cyprus
[9]SINTEF, NO
[10]TellU, NO
[11]CA Technologies, SP
[12]Department of Information and Communication Engineering, University of Murcia, Spain
[13]Department of Business Development and Technology, Aarhus University, Denmark
[14]Department of Corporate Technology, SIEMENS, Austria
[15]Odin Solutions S.L, Spain
[16]Department of Electrical and Electronic Engineering, University of Surrey, United Kingdom
[17]University of Applied Sciences Osnabrück, Germany
[18]AGT International, Germany
[19]NEC Laboratories Europe, Germany
[20]Athens Information Technology, Greece
[21]Intrasoft International, Luxembourg
[22]UBITECH LTD, Greece
[23]FUJITSU Europe, Germany
[24]Foundation for Research and Technology – Hellas (FORTH), Greece
[25]Siemens AG, Germany
[26]BlueSoft SP. z o.o., Poland
[27]Iquadrat, Spain
[28]Telecommunications Technological Centre of Catalonia (CTTC), Spain
[29]University of Passau, Germany
[30]Engineering Ingegneria Informatica S.p.A., Italy
[31]Sphynx Technology Solutions AG, Switzerland
[32]ST Microelectronics Srl., Italy
[33]Imperial College London, Great Britain & IITiS PAN, Poland
[34]IITiS PAN, Poland
[35]ITI-CERTH, Thessaloniki, Greece
[36]Aalto University, Finland
[37]Synelixis Solutions S.A., Greece
[38]Athens University of Economics and Business, Greece

Abstract

The chapter presents an overview of the eight that are part of the European IoT Security and Privacy Projects initiative (IoT-ESP) addressing advanced concepts for end-to-end security in highly distributed, heterogeneous and dynamic IoT environments. The approaches presented are holistic and include identification and authentication, data protection and prevention against cyber-attacks at the device and system levels. The projects present architectures, concepts, methods and tools for open IoT platforms integrating evolving sensing, actuating, energy harvesting, networking and interface technologies. Platforms should provide connectivity and intelligence, actuation and control features, linkage to modular and ad-hoc cloud services, The IoT platforms used are compatible with existing international developments addressing object identity management, discovery services, virtualisation of objects, devices and infrastructures and trusted IoT approaches.

7.1 BRAIN-IoT

7.1.1 BRAIN-IoT Project Vision

In line with the optimistic forecasts released in last years, Internet of Things (IoT) products and services are being more and more deployed in mass-market and professional usage scenarios, becoming a reality in our day-by-day life. Commercial and pilot deployments world-wide are progressively demonstrating the value of IoT solutions in real conditions, but also rising some concerns with respect to dependability, security, privacy and safety constraints.

The IoT technology and market landscape will become increasingly complex in the longer term i.e. 10+ years from now, especially after IoT technologies will have proven their full potential in business-critical and privacy-sensitive scenarios. An important shift is expected to happen as technology evolutions will allow to safely employ IoT systems in scenarios involving actuation and characterized by stricter requirements in terms of dependability, security, privacy and safety constraints, resulting in convergence between IoT and Cyber Physical Systems (CPS). Attracted by the trend, several organizations have started studying how to employ IoT systems also to support tasks involving actuation and control in business-critical conditions, resulting in a demand for more dependable and "smart" IoT systems. However, in order to turn such vision in reality, many issues must still be faced, including:

- Heterogeneity and (lack of) interoperability: a wide number of IoT platforms exist on the market, both cloud- based and locally hosted. Standardization and open-source initiatives are facilitating convergence among available platforms, which now employ similar usage patterns and increasingly converging sets of protocols, APIs, device models and data interchange formats. Nevertheless, full interoperability across platform still needs to be tackled on a case by case, platform by platform basis, due the wide amount of possible applications, design choices, customization options, formats and configurations that can be adopted by IoT developers and adopters.
- Difficulty of implementing "Smart Behaviours" in open collaboration context: while Machine Learning (ML) and Artificial Intelligence (AI) techniques are rapidly evolving to provide smart behaviours and solutions to increasingly complex problems, it is intrinsically difficult to generically "bind" such solutions to generic concrete IoT and CPS platforms and to make them collaborate for common tasks, since possible interactions between platforms remain unforeseen a priori.
- Security and safety: the distributed nature of IoT makes enforcement of good security practices intrinsically challenging. The market asks for IoT solutions suitable to safely support business-critical tasks, which can be deployed rapidly and with low costs. The emerging availability of actuation features in IoT systems calls for stricter security requirements. Nevertheless, many of today's IoT-based products are implemented with low awareness of potential security risks. As a result, many IoT products lack even basic, state-of-the-art security mechanisms, resulting in critical effects when such flaws deployed to mass-market scenarios.
- Enforcement of Privacy and Data Ownership policies: as IoT products are increasingly purchased and deployed by corporate and private users in their homes, work places, factories and commercial areas, privacy issues and violations become more frequent. While policies are quickly catching up by enforcing a suitable framework of rules within the EU, a comprehensive solution able to give back control of privacy aspects to users is still missing – creating significant issues when unaware users accept that their data is moved in foreign countries, outside the safe shield provided by EU regulations.
- Business models colliding with long-term resilience and survivability of IoT services: many IoT solutions on the market adopt fully centralized, cloud-oriented approaches. This is often done e.g. to ensure

that customers' devices are forced to use forever a single commercial back-end service. Such lock-in approaches create artificial monopolies, negatively affecting user rights and the overall market competitiveness. This practice introduces singular point of failures in IoT systems, making survivability and resiliency features difficult to be granted in the long term, therefore sometimes resulting in negative experiences for end users.

- Market Fragmentation and incumbency of large players: the current market of IoT platform solution is still affected by fragmentation among the many IoT platforms available each focused in specific application domain or associated technology stacks. Moreover, some market segments (i.e. the cloud-based IoT platforms market) are notably dominated by few dominant players – often based outside the EU, thus hampering the potential business opportunities for EU companies.

While EU-based initiatives and policies are doing significant amount of work to tackle such issues, often with very positive results, solutions suitable to tackle challenges arising for futuristic IoT usage scenarios are still missing. Future critical issues may be hiding under the hood already now and be ready to appear in the close future, putting at stake user acceptance and the credibility of the whole eco-system of IoT solutions vendors, integrators and adopters and hindering wider adoption of IoT solutions in potentially valuable markets.

7.1.2 Objectives

In order to tackle the aforementioned challenges, the BRAIN-IoT (*model-Based fRamework for dependable sensing and Actuation in INtelligent decentralized IoT systems*) project focuses on complex scenarios, where actuation and control are cooperatively supported by populations of heterogeneous IoT systems. In such a complex context, many initiatives fall into the temptation of developing new IoT platforms, protocols, models or tools aiming to deliver the ultimate solution that will solve all the IoT challenges and become "the" reference IoT platform or standard. Instead, usually they result in the creation of "yet-another" IoT solution or standard.

BRAIN-IoT will establish the principle that future IoT applications should *never* be supported by a single, unique, irreplaceable IoT platform. Rather future IoT services should exist within a federated/evolving environment that not only leverages current Industry Standards but is also capable of adapting to embrace future unforeseen industry developments. BRAIN-IoT aims at demonstrating that the lack of a single IoT standard and platform, which is generally recognized as the most notable weakness of IoT, can be turned into a strength and a guarantee for market competitiveness and user protection – if the proper framework for IoT dynamicity, security and privacy is in place.

The breakthrough targeted by BRAIN-IoT is to establish a practical framework and methodology suitable to enable smart cooperative behaviour in fully de-centralized, composable and dynamic federations of heterogeneous IoT platforms. BRAIN-IoT builds on model-based approaches and open industry standards and aims at supporting rapid development and deployment of applications and services in professional usage scenarios characterized by strict constraints in terms of dependability, safety, security and privacy. The BRAIN-IoT vision is realized through seven Technical Objectives (TOs), as described in Table 7.1.

Table 7.1 BRAIN-IoT technical objectives

Technical Objective (TO)	Description
TO1: to enforce interoperability across heterogeneous IoT devices autonomously cooperating in complex tasks.	BRAIN-IoT approach to interoperability is based on the adoption of shared semantic models, dynamically linked to concrete IoT devices (sensors, actuators, controls, etc.) operating autonomously in complex scenarios. Binding of models to concrete implementations leverages mapping to open industry standards for semantic device description.
TO2: to enable dynamic smart autonomous behaviour involving actuation in IoT scenarios	Building upon shared models (TO1) BRAIN-IoT facilitates the deployment of smart cooperative behaviour, realized by means of modular AI/ML features which can be dynamically deployed to heterogeneous IoT devices in mixed edge/cloud IoT environments. Smart behaviour features are enriched by distributed data processing, federated learning, virtualization/aggregation of data/events/objects, resolution of mixed-criticality situations and conflicts, verification and context-aware self-adaptation of connectivity and real-time event-oriented, reactive approaches.

(*Continued*)

Table 7.1 Continued

Technical Objective (TO)	Description
TO3: to enable the emergence of highly dynamic federations of heterogeneous IoT platforms able to support secure and scalable operations for future IoT use cases	This is achieved by leveraging fully de-centralized peer-to-peer approaches providing linkage between modular, ad-hoc IoT self-hosted and cloud-based services through existing open standards.
TO4: to establish Authentication, Authorization and Accounting (AAA) in dynamic, distributed IoT scenarios	BRAIN-IoT introduces a holistic end-to-end trust framework for IoT platforms suitable to be employed in scenarios characterized by strict security and safety requirements, associated with actuation and semi-autonomous operations, and by special needs for secure identification, authentication of data and devices, encryption, non-deniability, as well as detection of cyber-attacks and protection against them. This is done by adopting established security protocols, joint with distributed security approaches derived by peer-to-peer systems e.g. block-chain.
TO5: to provide solutions to embed privacy-awareness and privacy control features in IoT solutions	BRAIN-IoT develops new patterns for interaction between users and IoT solutions, leveraging semantic mapping of privacy requirements towards data and service models in use in each specific use case, introducing privacy-related APIs and models. This enables the possibility to programmatically inform users about privacy policies in place, as well as enabling them to exercise fine-grained privacy controls.
TO6: to facilitate rapid model-based development and integration of interoperable IoT solutions supporting smart cooperative behaviour	BRAIN-IoT provides tools to ease rapid prototyping (development, integration) of smart cooperative IoT systems. This is achieved by extending available tools for development, integration, commissioning and management of IoT and Cyber-Physical systems.
TO7: to enable commissioning and reconfiguration of decentralized IoT-based applications	BRAIN-IoT enables end-users to dynamically commission and reconfigure their modular IoT instances, choosing among the available platforms, modules implementations and services. This is achieved by extending existing open marketplace of IoT services and data jointly with available catalogues providing open IoT enablers and integrating them with its federation framework.

7.1.3 Technical Approach

The overall BRAIN-IoT concept is depicted in Figure 7.1 following the reference model proposed by Recommendation ITU-T Y.2060. BRAIN-IoT looks at heterogeneous IoT scenarios where instances of IoT architectures can be built dynamically combining and federating a distributed set of IoT services, IoT platforms and other enabling functionalities made available in marketplaces and accessible by means of open and standard IoT APIs and protocols.

At the bottom of the conceptual architecture, the IoT Devices and Gateways layer represents all physical world IoT devices with sensing or actuating capabilities, computing devices and includes complex subsystems such as autonomous robots and critical control devices. It is worth observing that BRAIN-IoT specifically aims to support the integration into an IoT environment of devices and subsystems with actuation features that could possible give rise to mixed-criticality situations and require the implementation of distributed processing approaches. The BRAIN-IoT Management capabilities includes all the features needed to support the envisioned fully decentralized scenario dynamically integrating heterogeneous IoT Devices and Gateways as well as:

- IoT Services – third party services accessible through open interfaces and offering data or various functionalities including data storage, data statistics and analytics, data visualization;

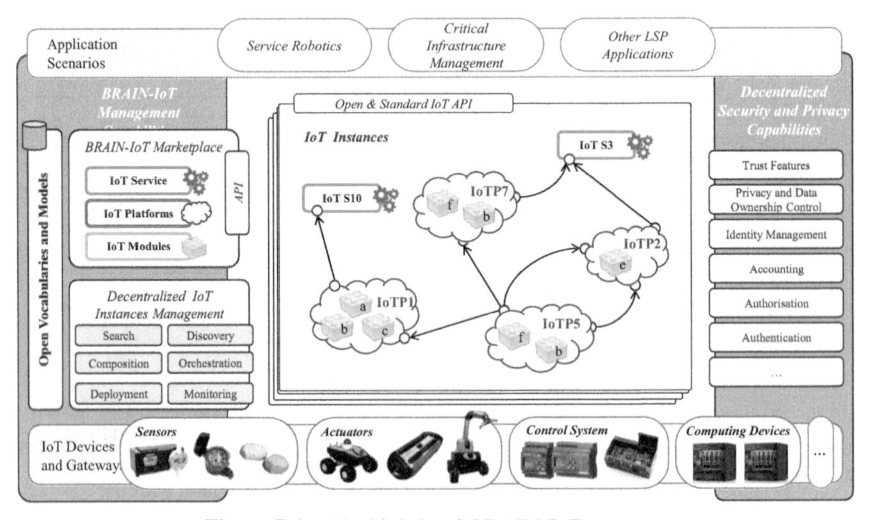

Figure 7.1 The high-level BRAIN-IoT concept.

- IoT Platforms – instances of open IoT platforms whose configuration and functionalities can be dynamically updated;
- IoT Modules – enabling functionalities (e.g., smart control features, data processing, data storage) that can be associated to a specific IoT platform instance and composed in order to meet given functional requirements.

Concerning the IoT Modules, the ones supporting smart control features are particularly relevant for the BRAIN-IoT challenging scenarios encompassing heterogeneous sensors and actuators autonomously cooperating in complex, dynamic tasks, possibly across different IoT Platforms. BRAIN-IoT will then develop a library of IoT modules implementing algorithms promoting collaborative context-based behaviours, control solutions based on Machine Learning Control, real-time data analysis and knowledge extraction techniques. Concerning the IoT Platforms, BRAIN-IoT will support different existing IoT solutions including e.g., FIWARE and SOFIA. All the above IoT building blocks can be described by a set of open and extendable vocabularies as well as semantic and behavioural models. This actually allows moving forward an easier, automated and dynamic integration within the BRAIN-IoT environment of new and existing IoT Services, Platforms and Modules available for traditional IoT applications. In fact, BRAIN-IoT defines a new meta-language, namely the IoT Modelling Language (IoT-ML), which uses the above set of vocabularies and models to formally describe an IoT Instance i.e., how a given set of IoT services and Platforms are interconnected with each other and federated and which IoT Modules are associated to the considered IoT Platforms. IoT-ML will base on existing solutions provided by OMG and W3C. The Decentralized IoT Instances management is instead in charge of offering the capabilities needed to support the dynamic composition of a given set of IoT building blocks into a specific IoT Instance. The vision is to progress from the fog computing paradigm and create distributed IoT Micro-cloud environments hosting IoT Platforms and IoT Modules and advertising their runtime capabilities. The resulting Micro-cloud environments are enhanced with management capabilities that allow search and discovery operations and their dynamic federation to form a specific IoT instance. These capabilities pave the way toward highly dynamic scenarios where IoT Modules and relevant functionalities can be composed and migrated runtime from one IoT Platform to another, complex tasks can be dynamically distributed between the edge and the cloud IoT Platforms depending on variable requirements and where IoT Instances can be fully reconfigured adding/removing runtime new IoT building blocks from the federation. BRAIN-IoT will also provide peculiar management strategies and techniques permitting the dynamic deployment/transfer of Smart Control IoT Modules across mixed edge and cloud environments. The Decentralized IoT Instances management also handles advanced IoT Instances configurations, properly orchestrating external IoT services with other IoT building blocks active in the resulting BRAIN-IoT fog environment. Finally, monitoring components allow to continuously supervise the overall IoT Instance and relevant composite application. In this way, it is possible to check the status of the federated building blocks, provide alerting, reporting and logging mechanisms and, if needed, trigger an IoT Instance reconfiguration e.g., because of a failure in one of the adopted IoT Modules, Platforms or Services. All the described management capabilities will base on relevant industry standards i.e., W3C Web of Things and OSGi, and will be extended to support agile composition and orchestration. The scalability aspects will be taken into careful consideration to support effective discovery and search of a potential high number of IoT building blocks. The orchestration process is conceived in such a way that it is possible to import/link IoT Modules, Platforms and Services made available from a BRAIN-IoT Marketplace characterized by a relevant set of open APIs. One of the most peculiar aspects being considered in BRAIN-IoT is the management of actuation capabilities in the considered Fog environment. In this context, the possibility to easily develop the previously introduced smart control features is pretty relevant. To this aim, BRAIN-IoT will evolve from already existing solutions, such as Eclipse Papyrus, and develop Model Binding and Synthesis tools extended to support the BRAIN-IoT open vocabularies and models, the IoT-ML and other IoT related standards. The resulting toolset will be used to develop novel Smart Control Features that could be possibly published as IoT Modules in the BRAIN-IoT Marketplace, as depicted in Figure 7.2.

Finally, Figure 7.3. summarizes the above description of the BRAIN-IoT environment offering a view of possible configurations of an IoT Instance with different distribution of the IoT building blocks between edge and cloud.

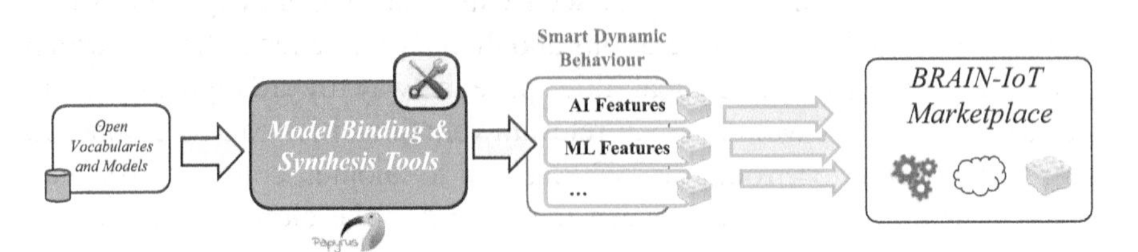

Figure 7.2 BRAIN-IoT development concept.

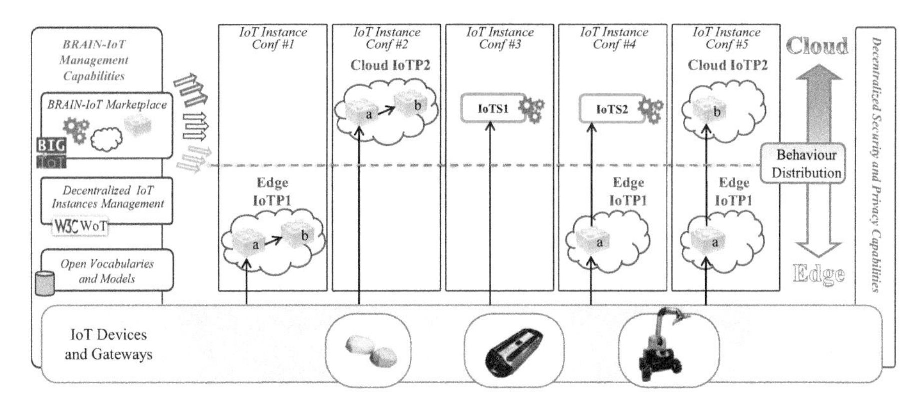

Figure 7.3 BRAIN-IoT deployment concept.

7.1.4 Security Architecture Concept

From the security and privacy perspective, IoT currently presents two main inherent weaknesses:

- Security is not considered at the design phase,
- As of today, no solution is offering a complete end-to-end security approach for any kind of devices (from the temperature sensor, the smoke detector, to the robot).

Existing systems don't apply the "secure-by-design" concept where security is seen as one of the major constraint of the system. To provide secure IoT solutions, modelling and analysis need to be integrated in the design and validation of application scenarios and IoT architectures. If the focus moves to a scenario where different heterogeneous building blocks are dynamically composed, additional security and privacy concerns arise. As a consequence, BRAIN-IoT provides a methodology to address security in the considered fog environment, based on an iterative process, allowing to take into account new scenarios. More specifically, BRAIN-IoT extend the successful methods of attack tree modelling and quantitative analysis to support secure composable IoT systems. This extension enables transparent risk assessment of IoT security architectures, i.e., it will address the needs and potential risks involved in an IoT environment specifying when and where to apply security controls in an understandable way thus raising user-awareness and trustworthiness. The results of the analysis are specific technical requirement to implement for each use case/scenario in order to reach the targeted security level.

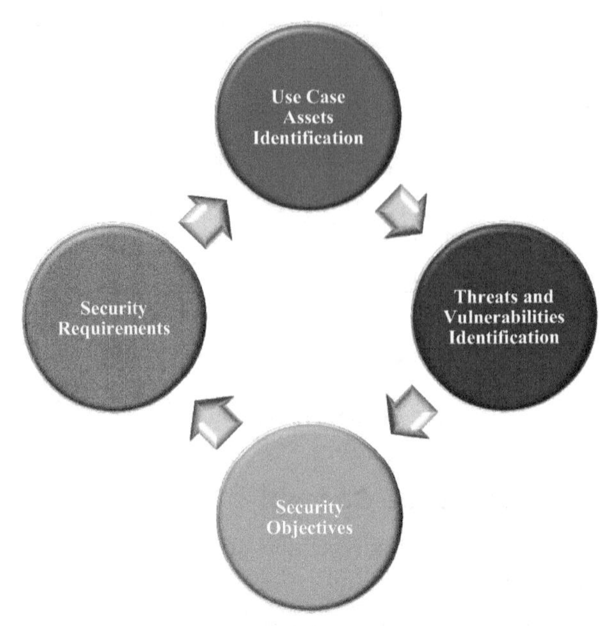

Figure 7.4 Iterative risk analysis methodology.

Second, existing security solution for IoT have many weaknesses, such as:

- Lot of flow disruption (with network component accessing data in clear text)
- Some protocol chooses to downgrade security algorithm to fit performance constraints,
- State-of-the-art solution are complex to set up in decentralized environment.

In order to provide a new approach, BRAIN-IoT integrates innovative Decentralized Security and Privacy Capabilities including Authentication, Authorization and Accounting for the overall distributed fog environment and end-to-end security for IoT data-flows. This security layer is based on a combination of well-established standards, such as PKI, with more innovative solution, stateless oriented, to fit the constraints of any kind of IoT (low power, low bandwidth, etc.)

A cross-platforms framework facilitating the adoption of privacy control policies is also hosted in the BRAIN-IoT environment. The objective is to provide end users with the means to easily monitor and control which data to – collect and to who make it available.

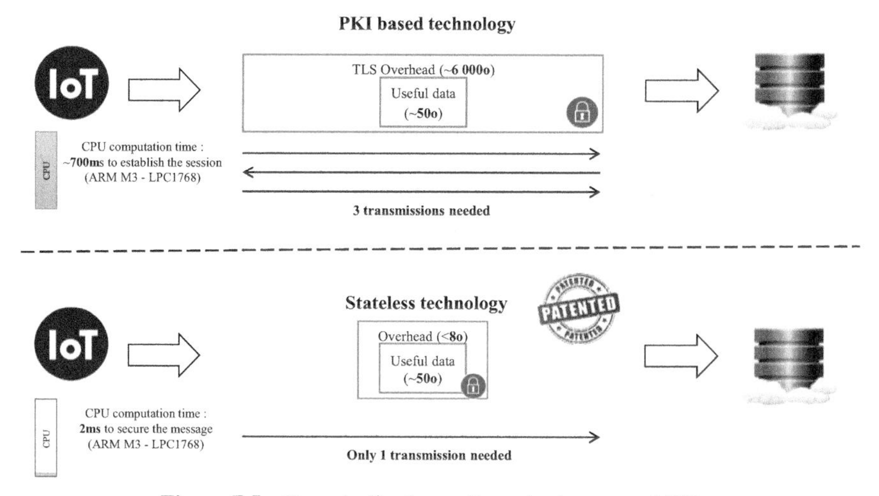

Figure 7.5 Decentralised security and privacy capabilities.

7.1.5 Use Cases and Domain Specific Issues

The overall depicted concept draws requirements and challenging use cases from IoT applications in two usage scenarios, namely Service Robotics and Critical Infrastructure Management, which provide the suitable setting to reflect future challenges in terms of dependability, need for smart behaviour, security and privacy/data ownership management which are expected to become more significant and impacting in the long-term (10+years).

7.1.5.1 Service robotics

The Service Robotics use case will involve several robotic platforms, like the open-source Robotics Operating System (ROS), which need to collaborate to scan a given warehouse and to assist humans in a logistics domain. The term Service Robotics is generally related to the use of robots to support operations done by humans and the logistics domain is one of the more interesting, for the presence of several tasks, where the robots can help the workers, making their tasks easier and safer. As example, they can cooperate to move a heavy object from one place to another. At the same time, robots involved in the scenario should scan the whole warehouse and update in real-time informative interfaces for the managers and the workers (e.g. warehouse's map), sharing the collected information. In addition, to the information related to the maps of the whole monitored area, the connected robots will also be equipped with a set of sensors, which will allow collecting interesting info, like room temperature, presence of humans or presence of obstacles in the robot path. Since several robots collaborate to collect the information, they can keep the status of the area updated in real-time and balance the effort required among them. At the beginning, the robots are configured with some default information, like the map of the warehouse. Then, this information is updated in real-time, while the robots perform their main tasks. The demonstration use-case proposed will include both real-time collection of data and control of the included robots. Particularly, the actuation of these robots will be an interesting test-bed for the platform, to demonstrate how the solutions developed by BRAIN-IoT allow to control remotely, in a standard way, the complex devices involved in this scenario.

The BRAIN-IoT solution enables the service robotics scenario, demonstrating how the tool-chain and marketplace developed by the project can be used to enable the cooperation of the different robots. In the envisioned scenario, the BRAIN-IoT toolkit will be leveraged to design and test all the aspects of the use case: the behaviours of the robots, the interactions with humans and the cooperation of involved robots, to do specific tasks. The use of the BRAIN-IoT toolkit will enable to limit the development of new ad-hoc software components, indeed, where possible, the solution will be based on open-source components and services already developed in other IoT platforms, provided to the developers through the BRAIN-IoT's marketplace and interconnected using the services and tools developed in the project. This scenario described will involve also several security aspects. Mechanisms of encryption and authentication will be adopted in the whole final solution, to guarantee the protection of the data exchanged and of the users' privacy. To avoid inappropriate use of the robots by malicious users and to avoid possible incidents due to remote control (i.e. authorized workers that try to control the robots remotely, without a correct visual of what is happening in the warehouse), techniques of indoor localization are considered to guarantee that only workers located in specific zones near the robots can control them. Furthermore, the solution will protect the privacy of users, anonymizing the data collected, to avoid sharing users' info, also if the data are stolen by malicious users.

7.1.5.2 Critical infrastructure management

The Critical Water Infrastructure Monitoring and Control use case focuses on the management of the water urban cycle in metropolitan environment of Coruña. The base of this system will be made of a complex portfolio of probes, meters, sensors, devices and open-data sources deployed on the field, including: water, flow and pressure meters on the water mains; smart devices, which measure the main chemical-physical characteristics of water; pluviometers, which can monitor the level of rainwater in a specific zone, water circular pumps, which can be used to control the flow of liquids in heating systems. These devices will be geographically distributed, heterogeneous and will be provided by different owners: directly by the water utility, by end-users themselves or by third-party service providers, like, SMEs providing ancillary water services. For this reason, there will be many different data involved in these scenarios: meteorological open data, reservoir water level data, purification data, distribution data in the various subsystems, customer data in urban water supply processes, sewage collection and sewage treatment data in different subsystems. The collection of all these data will allow to provide value-added services, like showing to the client, commercial or not, the quality of the water provided to them or the possibility to react quickly to critical situation and to do predictive maintenance, through the ability to detect anomalous behaviours and to fix them, before they become an issue difficult to be fixed.

The BRAIN-IoT solution will enable this scenario, allowing to collect data from all the different domains and to actuate the devices where needed. The WoT-based approach used for the design of the platform, will be leveraged to collect the data, provided by different public and private IoT platforms, using heterogeneous protocols and data formats. Furthermore, BRAIN-IoT focuses particularly to design and develop ways to control devices abstracted by these solutions. For example, the system needs to allow the managers to control the circulator pumps, regulating the fluid flow in heating system, to avoid problems or to react to some critical situation detected through the monitoring sensors. The collection of the data about water consumption from different sources generates risks about privacy protection. Indeed, the data can be shared with public entities or third-party services providers for several purposes, like statistical measures. To do this, the data need to be associated with all the potentially interesting contextual information (i.e. Position of the data, timeslot when the data have been measured and so on) but removing the association with all the personal info of the entity, related with that data. Finally, security mechanisms will be used for the actuation of devices that is potentially a dangerous task, which must be executed only by expert personal that need to know well what they are doing and the context in which the device is operating. For this reason, mechanisms of accounting, authentication and authorization will be used to guarantee that only authorized expert users are able to do these tricky operations.

7.2 Cognitive Heterogeneous Architecture for Industrial IoT – CHARIOT

7.2.1 Introduction

Recently, cloud Computing as well as Internet of Things (IoT) technologies are rapidly advancing under the concept of future internet. Numerous IoT systems and devices are designed and implemented following industrial domain requirements but most of the times not considering recent risk relating to openness, scalability, interoperability as well as application independence, leading to a series of new risks relating to information security and privacy, data protection and safety. As a result, securing data, objects, networks, infrastructure, systems and people under IoT is expected to have a prominent role in the research and standardization activities over the next several years. CHARIOT EC co-funded, research project, clearly recognises and replies to this challenge, identifying needs and risks and implementing a next generation cognitive IoT platform that can enable the creation of intelligent IoT applications with intelligent shielding and supervision of privacy, cyber-security and safety threats, as well as complement existing IoT systems in non-intrusive ways and yet help guarantee robust security by placing devices and hardware as the root of trust. The scope of this article is to provide a detailed overview of the CHARIOT vision, technical objectives and overall solution, a high-level presentation of the system architecture as the project approaches in the design of the CHARIOT solution and platform.

7.2.2 Business Challenge and Industrial Baselines

The CHARIOT project activities are aligned with actual business and industrial requirements on the recent needs on data safety, security and privacy over modern IoT systems following demands of highly increasing numbers of IoT devices. It is expected that by 2025, there will be 75 Billion IoT-connected devices World Wide while spending on IoT devices and services reached $2 trillion in 2017, with China, North America, and Western Europe accounting for 67% of all devices [8]. This growth in connected devices is anticipated accelerate due to a rise in adoption of cross-industry devices (LED lighting, HVAC systems, physical security systems and lots more). On top of this CHARIOT also recognizes various IoT security breaches that have been dominating headlines, while 96% of security professionals expect an increase in IoT breaches this year [13]. In the direction of a more secure IoT infrastructure, there have been some requests for government regulation of the

IoT, asserting that IoT manufacturers and customers are not paying attention to the security of IoT devices [14]. CHARIOT has clearly recognized the above requirements and has an aligned set of objectives towards increase of security, privacy and safety of industrial IoT networks and components.

7.2.3 The CHARIOT EC, Research Project – Vision and Scope

CHARIOT (Cognitive Heterogeneous Architecture for Industrial IoT) is an EC, co-funded, research project granted under the IoT-03-2017 – R&I on IoT integration and platforms as a Research and Innovation (RIA) EC topic. The CHARIOT consortium consists of research and innovation organisations from major research streams all merged into the CHARIOT solution providing the competence to deliver a 'holistic approach addressing Privacy, Security and Safety of IoT operation in industrial settings with safety critical elements'. The consortium includes competences in the fields of Project management and IoT governance (INLECOM, UK), Cognitive Architectures & Platforms for IoT (IBM, Ireland), Static source code analysis tools (CEA, France), Analytics Prediction models and Dashboard development (EBOS, Cyprus) as well as IoT deployment architectures, cloud/fog technologies (VTLN, Belgium, TELCOSERV, Greece), security including cybersecurity (ISC, ASPISEC, Italy) and integration aspects (CLMS, Greece).

CHARIOT provides a design method and cognitive computing platform supporting a unified approach towards Privacy, Security and Safety (PSS) of IoT Systems including the following innovations summarised below:

- A **Privacy and security protection method** building on state of the art Public Key Infrastructure (PKI) technologies to enable the coupling of a pre-programmed private key deployed to IoT devices with a corresponding private key on a Blockchain system. This includes the implementation of security services utilising a cryptography-based approach and IoT security profiles all integrated to the CHARIOT platform.
- A **Blockchain ledger** in which categories of IoT physical, operational and functional changes are both recorded and affirmed/approved by the various run-time engines of the CHARIOT ecosystem while leveraging existing blockchain solutions in innovative ways.
- **Fog-based decentralised infrastructures** for Firmware Security integrity checking leveraging Blockchain ledgers to enhance physical, operational and functional security of IoT systems, including actuation and deactivation.

- An **accompanying IoT Safety Supervision Engine** providing a novel solution to the challenges of securing IoT data, devices and functionality in new and existing industry-specific safety critical systems.
- A **Cognitive System and Method** with accompanying supervision, analytics and prediction models enabling high security and integrity of Industrials IoT.
- **New methods and tools for static code analysis** of IoT devices, resulting in more efficient secure and safer IoT software development and V&V.

CHARIOT is closely following a business and industrially driven approach to align the developed technologies and outcomes to actual industrial needs in the fields of transport, logistics etc and in general domains of IoT applications. With this vision, CHARIOT, will apply its outputs and recent developments to three living labs in order to demonstrate its realistic and compelling heterogeneous solutions through industry reference implementations at representative scale, with the underlying goal of demonstrating that Secure, Privacy Mediated and Safety IoT imperatives are collectively met, in turn delivering a key stepping stone to the EU's roadmap for the next generation IoT platforms and services. The actual living labs will be implemented in the industrial framework of TRENITALIA (rail), Athens International Airport (transport) and IBM Ireland (smart buildings) [9, 10].

7.2.4 CHARIOT Scientific and Technical Objectives

We present below a summary of the CHARIOT scientific and technical objectives as the main scope and outcomes of the CHARIOT unified design method and cognitive computing platform supporting a unified approach towards Privacy, Security and Safety (PSS) of IoT Systems, that places devices and hardware at the root of trust, in turn contributing to high security and integrity of industrial IoT.

- **Objective 1:** Specify a Methodological Framework for the Design and Operation of Secure and Safe IoT Applications addressing System Safety as a cross cutting concern. The CHARIOT design method will bridge the systems engineering gaps that currently exists between a) the formal safety engineering techniques applied in the development and testing of safety critical systems and b) the rapidly evolving and ad-hoc manner in IoT devices are developed and deployed. This includes classification and usage guidelines of relevant standards and platforms, introduction of new concepts and methods for coupling pre-programmed private security keys on the IoT device with a Blockchain system and ledger to enhance its security and privacy protection and guarantee that only authorised entities who have a matching key can influence operation, function and change, thereby invalidating the potential for a substantial spectrum of cyber-attacks and significantly before they become actual exploits. Developments will also include a specialized static source code analysis tool and cross-compiler to help avoid safety defects and add some meta-data into the binary permitting that binary executable to be suitably "filtered" or "authenticated" by gateways and, in turn, shielding against cyber-attacks while consolidate all the above into the CHARIOT IoT Design Method.

- **Objective 2:** Develop an Open Cognitive IoT Architecture and Platform (the CHARIOT Platform), that exhibits intelligent safety behaviour in the diverse and complex ways in which the safety critical system and the IoT system will interact in a secure manner. This includes the creation of an open IoT Cognitive Architecture for a "Web-of-Things" like environment, supporting a range of solutions and applications interacting with highly distributed, heterogeneous and dynamic IoT and critical safety system environments. Under this objective, CHARIOT will also provide interfacing to a topological representation and functional behaviour models of IoT system components and safety profiles as well as a integrated IoT Platform by enhancing the existing state of the art in cognitive computing platforms and build the additional CHARIOT safety and privacy features through open APIs and including security services utilising the Blockchain technology, the IoT security profiles and fog computing services.
- **Objective 3:** Develop a runtime IoT Privacy, Security and Safety Supervision Engine (IPSE) which will act continuously to understand and monitor the cyber-physical ecosystem made up of the IoT devices, safety critical systems and a PSS policy knowledge-base in real-time. This cognitive engine will ensure that potentially endangering behaviours of the IoT system are predicted and avoided and, where that is not possible, handled in an agreed manner in conjunction with safety critical systems runtime environments to avoid a breach of the safety constraints. IPSE will include four innovative cognitive applications: A Privacy Engine based on PKI and Blockchain technologies, a Firmware Security integrity checking, an IoT Safety Supervision Engine (ISSE) and an Analytics Prediction models and Dashboard.

- **Objective 4:** Test and validate against Industrial IoT safety in three Living Labs (LLs) addressing different industrial areas in IoT safety: in transport (rail and airports) and in buildings. The LLs will be used to demonstrate the capabilities of the proposed approach and provide compelling and representative industry use cases with associated test data that will effectively demonstrate an integrated end-to-end application for how the broader CHARIOT approach to security, privacy and safety will be applied in different industry-representative contexts at enterprise scale.
- **Objective 5:** Ensure large outcomes scale up through wide dissemination, exploitation actions and a Capacity Building Programme aiming at infrastructure sustainability, organisational development, and human capital development through training on the practical use of the CHARIOT Concepts, Capabilities, Services and Platform Offering.

7.2.5 Technical Implementation

The technical implementations in CHARIOT will be performed in a series of phases, perfectly aligned to the project scientific objectives presented above. These include the design, development, integration and testing of several key-components as will be presented in the chapters that follow.

7.2.5.1 The CHARIOT Open IoT cognitive cloud platform

The CHARIOT cognitive platform comprises of a set of functions, logical resources and services hosted in a cloud data centre supporting a range of cognitive solutions and application interacting with an ecosystem of highly distributed, heterogeneous and dynamic IoT and critical safety system environments. This module provides connectivity and intelligence, supporting actuation and control features as required by the final applications. It takes advantage of an existing IoT platform (IBM's Watson IoT [15]) to demonstrate concept and capability and will also support integration with other safety, privacy and machine-learning cloud services via relevant open APIs, thus supporting third party integration and innovation. Through such interfaces, the CHARIOT platform will subsequently be compatible with existing international developments, addressing object identity management, fog, discovery services, virtualisation of objects, devices and infrastructures and trusted IoT approaches. The CHARIOT platform is being designed respecting open principles.

While the open nature of the architecture does not preclude the adoption of specific vendor technologies in the initial platform Proof-of-Concept (PoC) implementation for the living labs, the architecture will be intentionally designed with open interfaces such that individual middleware and components can be easily substituted with alternatives in future implementations. The platform will also explore the development and deployments of probes to provide methods of collecting information on the IoT devices and on the safety-critical-systems in real-time, in turn facilitating the creation of a topological representation and functional behaviours of the IoT systems by the Safety Supervision Engine.

The cognitive engine will be used to test the concept of adapting autonomously, instructing the "system" to behave in intended ways and perform required updates and changes through authorised actors. Based on a pattern of events evidenced in ledgers, the cognitive system will adapt/instruct the IoT system(s) to adapt in appropriate ways based on leveraging innovative machine learning and data mining approaches.

PKI and Blockchain Technologies

Leveraging existing blockchain technologies along with traditional PKI schematics enables CHARIOT to revolutionize the field of identity management and access control. Blockchain acts as the backbone of the system by enabling trust between the various CHARIOT services as well as between the gateways and the IoT sensors within the network. The implementation will be based on a permissioned blockchain that will become the mediator of any communications occurring within the network.

7.2.5.2 Static code analysis and firmware security tool

A significant component of the CHARIOT overall solution is the development and enhancement of a free software cross-compilation toolset – leveraging on existing open source technologies – for IoT engineers designing IoT systems and developing source code running on them. Strong highly safety-critical IoT software requires a costly, but extensive, formal methods approach [11], in which developers agree to put a lot of efforts in formally specifying then analysing their source code and using proof assistants to ensure lack of bugs (w.r.t. some explicitly, detailed and formalized specification). But the CHARIOT project aims to help less life-critical IoT software developers by providing them with a tool to help them in developing IoT software and better use of existing free software IoT frameworks. This will be an open software toolset that assists IoT software developers, particularly as not experts in computer science but a competent engineer in a specific industrial domain (railroad, automotive, smart building, maritime, etc.), so even heuristic source code analysis techniques (leveraging above some formal methods approaches) can improve his/her coding productivity. This tool will be developed as part of the CHARIOT solution and a plugin/extension module for GCC based compilers that the software industry is currently using and will be executed at compilation/linking stage and will use meta-programming techniques to foster "declarative" high-level programming styles. This will enable the developers (as the IoT device firmware developers) to identify most safety critical functions executed at the IoT device or gateway level. Also, firmware compiled with that toolset will carry some cryptographic signature to enable filtering of firmware updates in the gateway.

7.2.5.3 Integrated IoT privacy, security and safety supervision engine

This engine is a set of novel runtime components which act in concert to understand and monitor the cyber-physical ecosystem made up of the IoT gateway and devices, the safety critical systems and safety/security policy knowledge-base. The Privacy Engine utilises existing security protocols and technologies such as Blockchain to provide a strong foundation for the trusted interchange of information about and between the participants in the system-of-systems. The Safety Engine also analyses the IoT topology and signal metadata relative to the relevant safety profiles and applies closed-loop machine-learning techniques to detect safety violations and alert conditions. The objective of this engines is to develop a cognitive engine that will leverage the Cyber-Physical topological representation of the system-of-systems combined with the security/safety-polices to provide a real-time risk map will allow for both static analysis and continuous monitoring to assess safety impact and appropriate response actions.

The supervision engine will be responsible for interacting with the CHARIOT IoT platform, providing the centralised intelligence and control functionality for applying the necessary privacy, security, and safety policies to all components in the IoT system of systems, monitor IoT devices and systems to detect abnormalities in their behaviour and analyse their causes, maintain an internal topological representation of the constantly evolving IoT system of systems and collect and represent PSS policies and the threat intelligence in the topology to provide a real-time risk map, impact assessment and triggering of appropriate response actions. The engine will also maintain safety, security and privacy even when unknown devices and sensors are connected to the network, ensuring that they do not interfere to the normal operation of existing IoT components, assess the topology to detect whether the IoT ecosystem has entered or is predicted to be advancing towards an abnormal (unsafe/insecure) state, and automatically activate a safety remediation in response to this unsafe state, to reduce the impacts on users and other IoT components and restrict abnormal operations and allow operations of safe functions to maintain at reduced level the operation of the controlled system.

7.2.5.4 Analytics prediction models user interface

This system component is an innovative cognitive web application, which constitutes together with other relevant components – such as the Privacy and the IoT Safety Supervision Engine – the IPSE. The application collects the data received by the various IoT gateways and sensors in the fog network and using appropriate algorithms, Analytics Prediction models will be created and presented through a user friendly configurable dashboard.

This module will be the advanced-intelligence dashboard for both understanding of the IoT ecosystem topology and for post data analytical purposes to assist in the refinement and improvements of PSS policies while at the same time act as the interface between the CHARIOT platform and the system operator/user.

7.2.6 System Demonstration, Validation and Benchmarking

The overall system operation will be demonstrated and validated via full integration to the actual operating environments and infrastructures of three industrial sites over precise key-performance-indicators that contribute to the separate business environment and value. The three key selected sites (living labs, LLs) will be: a) Trenitalia (transport – rail) b) IBM Ireland business campus (smart buildings) and c) Athens international airport (transport – airport). Details on the three separate cases have been included below:

7.2.6.1 Living lab 1: Trenitalia

The primary objective in this LL is to enhance the safe operation of the Italian railways service. This includes, reduction of risk to passengers and personnel, compliance with appropriate regulations, and creation of a safe and efficient operating environment in the railways. At the same time this use case will focus on utilizing the feed from IoT used to monitor electrical and mechanical components dedicated on assessing energy consumption and dispatch them to the on-board control servers and the land-based central control system. The application of the CHARIOT tool will facilitate the timely recognition of sensors malfunction, along with prediction of maintenance requirements.

7.2.6.2 Living lab 2: IBM business campus

In this LL, the objective will be to enable the continued IoT evolution of the IBM technology campus from a set of individuals "automated/smart" buildings into to a truly cognitive IoT environment that provides a safer and more efficiently managed working environment for all IBM staff, customers and visitors and also to use the knowledge gained to help drive advancements in Cognitive IoT to a global scale by reflecting it in IBM products and services.

7.2.6.3 Living lab 3: Athens international airport

The application of CHARIOT in this Living Lab will address safety of airport Infrastructures, enhance protection of Athens airport's facilities from physical and cyber threats. To achieve this, CHARIOT will enhance airports capability on early detection/prediction of hazardous situations, in parallel with reduction in false positive alarms that disrupt airport operations.

7.2.7 Summary and Discussion

This chapter provides the overall concept of the CHARIOT project and business orientation. It summarizes the project scope and business value as derived from actual industrial needs in the framework of safety, security and privacy of industrial IoT. CHARIOT started in January 2018 and it currently in the stage of requirements extraction and definition of the system overall architecture as this is aligned with the project end-users (living labs) that drive and validate the technological developments. Currently, CHARIOT is also defining the technical and methodological framework of the overall solution adapted for the cases of the three living labs that is going to evolve into the concise implementations for the next project phases, in a systematic approach to Privacy, Security, and Safety in Industrial IoT environments, using a strategic/objectives driven systematic way, in a process of continuous improvement. CHARIOT intends to have a first implementation of the system within the first months of 2019 and will integrate this to all infrastructures involved and as planned. This project has received funding from the

European Union's Horizon 2020 research and innovation programme under grant agreement No 780075". The authors acknowledge the research outcomes of this publication belonging to the CHARIOT consortium.

7.3 ENACT: Development, Operation, and Quality Assurance of Trustworthy Smart IoT Systems

Until now, IoT system innovations have been mainly concerned with sensors, device management and connectivity, with the mission to gather data for processing and analysis in the cloud in order to aggregate information and knowledge [16]. This approach has conveyed significant added value in many application domains, however, it does not unleash the full potential of the IoT [82]. The next generation IoT systems need to perform distributed processing and coordinated behaviour across IoT, edge and cloud infrastructures [17], manage the closed loop from sensing to actuation, and cope with vast heterogeneity, scalability and dynamicity of IoT systems and their environments. Moreover, the function and correctness of such systems has a range of criticality from business critical to safety critical. Thus, aspects related to trustworthiness such as security, privacy, resilience and robustness, are challenging aspects of paramount importance [16]. Therefore, the next generation of IoT systems must be trustworthy above all else. In ENACT, we will call them trustworthy smart IoT systems, or for short; trustworthy SIS.

Developing and managing the next generation trustworthy SIS to operate in the midst of the unpredictable physical world represents daunting challenges. Challenges, for example, that include that such systems always work within safe operational boundaries [18] by controlling the impact that actuators have on the physical world and managing conflicting actuation requests. Moreover, the ability of these systems to continuously evolve and adapt to their changing environments are essential to ensure and increase their trustworthiness, quality and user experience. DevOps is a philosophy and practices that covers all the steps from concept to delivery of a software product. In ENACT we see DevOps advocating a set of software engineering best practices and tools, to ensure Quality of Service while continuously evolving complex systems, foster agility, rapid innovation cycles, and ease of use [19]. DevOps has been widely adopted in the software industry. However, there is no systematic DevOps support for trustworthy smart IoT systems today [18–20]. The aim of ENACT is to enable DevOps in the domain of trustworthy smart IoT systems.

7.3.1 Challenges

The key research question of ENACT is thus the following: "*how we can tame the complexity of developing and operating smart IoT systems, which (i) involve sensors and* ***actuators*** *and (ii) need to be* ***trustworthy****?*". Our fundamental approach is to evolve DevOps methods and techniques as baseline to address this issue. We thus refine the research question as follows: "*how we can apply and evolve the DevOps tools and methods to facilitate the development and operation of trustworthy smart IoT applications?*".

Challenge 1: Support continuous delivery of trustworthy SIS. Currently there is little effort spent on providing solutions for the delivery and deployment of application across the whole IoT, edge and cloud space. In particular, there is a lack of languages and abstractions that can be used to support the orchestration of software services and their continuous deployment on heterogeneous devices [21] together with the relevant security mechanisms and policies.

Challenge 2: Support the agile operation of trustworthy SIS. The operation of large-scale and highly distributed IoT systems can easily overwhelm traditional operation teams. Other management models such as NoOps and Serverless Computing are evolving to solve this problem. Whatever the operations management model the major challenges will be to improve efficiency and the collaboration with development teams for rapid and agile evolution of the systems. Currently, there is a lack of mechanisms dedicated to smart IoT systems able to *(i)* monitor their status, *(ii)* indicate when their behaviour is not as expected, *(iii)* identify the origin of the problem, and *(iv)* automate typical operation activities. Furthermore, the impossibility of anticipating all the adaptations a system may face when operating in an open context, creates an urgent need for mechanisms that will automatically maintain the adaptation rules of a SIS.

Challenge 3: Support continuous quality assurance strengthening trustworthiness of SIS. Maintaining quality of service is a complex task that needs to be considered throughout the whole life-cycle of a system. This complexity is increased in the smart IoT system context where it is not feasible for developers and operators to exhaustively explore, anticipate or resolve all possible context situations that a system may encounter during its operation. This is due to the open context in which these systems operate and as a result can hinder their trustworthiness. Quality of Service is particularly important when the system can have an impact on the physical world through actuators. In addition, testing, security assurance as well as the robustness of such systems is challenging [20].

7.3.2 The ENACT Approach

DevOps seeks to decrease the gap between a product design and its operation by introducing software design and development practices and approaches to the operation domain and vice versa. In the core of DevOps there are continuous processes and automation supported by different tools at various stages of the product life-cycle. In particular, the ENACT DevOps Framework will meet the challenges below and support the DevOps practices during the development and operation of trustworthy smart IoT systems. ENACT will provide innovations and enablers that will feature trustworthy IoT systems built by implementing the seven stages of the process as depicted in Figure 7.6.

Plan: The ENACT approach is to introduce a new enabler to support the risk-driven and context-aware planning of IoT systems development, including mechanisms to facilitate the selection of the most relevant and trustworthy devices and services to be used in future stages.

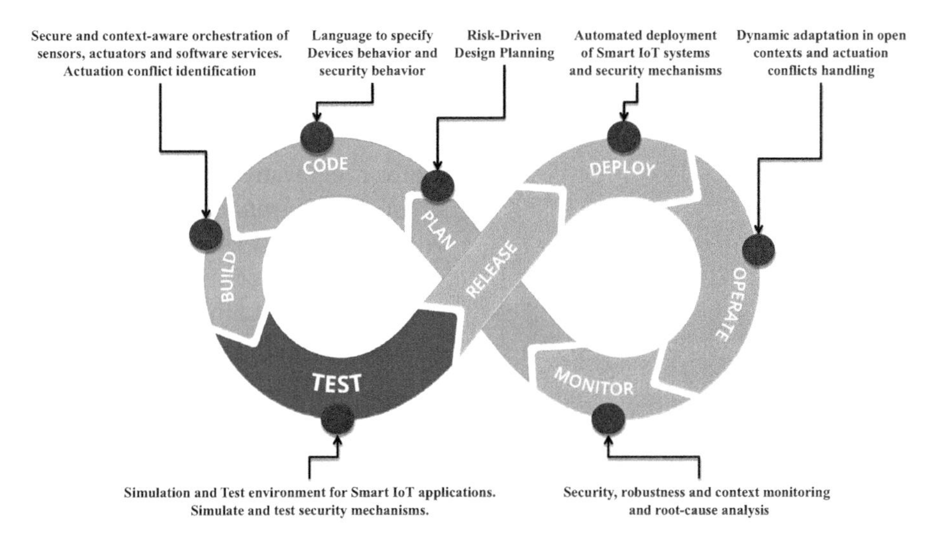

Figure 7.6 ENACT support of DevOps for trustworthy smart IoT systems.

Code: The ENACT approach is to leverage the model-driven engineering approach and in particular to evolve recent advances of the ThingML [21] language and generators to support modelling of system behaviours and automatic derivation across vastly heterogeneous and distributed devices both at the IoT and edge layers.

Build and Deploy: The ENACT approach is to provide a new deployment modelling language to specify trustworthy and secure orchestrations of sensors, actuators and software components, along with the mechanisms to identify and handle potential actuation conflicts at the model level. The deployment engine will automatically collect the required software components and integrate the evolution of the system into the run-time environment across the whole IoT, Edge and Cloud space.

Test: ENACT enablers will allow continuous testing of smart IoT systems in an environment capable of emulating and simulating IoT and edge infrastructure by targeting the constraints related to the distribution and infrastructure of IoT systems. This system is intended to be able to simulate some basic attacks or security threats.

Operate: The ENACT approach will provide enablers for the automatic adaptation of IoT systems based on their run-time context, reinforced by online learning. Such automatic adaptation will address the issue of the management complexity. The complexity of open-context IoT systems can easily exceed the capacity of human operation teams. Automatic adaptation will improve the trustworthiness of the smart IoT system execution.

Monitor: The ENACT approach is to deliver innovative mechanisms to observe the: status, behaviour, and security level of the running IoT systems. Robust root cause analysis mechanisms will also be provided.

In addition to the DevOps related contributions identified above, the ENACT DevOps Framework will provide specific cross-cutting innovations related to trustworthiness, which can be seamlessly applied, in particular based on the following ENACT concepts:

Resilience and robustness: The ENACT approach is to provide novel solutions to make the smart IoT systems resilient by providing enablers for diversifying IoT service implementations, and deployment topologies (e.g., implying that instances of a service can have a different implementation and operate differently, still ensuring consistent and predictable global behaviour). This will lower the risk of privacy and security breaches and significantly reduced impact in case of cyber-attack infringements.

Security, privacy and identity management: The ENACT approach is to provide support to ensure the security of trustworthy SIS. This not only includes smart preventive security mechanisms but also the continuous monitoring of security metrics and the context with the objective to trigger reactive security measures.

7.3.3 ENACT Case Studies

Three use cases from the Intelligent Transport Systems (Rail), eHealth and Smart Building application domains will guide, validate and demonstrate the ENACT research.

7.3.3.1 Intelligent transport systems

This use case will assess the feasibility of IoT services in the domain of train integrity control, in particular for the logistics and maintenance of the rolling stock and on-track equipment. In this domain, the infrastructure and the resources that should be used are usually expensive and require a long-time in planning and execution. Therefore, the usage of the rail systems must be optimised at maximum, following security and safety directives due to the critical and strategic characteristics of the domain. This use case will involve logistic and maintenance activities. Within the ENACT scope, it will be focused on the logistics activities.

A logistic and maintenance scenario will be defined with the aim to provide information about the wagons that form the rolling stock. This scenario will cover not only optimizing cargo storing and classification, but also providing the appropriate resources to assure the correct functioning of the system. These will be only possible if the train integrity is confirmed when the different wagons are locked and moving together. This situation will assure the proper transportation of cargo or passengers, avoiding possible accidents. This use case will involve an infrastructure consisting of large sets of on-board sensors (e.g., Integrity Detector, Asset data info, Humidity and temperature sensors) and multiple gateways interacting with cloud resources.

7.3.3.2 eHealth

The eHealth use case will develop a digital health system for supporting and helping various patients staying at home to the maximum extent possible either during treatment or care. Elderly people are one type of subject in this case study. The Digital health system will feature elderly care to allow the subjects to live at home as long as possible. Another type of patients that we consider is Diabetes patients that need to follow their glucose level and regularly be followed up by health personnel.

The digital health system will both control equipment normally present in smart homes to make life comfortable (automatic light control, door locks, heater control, etc.), and control various types of medical devices and sensors. These devices and sensors support the care and wellness for the specific patient and consist of a wide variety of types, including: blood pressure meter, scales, fall detection sensors, glucose meter, video surveillance, medicine reminder, indoor and out-door location etc). In addition, the system needs to integrate with other systems to provide information or alarms for example to response centres, care-givers, physicians, next of kin etc., and to feed information to medical systems such as electronic patient journals (EPJ). The pivotal role of the system's Edge Computing will be what we denote "the medical gateway" which integrates sensors and devices, controls the edge and ensures the right data are provided to the various stakeholders and to integrated systems such as EPJ.

7.3.3.3 Smart building

This use case will make use of smart building sensors, actuators and services. To this aim two sets of applications covering Smart Energy Efficiency and Smart Elderly Care will be developed within a Care Centre environment. Energy efficiency of new and existing buildings is crucial to achieve carbon emission reduction, and as we increasingly spend more time indoors, adequate levels of user comfort need to be guaranteed by the smart buildings. This implies a trade-off between energy use and the different aspects of users' comfort. They will be tested in the KUBIK, a smart building especially designed for testing new solutions for sustainable buildings. The use case will simulate a care centre consisting of small apartments where a group of elderly people live together. This care centre use case includes sensors and actuators that monitor and control the environment in order to ensure the safety of the facilities, to perform energy efficiency measures and also to support the care-takers in monitoring the wellbeing of users.

The trend for smart buildings is to provide an increasing range of services supported by an increasing number of IoT sensors and actuators. Example of such services or applications include thermal comfort, visual comfort, energy efficiency, security, etc. Applications in this space need to share building infrastructure and may have conflicting objectives. The solution requires a clear hierarchy between the different actuation scenarios.

7.4 Search Engines for Browsing the Internet of Things – IoTCrawler

Efficient and secure access to Big IoT Data will be a pivotal factor for the prosperity of European industry and society. However, today data and service discovery, search, and access methods and solutions for the IoT are in their infancy, like Web search in its early days. IoT search is different from Web search because of dynamicity and pervasiveness of the resources in the network. Current methods are more suited for fewer (hundreds to millions), static or stored data and services resources. There is yet no adaptable and dynamic solution for effective integration of distributed and heterogeneous IoT contents and support of data reuse in compliance with security and privacy needs, thereby enabling a true digital single market. Previous reports show that a large part of the developers' time is spent on integration. In general, the following issues limit the adoption of dynamic IoT-based applications:

- The heterogeneity of various data sources hinders the uptake of innovative cross-domain applications.
- The large amount of raw data without intrinsic explanation remains meaningless in the context of other application domains.
- Missing security and neglected privacy present the major concern in most domains and are a challenge for constrained IoT resources.
- The large-scale, distributed and dynamic nature of IoT resources requires new methods for crawling, discovery, indexing, physical location identification and ranking.
- IoT applications require new search engines, such as bots that automatically initiate search based on user's context. This requires machine intelligence.
- The complexity involved in discovery, search, and access methods makes the development of new IoT enabled applications a complex task.

Some ongoing efforts, such as Shodan and Thingful provide search solutions for IoT. However, they rely mainly on a centralised indexing and manually provided metadata. Moreover, they are rather static and neglect privacy and security issues. To enable the use of IoT data and to exploit the business potential of IoT applications, an effective approach needs to provide:

- An adaptive distributed framework enabling abstraction from heterogeneous data sources and dynamic integration of volatile IoT resources.
- Security, privacy and trust by design as integral part of all the processes from publication, indexing, discovery, and subscription to higher-level application access.

- Scalable methods for crawling, discovery, indexing and ranking of IoT resources in large-scale cross-platform and cross-disciplinary systems and scenarios.
- Machine initiated semantic search to enable automated context dependent access to IoT resources.
- Monitoring and analysing the Quality of Service (QoS) and Quality of Information (QoI) to support fault recovery and service continuity in IoT environments.

IoTCrawler is an EU H2020 project that addresses the above challenges by proposing efficient and scalable methods for crawling, discovery, indexing and ranking of IoT resources in large-scale cross-platform and cross-disciplinary systems and scenarios. It develops enablers for secure and privacy-aware discovery and access to the resources, and monitors and analyses QoS and QoI to rank suitable resources and to support fault recovery and service continuity. The project evaluates the developed methods and tools in various use-cases, such as Smart City, Social IoT, Smart Energy and Industry 4.0. The key elements of IoTCrawler are shown in Figure 7.7.

The project aims to create scalable and flexible IoT resource discovery by using meta-data and resource descriptions in a dynamic data model. This means, for example, that if a user is interested in measuring temperature in a certain location, the result (e.g. list of sensors) should only contain sensors

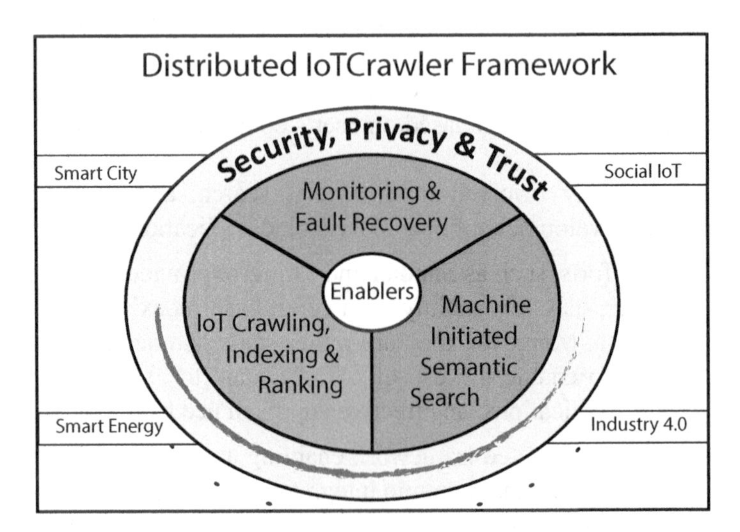

Figure 7.7 Key concepts of the IoTCrawler proposal. [40] ©2018 IEEE.

that can measure temperature, but the user may accept sensors that closely fulfil her/his application requirements even though all other characteristics may not be favourable (e.g. cost of acquisition may be high and sensor response time may be slow). For this reason, the system should understand the user priorities, which are often machine-initiated queries and search requests, and provide the results accordingly by using adaptive and dynamic techniques.

7.4.1 Architecture of IoTCrawler

IoTCrawler provides novel approaches to support an IoT framework of interoperable systems including security and privacy-aware mechanisms, and offers new methods for discovery, crawling, indexing and search of dynamic IoT resources. It supports and enable machine-initiated knowledge-based search in the IoT world. Figure 7.8 depicts the IoTCrawler framework and highlights its key components, which are detailed next.

7.4.1.1 IoT framework of interoperable (distributed) systems

The diversity of the market has resulted in a variety of sophisticated IoT platforms that will continue to exist. However, to evolve and enable the full benefits of IoT, these platforms need access to data, information and services across various IoT networks and systems within an integrated ecosystem of IoT resources. IoTCrawler envisions a cooperation of platforms and systems to provide smart integrated IoT based services. Nevertheless, instead of defining an overarching hyper-platform on top, the integration proposed by IoTCrawler is carried out by the definition of a common interface, enabling this way cooperation and interconnection of various platforms by making their data and services discoverable and accessible to other applications and services. An IoTCrawler-enabled platform can internally be implemented in different ways, since it only has to support the common and open interfaces to join the ecosystem. The open IoT interfaces are split in two planes that are called control and data planes. The control plane will coordinate and control the data and information processing in the platforms (monitoring and quality analysis). The data plane will allow for IoT data flow exchange between platforms (crawling, indexing and search).

7.4.1.2 Holistic security, privacy and trust

An ecosystem of IoT platforms brings immense benefits but also potential risks for users and stakeholders. The very principle that makes the IoT so

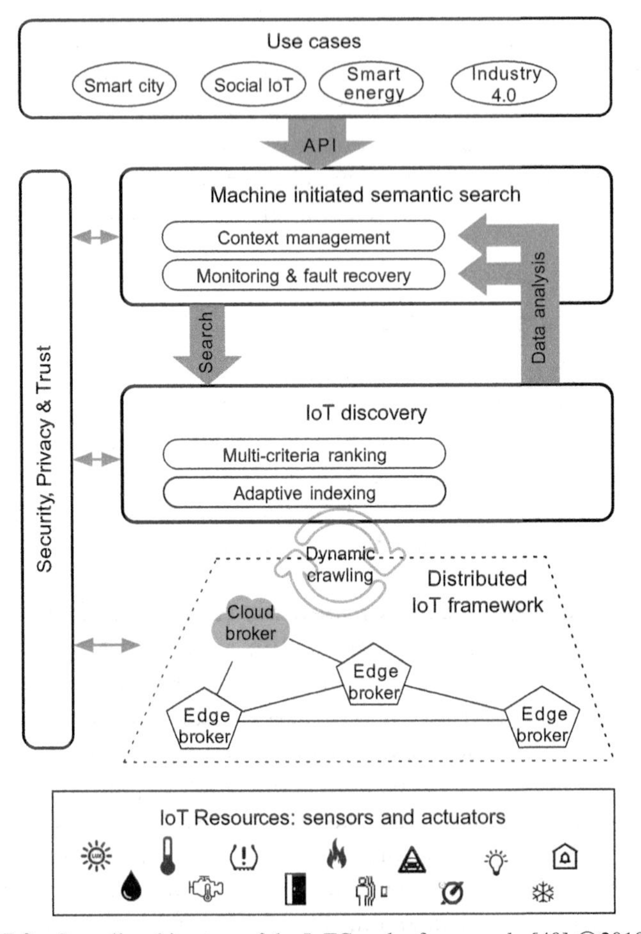

Figure 7.8 Overall architecture of the IoTCrawler framework. [40] ©2018 IEEE.

powerful – the potential to share data instantly with everyone and everything – creates huge security and privacy risks. Since IoT systems are, by their nature, distributed and operate often in unprotected environments, the maintenance of security, privacy, and trust is a challenging task. IoTCrawler addresses quality, privacy, trust and security issues by employing a holistic and end-to-end approach to the data and service publication to search and access workflow. Device and connectivity management will ensure that the end devices only connect to trusted access networks. IoTCrawler develops solutions for mitigating privacy intrusion and data correlation based on data collected from multiple sources. Both technical and information governance procedures and guidelines are defined and implemented. This makes sure that the technical solutions are in place for avoiding the security and privacy risks, and also appropriate information governance procedures and best practices and measures are followed in development, deployment and utilisation of the use-cases and third-party applications.

7.4.1.3 Crawling, discovery and indexing of dynamic IoT resources

Information access and retrieval on the early days of the Internet and the Web mainly relied on simple functions and methods. For example, Yahoo's first search engine was simply based on the "grep" function in Unix or the AltaVista search engine initially did not have a ranking mechanism. The Internet and the Web have gone a long way in the past two decades to improve the way we access the information on the Web. While the current information and search retrieval on the Web is far from ideal, there are several sophisticated methods and solutions that provide crawling, indexing, ranking and search and retrieval of extremely large volumes of information on the Internet. The new generations of Web search engines have now focused on information extraction, personalised and customised knowledge and extraction techniques and solutions. Some early works are demonstrated by Google's knowledge graph, Wolfram Alpha and Microsoft Bing. The current information access and retrieval methods on the IoT are still at the same stage that the Web and the Internet were in their early days. Information retrieval on the large-scale IoT systems is currently based on the assumption that the sources are known to the devices and consumers or it is assumed that opportunistic methods will send discovery and negotiation messages to find and interact with other relevant resources in their outreach (e.g. Google's recent Physical Web project is designed based on this assumption). Overall, IoT systems have more ad-hoc resources that do not comply with document and URL processing and indexing norms; the resources, such as mobile phones and sensing devices, can publish data and then move to another location or disappear. Service and data crawling and discovery for smart connected devices and services will also involve automated associations and integration to provide an extensible framework for information access and retrieval in IoT. IoTCrawler focuses on providing reliable, quality and resource-aware and scalable mechanisms for data and services publishing, crawling, indexing in very large-scale distributed dynamic IoT environments.

7.4.1.4 Machine-Initiated semantic search

In the past, search engines were mainly used by human users to search for content and information. In the newly emerging search model, information is provided depending on the users' (human user or a machine) context and requirements (for example, location, time, activity, previous records, and profile). The information access can be initiated without the user's explicit query or instruction but used on its necessity and relevance (context-aware search). This will require machine interpretable search results in semantic forms. Moreover, social media, physical sensors (numerical streaming values), and Web documents must be better integrated, and the search results should become more machine interpretable information rather than remaining as pure links (e.g. the Web search engines mainly return a list of links to the pages as their results; with some exceptions on popular questions and topics).

IoTCrawler enables context-aware search and automated processing of data by semantic annotation of the data streams, thus making their characteristics and capabilities available in a machine processable way. There are several existing works that provide methods and techniques for semantic annotations and description of the IoT devices, services and their messages and data. However, most of these methods rely on centralised solutions and complex query mechanisms that hinder their scalability and wide scale deployment and use for the IoT. IoTCrawler supports an ecosystem of multiple platforms and develops dynamic semantic annotation and reasoning methods that will allow continuous and seamless integration of new devices and services by exploiting and adapting existing annotations based on similarity measures.

The automatic discovery has to consider the current context. Context-awareness requires the integration and analysis of social, physical and cyber data. IoTCrawler develops enablers for context-aware IoT search. Hence the requirements of the different applications are mapped to the solutions by selecting resources considering parameters such as security and privacy level, quality, latency, availability, reliability and continuity. IoTCrawler improves reliability and robustness by fault recovery mechanisms and mitigation of malfunctioning devices using device activation/deactivation in the associated area. The fault recovery also requires mechanisms to support communication among networked IoT resources located in diverse locations and across different platforms, and to provide secure and efficient re-distribution of information in case of failure.

7.4.2 Use Cases

IoTCrawler is currently evaluating its technologies in four real world use-cases: Smart Cities, Social IoT, Smart Energy, and Industry 4.0 (see Figure 7.9). Further use-cases will be identified and ranked in co-creation workshops with the relevant stakeholders within the project.

7.4.2.1 Smart city

The city of Aarhus has been considered as a target for smart city deployment in the project. IoTCrawler helps to overcome the negative perceptions of Internet of Things and Smart Cities by developing smart city experimentation

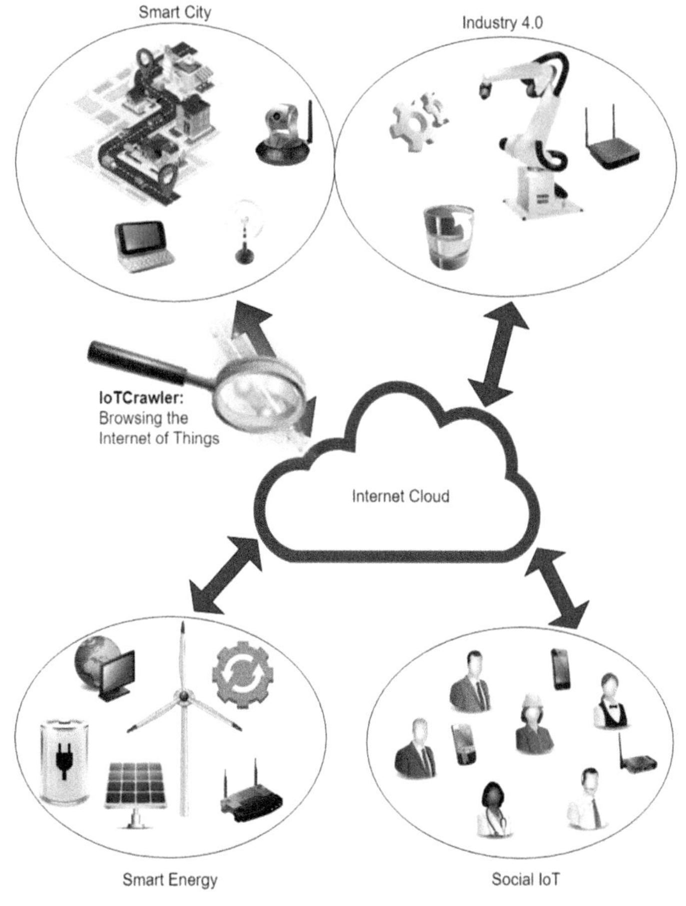

Figure 7.9 IoTCrawler use cases at a glance.

tools for Aarhus' City Lab that can make citizens and companies engaged and be curious about smart city solutions. IoTCrawler also provides the enabling technologies to discover new data sources in Aarhus for Open Data platforms and has the potential to become a reference platform supporting IoT data and service sharing as part of the sharing economy. To track the performance of a smart city, IoTCrawler develops enablers for monitoring activity and quality of the sensors. This can be used to set up KPI's for City Labs and to track its performance. The smart city deployment of Murcia is also considered in IoTCrawler, exploiting the large sensor platform installed.

7.4.2.2 Social IoT

Social IoT relates to using sensors deployed at sports and entertainment events in order to quantify the performance of professionals or experience of participants. This enables participants to engage in events beyond simply watching, thus creating a unique personal record of their experience, and in combination with social and digital media allows event manager to create new insights and content for their audience. IoTCrawler has access to over 800 events, including fashion events (e.g. New York Fashion week), culinary events, sports events (e.g. Basketball Final Four), or events such as Miss Universe. For each event, sensors are deployed at local venues and participants and spectators are equipped with wearable devices. This results in a range of diverse data sets that are collected, analysed, stored, and used, e.g. for content creation. Discovering and semi-automatically describing existing sensors, data sets and streams using IoTCrawler technologies has the potential to significantly increase the overall value of the dataset access and their integration, making it accessible to a larger group of people and enabling new applications. As described above, the data sets include raw sensor data and processed analytic results. However, data processing often involves data from other third-party sources. For this reason, play-by-play data is used to correlate analytical results to match events, and social media sources can be used to link to user generated content. IoTCrawler's discovery, indexing and search enablers have the potential to significantly reduce the effort associated with the integration of sensor technologies, and other external data sources.

7.4.2.3 Smart energy

Smart Buildings play an important role in distributed energy systems as they turn from energy consumers to the so-called "energy prosumers". In future energy systems, Smart Buildings actively interact with the Smart

Grids in order to stabilise them or participate in energy trading as well as for structural condition monitoring and proactive maintenance. For this purpose, buildings offer semantically annotated properties of the technical equipment within especial energy flexibilities (i.e. for shifting electrical and thermal loads). In this frame, this use-case employs the technologies developed in IoTCrawler to dynamically discover the flexibilities of Smart Buildings and analyse their potential as well as their demand for applications that are necessary to manage and offer energy to the Smart Grid or the energy market. This information can be used by energy retailers or grid operators to deploy best fitting applications to individual buildings. The project uses semantic enrichment of grid data and data analytics to enhance smart grid applications and reduce the need for manual engineering and setup of systems.

7.4.2.4 Industry 4.0

Industry 4.0 includes advances such as predictive maintenance, energy prediction, or human-robot collaboration. The results of IoTCrawler will be used to improve predictive maintenance planning for horizontal machining centres in aerospace and Die&Mould industries. Currently, data integration consumes more than 80% of the time in the industry. IoTCrawler has the potential to significantly accelerate the development and deployment of Industry 4.0 analytics solutions, by discovering and semi-automatically integrating machine metadata, sensor data provided by the machines and information stored in related enterprise databases. Extending the discovery to actuator services (e.g. air conditioning, heating, and machine operation) allows to link actions for avoiding load peaks to energy analytics pipeline. IoTCrawler also increases workers' safety by identifying critical conditions (e.g. gas exposition) in the permanent sensor data stream of drones, and forward such condition markers to monitoring teams and production management subsystems.

7.4.3 Main Innovations in the Areas of Research

The literature within key areas of the IoTCrawler proposal is reviewed next, indicating the main innovations of the work within the general framework described above.

7.4.3.1 Search and discovery

Being essential for any network architecture, one of the key components of the proposed architecture is the search and discovery operation. Distributed

Hash Table (DHT) is used to provide a high scalability in storage and a flexible support for query and update operations. DHT is a totally decentralised system that stores data objects for easy and quick access (query) and update (store). DHTs are built on top of overlay networks into which network objects are spread and identified with unique keys, e.g. the well-studied overlay network and DHT Chord mechanism [22], which is the direct ancestor of Kademlina [23] (BitTorrent's DHT). Overlay networks and DHTs are well suited to form the basement of a proper discovery mechanism, such as the Overlay Management Backbone (OMB) approach [24]. To add suitable schema evolution to the information/content discovery, description mechanisms such as the Resource Description Framework (RDF) and JSON-LD [25] are needed. Combining a DHT mechanism with RDF, the work in [26] proposes to use an adapted version of RDQL [27] to perform the queries. The main problems of this approach are that it consumes a lot of storage space and that it is not efficient for simple searches. SPARQL [28] is the de facto query language for RDF, by providing a coherent and simple search mechanism.

The IoTCrawler approach exploits the remarkable qualities of the overlay network and DHT described above to build a distributed discovery infrastructure. However, the nodes are deployed in separate domains to distribute both the storage/finding load and the management of information access.

7.4.3.2 Security for IoT

In spite of the emergence of different cross-world initiatives in recent years (IERC, ITU-T SG20, IEEE IoT Initiative4 or IPSO Alliance are just some of them), there is a lack of a unified vision on security and privacy considerations in the IoT paradigm, which embraces the whole lifecycle of smart objects that are making up the digital landscape of the future. In the IoT, data confidentiality and authentication, access control within the networks, privacy and trust among users and things are among some of the key issues [29].

IoTCrawler explores the use of advanced cryptographic techniques based on Attribute-Based Encryption (ABE). Specifically, it analyses the application and extension of the Ciphertext-Policy Attribute-Based Encryption (CP-ABE) as a flexible and promising cryptographic scheme in order to enable information to be shared while confidentiality is still preserved. In CP-ABE, the cipher-text embeds the access structure to describe which private-keys can decrypt it, and the same private-key is labelled with descriptive attributes. IoTCrawler addresses the integration of CP-ABE with different signatures schemes to provide end-to-end integrity to the information that is shared for anticipatory purposes. Users are given means to define how their personal information is shared and under which circumstances using a policy-based approach. Additionally, IoTCrawler investigates the integration of this solution within the search and discovery process for IoT.

The Blockchain paradigm [30] is also included in IoTCrawler. A Blockchain is a distributed database that maintains a continuously growing set of transactions in a way that is designed to be secure, transparent, highly resistant to outages, auditable, and efficient, at the same time it is distributed. However, despite the benefits that Blockchain technologies offer, we still need to overcome two major challenges in IoTCrawler. First, privacy, since transactions tend to be public, and encryption to protect transactions' contents is not enough because it still allows the remaining nodes in the system to learn about the occurrence of a particular exchange in the system; and, second, scalability, because existing permission-less blockchains (e.g. Bitcoin) are only able to scale to a considerable number of nodes at the expense of attained throughput, e.g. Bitcoin's throughput is about seven transactions per second.

Moreover, IoTCrawler will leverage Trusted Execution Environments (TEEs) to enhance the security primitives deployed in the proposed framework, given that existing TEEs suffer from a number of shortcomings, especially with respect to their security and privacy provisions.

In the area of Authentication, Authorisation and Accounting (AAA), IoTCrawler proposes a lightweight access control scheme based on Capability Tokens for IoT as presented in [31, 32], where these tokens act as a proof of possession providing a straightforward validation mechanism without requesting a third party. We propose a mechanism for interoperability of different authentication and authorisation solutions based on a bridge to third party elements, such as the standard stacks as LDAP and FIWARE Service Enablers to support a lightweight federation-like approach.

7.4.3.3 Data validation and quality analysis

The assessment of Quality of Data can basically be evaluated in five common dimensions: Completeness, Correctness, Concordance, Plausibility and Currency. In [33] the authors provide a table of different terms used to describe one of the dimensions of data quality. Furthermore, they provide a mapping between data quality dimensions and data quality assessment methods. In [34] Sieve is introduced, a framework to flexibly express quality assessment methods and fusion methods. The STAR-CITY project [35] describes a system for semantic traffic analytics. Based on various heterogeneous data sources (e.g., Dublin bus activity, events in Dublin city), their system is able to predict future traffic conditions with the goal to make traffic management easier and to support urban planning.

One of the major challenges in the assessment of quality metrics to sensory IoT data is the lack of ground truth. The authors of [36] and [37] developed and evaluated a concept for the assessment of node trustworthiness in a network based on data plausibility checks. They propose that every node performs a plausibility check to identify malicious nodes sending faulty data. Similar to this work, they use data sources in order to find "witnesses" for a given sensor reading. The authors in [38] propose three different approaches to deal with a missing ground truth in social media: spatiotemporal, causality, and outcome evaluation. Their concept to use spatiotemporal evaluation to predict future behaviour of humans is like the proposed IoTCrawler approach, disregarding that we evaluate past events. Prior work of the authors emphasised the importance of an appropriate distance model reflecting infrastructure, e.g., roads, and physics, i.e. traffic or air movements [39]. The approach in IoTCrawler refines the state of the art by utilising sensor and domain independent correlation and interpolation models whilst incorporating knowledge of the city infrastructure to evaluate data stream plausibility.

7.4.4 Conclusion

This part presents the key ideas and the architecture of a crawling and discovery engine for the Internet of Things resources and their data. We describe our work in the context of the H2020 IoTCrawler project, which proposes a framework to make possible the effective search over IoT resources. The system goes beyond the state of the art through adaptive, privacy-aware and secure algorithms and mechanisms for crawling, indexing and search in distributed IoT systems. Innovative technological developments are proposed as enablers to support any IoT scenario. We discuss four use cases of the platform, which are presented in the areas of Smart Cities, Social IoT, Smart Energy and Industry 4.0. The project is currently implementing the envisaged framework, at the same time the main interoperability issues are considered to support the real-life uses cases identified. This work has been sponsored by the European Commission, through the IoTCrawler project (contract 779852), and the Spanish Ministry of Economy and Competitiveness through the Torres Quevedo program (reference PTQ-15-08073).

7.5 SecureIoT: Multi-Layer Architecture for Predictive End-to-End Internet-of-Things Security

The proliferation and rising sophistication of Internet of Things (IoT) infrastructures and applications comes with a wave of new cybersecurity challenges. This is evident in several notorious security incidents on IoT devices and applications, which have occurred during the last couple of years. These include the "Lizard Stressor" attacks on home routers (January 2015), the 1.4 million cars that were recalled by Chrysler due to potential hacking of their control software (July 2015), Tesla's autopilot crash (July 2016), as well as the first large scale distributed denial of service (DDoS) attack based on IoT devices (October 2016). Most of these incidents are directly associated with the complexity, heterogeneity and dynamic behaviour of emerging IoT deployments, which poses security challenges, which can be hardly addressed by state of the art platforms. Some of the most prominent of these challenges, include:

- The fact that they provided limited support for end-to-end security, since they lack mechanisms that address IoT security at all levels, i.e. from the field and devices level to the edge and cloud levels. Moreover, existing security solutions tend to be framed within a single platform and ecosystem and cannot effectively operate in scenarios involving multiple platforms and ecosystems [41].
- Their inability to deal with very volatile and dynamic environments comprising networks of smart objects. State-of-the-art IoT platforms and their security mechanisms provide within cloud-based environments that ensure cybersecurity for large numbers of IoT devices. Nevertheless, they make only limited provisions for dynamic applications involving networks of smart objects (i.e. objects with (semi)autonomous behaviour). In the latter, IoT devices and smart objects are likely to join or leave, while security and privacy policies can also change dynamically and without prior notice. Hence, to support large scale interactions across multiple IoT platforms and networks of smart objects, there should be some means of predicting and anticipating the security behaviour and trustworthiness of an IoT entity (e.g., device, platform, groups of objects) prior to interacting with it.

SecureIoT is motived by the need to support cyber-security in scenarios involving cross-platform interactions and interactions across networks of smart objects (i.e. objects with semi-autonomous behaviour and embedded intelligence), which require more dynamic, scalable, decentralized and intelligent IoT security mechanisms. To this end, it introduces a multi-layer, data-driven security architecture, which collects and processes information from the field, edge and cloud layers of an IoT system, in order to identify security threats at all these layers and accordingly to drive notifications and early preparedness to confront them. Furthermore, SecureIoT foresees cross-layer coordination mechanisms and will employees advanced analytics towards a holistic and intelligent approach that will predict and anticipate secure incident in order to timely confront them. Also, SecureIoT introduces a range of security interoperability mechanisms in order to support cross-vertical and cross-platform cyber-security scenarios. The SecureIoT architecture serves as basis for the provision of security services to IoT developers, deployers and platform providers, including a risk assessment, a compliance auditing and a secure programming support service. In this context, the rest of this chapter is structured as follows: Section 2 introduces the SecureIoT architecture and its main principles. Section 3 discusses the security services to be offered by the project, while Section 4 presents some use cases that will be used to validate the project's results.

7.5.1 SecureIoT Architecture

7.5.1.1 SecureIoT architecture overview

Figure 7.10. provides a high-level overview of the security architecture of the project. The architecture provides placeholders for predictive IoT security mechanisms, which can be contributed by different security experts in order to protect IoT infrastructures and services. In the scope of SecureIoT the partners will specify and implement such mechanisms in the areas of security monitoring and predictive analysis, which will serve as a basis for supporting the project's use cases. Nevertheless, the project's architecture is more general and therefore able to accommodate additional algorithms and building blocks. The architecture complies with the reference architectures specified by the Industrial Internet Consortium (IIC) and the OpenFog consortium [42], as it specifies: (i) **The field level**, where IoT devices (including smart objects) reside; (ii) **The fog/edge level**, which controls multiple devices close to the edge of the network. Note that the fog/edge level might be the first security layer in an IoT application, especially when resource constrained devices are deployed; (iii) **The enterprise and platform levels**, which reside at the core and where application and platform level security measures are applicable. Moreover, the SecureIoT architecture will also specify:

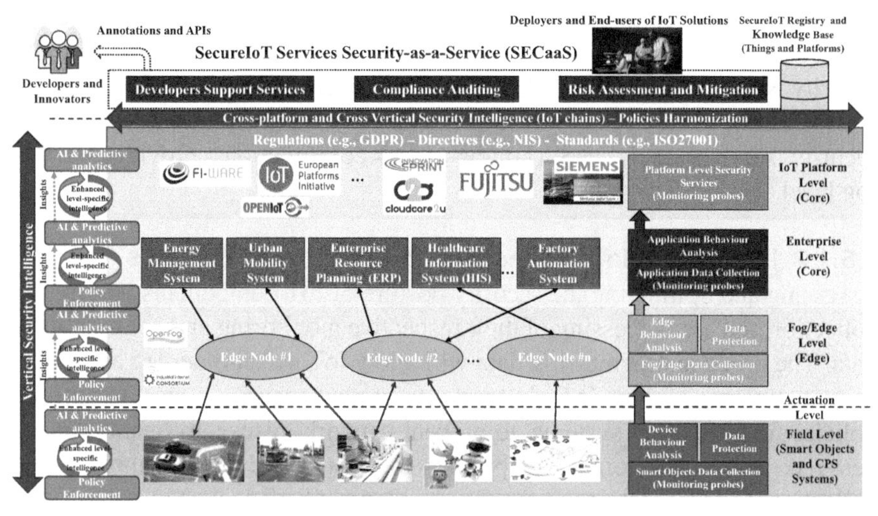

Figure 7.10 Overview of SecureIoT Architecture.

- Interfaces for (security) data collection at all levels of the security architecture, including monitoring probes that are deployed at all levels.
- Data analytics modules (including AI and predictive analysis) at all levels, which extract insights about the future security state of the IoT infrastructure and applications.
- Semi-automated Policy Enforcement Points (PEPs), which are driven by predictive insights and enforce policies at different levels. PEPs will provide the means for enforcing security and cryptographic functionalities, configuring IoT platforms and devices for enhanced security, as well as for distributing security sensitive datasets.
- Multi-level security mechanisms and measures, which combine security monitoring, analytics and insights from multiple levels.

Applicable policies and security measures are driven by regulations (e.g., GDPR), directives (e.g., NIS, ePrivacy) and standards (such as ISO27001 [43]). The ultimate goal of the architecture is to provide concrete services such as the SECaaS. The delivery of these services is facilitated by the development and maintenance of a security knowledge base, where metadata about IoT entities (i.e. objects platforms etc.) are registered along with knowledge collected and summarized based on multiple publicly available threat intelligence sources. Note that the security services of the architecture are offered as a service based on a Security-as-a-Service (SECaaS) paradigm. This however does not imply that the security services are solely deployed in the cloud. Rather, they can be offered based on a combination of cloud-based SaaS (Software-as-a-Service) security services and FaaS (Fog-as-a-Service) functions provided at the fog level.

7.5.1.2 Intelligent data collection and monitoring probes

Assessing and optimizing the security posture of IoT components require the collection and the processing of their respective monitoring and configuration data. The produced monitoring data will allow IoT stakeholders to assess the security posture of their IoT platforms, to predict security issues, to enforce policies for hardening systems, to prevent network misuse, to quantify business risk, and to collaborate with partners to identify and mitigate threats. The collection of these data requires the development of dedicated probes and monitoring layers at different levels of the deployed IoT platform (device, network, edge and core) to capture a comprehensive and a complete view of its operations and interactions. In SecureIoT, monitoring probes will be provided to support the collection of log data, including network flows and software configurations, at the component, services and network levels.

A key characteristic of SecureIoT's security monitoring infrastructure (and related probes) will be its built-in intelligence in the data collection and pre-processing mechanisms, which will be implemented over the SecureIoT monitoring probes that will interfaces to different IoT platforms. As part of this intelligence, the data collection mechanisms will ensure data quality, data filtering, as well as adaptive selection of the needed data sources based on dynamic changes to the configuration of the IoT platforms, applications and smart objects. In order to implement this intelligence, the monitoring probes will be enhanced with data streaming analytics mechanisms, which will be able to process security-related information sources on the fly (i.e. almost at real-time) in order to adapt the filtering and data collection accordingly. This data collection intelligence will facilitate fast processing, as well as the implementation of predictive analytics schemes.

7.5.1.3 SecureIoT systems layers and information flows

Figure 7.11. presents the layers of a SecureIoT compliant system, with emphasis on the flow of information from an IoT platform to the SecureIoT SECaaS services. The following layers are presented:

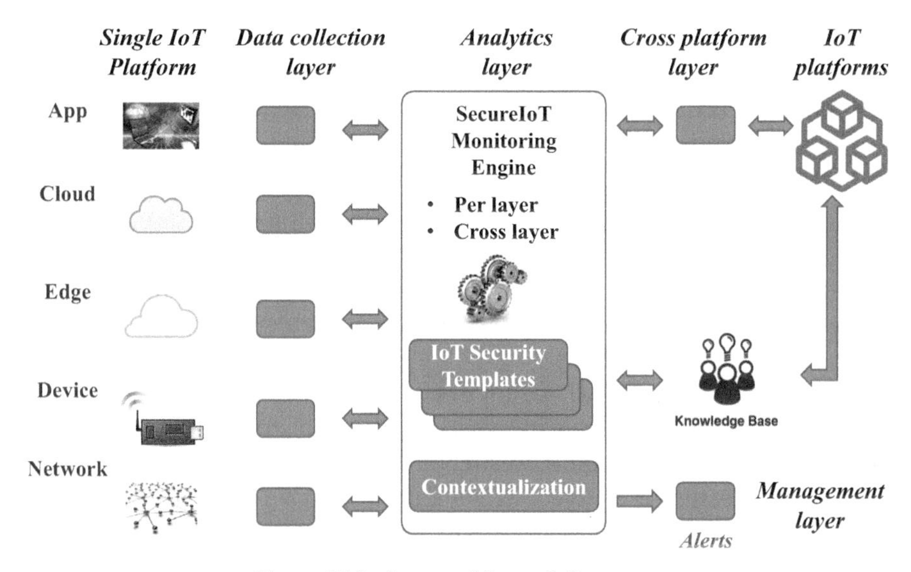

Figure 7.11 Layers of SecureIoT systems.

- **A layer of an individual IoT platform or system**, which typically comprises network, devices/field, edge/fog, cloud and application-level components. These components are usually part of the target IoT platform or systems that needs to be secured based on SecureIoT.
- **A data collection layer**, which comprises the above-mentioned security monitoring probes. Note that probes will be specified and developed for all parts and components of the IoT system i.e. from the network and devices components all the way up to the IoT applications' components.
- **A data analytics layer**, which is destined to process the data derived from the various probes. This layer is empowered by data analytics algorithms, but also by a range of cybersecurity templates, which specify rules and patterns of the security incidents that are to be identified. Taking network-level attacks as example, templates for specific types of network attacks will be specified such as TCP SYN attacks, UDP flood attacks, HTTP POST DoS (Denial of Service) attacks [43]. Each of the templates will comprise the rules and conditions under which the attacks will be identified. Likewise, templates for other types of attacks, including application specific ones will be specified and used. Along with these templates, the data analytics layer will comprise a contextualization

component, which will be used to judge whether the attacks indicators are abnormal for the given IoT platform and application context.
- **A cross-platform layer**, which is destined to aggregate and correlate information derived from multiple-IoT platforms. It will serve as a basis for identifying attack indicators in cross-platform scenarios.

All of the above layers and components will leverage the services of a knowledge base that will comprise information and knowledge about IoT-related cybersecurity attacks. It will be also used to drive the operation of the IoT security templates and the contextualization component.

7.5.1.4 Mapping to RAMI 4.0 layers

SecureIoT is destined to support cybersecurity scenarios in both consumer and industrial settings. In order to strengthen the industrial relevance of the project's architecture, the project will provide a mapping of the main building blocks of the SecureIoT architecture to the Reference Architecture Model Industry4.0 (RAMI 4.0) [45]. While this mapping is work in progress, the following associations and mappings are envisaged:

- **The SecureIoT field layer**, maps to the Field and Control Device hierarchy levels of RAMI4.0, as well as to its Asset Integration layer.
- **The SecureIoT edge layer**, maps tot eh Station and Workcenter hierarchy levels of RAMI4.0, as well as to its Asset, Integration and Communication layers.
- **The SecureIoT cloud layer**, maps to the Workcenter, Enterprise and Connected World hierarchy levels of RAMI4.0, as well as to its Information, Functional and Business layers.
- **The SecureIoT application layer**, maps to the Enterprise and Connected World hierarchy levels of RAMI4.0, as well as to its Business layer.
- **The SecureIoT data collection layer**, maps to the Field Device, Control Device, Station and Work Centers hierarchy levels of RAMI4.0, as well as to its Communication and Information layers.
- **The SecureIoT analytics layer**, maps to the Enterprise and Connected World hierarchy levels of RAMI4.0, as well as to its Information layer.
- **The SecureIoT management layer**, maps to the Enterprise and Connected World hierarchy levels of RAMI4.0, as well as to its Information, Functional and Business layers.

7.5.2 SecureIoT Services

Based on its architecture, the project will offer risk assessment, compliance auditing and programmers' support services as outlined in the following paragraphs.

7.5.2.1 Risk assessment (RA) services

The SecureIoT RA services will aim at an efficient balance between realizing opportunities for gains, while minimizing vulnerabilities and losses. They will strive to ensure that an acceptable level of security is provided at an affordable cost. The SecureIoT framework will quantify risks in terms of a "likelihood factor", which will be calculated based on combination of the probability and impact of any identified vulnerabilities. This "likelihood factor" will be appropriately weighted and ultimately normalized based on a risk calculation model in-line with NIST's Common Vulnerability Scoring System (CVSS). Special emphasis will be paid in evaluating the criticality of risks associated with the behaviour and the operation of smart objects, as well as of services spanning multiple platforms. SecureIoT will therefore formulate a formal methodology and an accompanying model that will produce risk quantifications based on the identified vulnerabilities, potential threats and the impact estimation per potentially successful exploitation. SecureIoT will develop a risk quantification engine based on an expert system, which will provide flexibility in implementing different rules and assign different rates to the various risks.

7.5.2.2 Compliance auditing services

This service will be delivered as a tool available to solution deployers, operators and end-users. Based on information collected through the security analytics, including the information of the IoT knowledge base. It will provide support for a set of security and privacy controls on the IoT infrastructures at multiple levels. The tool will be configured to support auditing of IoT infrastructures and services, against existing sets of security and privacy controls. The auditing will identify non-compliant behaviours and will provide recommendations about areas that require attention. Several prominent sets of security and privacy rules that will be supported concerning controls and measures specified in the scope of the GDRP regulation, NIS and ePrivacy directives.

7.5.2.3 Programming support services

This service will enable developers to secure applications as part of their programming efforts. In particular, it will enable them to: (a) Enforce Distributed Access Control; (b) Ensure the cryptographic protection of data; and (c) Physical distribute sensitive data for enhanced security. These activities will be supported based on programming annotations, which will specify distributed access control, cryptographic protection and physical data distribution activities. A series of source generation, bytecode transformation and runtime reflection actions will be undertaken at specified Policy Enforcement Points (PEPs), which will be implemented at various levels i.e. the device, edge, core and application layers of the SecureIoT architecture. To this end, along with the security monitoring probes, the SecureIoT architecture will provide the means for configuring elements at the PEPs.

7.5.3 Validating Use Cases

The project's architecture and services will be validated in three use cases, which are briefly discussed in the following paragraphs.

7.5.3.1 Industrial plants' security

The use case will focus on plant networks for operations and support – e.g. SCADA, MES, PLCs, etc. – and enterprise networks connected to IoT-platforms providing support for automation and supply chain collaboration. The technical approach of the industrial IoT use case is twofold as reliability and availability of real world production must not be brought at risk. The following security challenges will be addressed, based on the SecureIoT services:

- **Secure operations of connected factories with thread prediction**: The SecureIoT risk assessment service will be therefore used to predict security issues arising from deployed automation technologies in a multi-vendor environment. Furthermore, SecureIoT's prediction and mitigation services will enable the plant control to draw the right conclusions and prepare for attacks before they emerge.
- **Compliance and Protection of product/user data in a multi-vendor environment**: Factories need to protect product and user data sets. SecureIoT will be used in order to enforce privacy and data protection policies. Likewise, the compliance auditing SecureIoT service will be also used to identify and remedy gaps in the industrial IoT environment.
- **Predictive Maintenance and Avoiding Machine Break-Downs in "Human in the Loop" Scenarios**: Predictive maintenance requires

trustworthy exchange, storage and processing of sensor and asset management datasets. Security analytics of IoT application level entities will be exploited as part of the SecureIoT risk assessment service in order to proactively identify issues with transmission and protection of datasets involved in the predictive maintenance process, in order to ensure the reliability of the process and avoid damages/losses in scenarios where machines foretell their lifetime and initiative actions in the supply chain (e.g., ordering of spare parts, scheduling of maintenance).

7.5.3.2 Socially assistive robots

This use case will focus on security challenges associated with the integration of a socially assistive robot (i.e. QT robot from SecureIoT partner LuxAI) with a cloud-based IoT platform. This integration will target the delivery of personalized ambient assisted living functionalities, such as personalized rehabilitation and coaching exercises. In order to support these applications a dense IoT network, enable continuous interaction between IoT devices, robots, human users and the environment will be established. The integration challenge will however lie on keeping track of the state of the robot and the environment, as well as on implementing distributed task assignment strategies (such as the Consensus-Based Bundle Algorithm (CBBA)), which enable the distribution of application logic across different smart objects. The following security challenges will be addressed:

- **Network and message security**: The SecureIoT risk assessment and mitigation services will be used to identify threats associated with communications and network performance in order to appropriately adapt the operation of the application (e.g., stop the training if needed and deliver proper alerts to users).
- **Prediction and avoidance of dangerous/risky situations**: SecureIoT will monitor the robots' operation both at the software level (i.e. through information flow tracking) in order to identify possible hacking of the robot, but also at the application level in order to detect abnormal operation/behaviour that can lead at risk.
- **Secure programming environment for robotics missions**: The programming interfaces of the robot will be enhanced with SecureIoT programming model and annotations in order to enable the developer of a rehabilitation mission to enforce policies specified in some policy language such as XACML (eXtensible Access Control Markup Language).

- **Compliance to GDPR**: An analysis of the application for GDPR compliance will take place, including automated identification of non-compliance risks (based on the SecureIoT risk assessment) and subsequent implementation of GDRP compliant policies based on the secure programming XACML-based mechanisms.

7.5.3.3 Connected car use cases

This use case concerns security in connected cars scenarios, including: (i) **Usage Based Insurance** scenarios where vehicle data are analysed to assess driver behaviour and hence determine risks in order to better tailor insurance premiums for the customers; and (ii) **Warnings on traffic and road conditions**, that involve analysing data coming from multiple vehicles to understand the traffic conditions in different locations. From the point of view of cybersecurity for the usage-based insurance, it is important to ensure that the data transmitted is only accessible by the responsible organisation (privacy) and that the system cannot be corrupted such that a risky driver appears to be low risk. Moreover, the integrity of the data is a key requirement to ensure that insurance premiums are calculated fairly based on objectively assessed risk using accurate and trusted data. Likewise, for the traffic and road condition warnings it is vital that the data sent to the car is an accurate interpretation of the data provided from each vehicle. It This is because the system could be used maliciously to create congestion if the data is corrupted. Moreover, integrity of software running in the connected car is crucial. Recent attacks or security alarms raised has been focused on taking control over IoT devices and gateways. Over the air firmware update could be used as a countermeasure mechanism after an anomalous (or malware) detection.

To address these challenges, the SecureIoT risk assessment framework will be employed, including predictive risk assessment functionalities. In case of identified issues, preventive measures will be activated (i.e. enforcement of data protection policies, provision of alerts to end-users, instigation and scheduling of over the air updates).

7.5.4 Conclusion

SecureIoT is a first of a kind attempt to introduce a standards-based architecture for end-to-end IoT security. The project's architecture is aligned to recent standards for industrial IoT security, including standards of the Industrial Internet Consortium and the OpenFog consortium. It makes provisions for collecting and analysing data from all layers of an IoT platform, while at the same time catering from cross platform and cross layer security analysis. Moreover, the SecureIoT architecture provides the means for defining and executing security actions at specific PEPs, as a means of enforcing policies and instigating mitigation actions. Based on this architecture, the project will implement risk assessment, compliance and the programming support services.

SecureIoT is currently in its requirements engineering and specification phase, while it has also commenced its architecture specification activities. As part of the latter, the project will provide a mapping of its architectural concepts to the RAMI4.0 reference model. Moreover, the project will start the implementation of the data collection and data analytics mechanisms that will underpin the realization of the architecture and of its services. The project holds the promise to enhance the functionalities and lower the costs for securing IoT applications spanning multiple IoT platforms and smart objects. We will aspire to disseminate more detailed results through publications, presentations and other activities of the IERC cluster in the coming ten months. This work has been carried out in the scope of the H2020 SecureIoT project, which is funded by the European Commission in the scope of its H2020 programme (contract number 779899). The authors acknowledge valuable help and contributions from all partners of the project.

7.6 SEMIoTICS

7.6.1 Brief Overview

SEMIoTICS aims to develop a pattern-driven framework, built upon existing IoT platforms, to enable and guarantee secure and dependable actuation and semi-autonomic behaviour in IoT/IIoT applications. Patterns will encode proven dependencies between security, privacy, dependability, and interoperability (SPDI) properties of individual smart objects and corresponding properties of orchestrations involving them. The SEMIoTICS framework will support cross-layer intelligent dynamic adaptation, including heterogeneous smart objects, networks and clouds, addressing effective adaptation and autonomic behaviour at field (edge) and infrastructure (backend) layers based on intelligent analysis and learning. To address the complexity and scalability needs within horizontal and vertical domains, SEMIoTICS will develop and integrate smart programmable networking and semantic interoperability mechanisms. The practicality of the above approach will be validated using three diverse usage scenarios in the areas of renewable energy (addressing IIoT), healthcare (focusing on human-centric IoT), and smart sensing (covering both IIoT and IoT); and will be offered through an open Application Programming Interface (API). SEMIoTICS consortium consists of strong European industry (Siemens, Engineering, STMicroelectronics), innovative SMEs (Sphynx, Iquadrat, BlueSoft) and academic partners (FORTH, Uni Passau, CTTC) covering the whole value chain of IoT, local embedded analytics and their programmable connectivity to the cloud IoT platforms with associated security and privacy. The consortium is striving for a common vision of creating EU's technological capability of innovative IoT landscape both at European and international level.

7.6.2 Introduction

Global networks like IoT create an enormous potential for new generations of IoT applications, by leveraging synergies arising through the convergence of consumer, business and industrial Internet, and creating open, global networks connecting people, data, and "things". A series of innovations across the IoT landscape have converged to make IoT products, platforms, and devices, technically and economically feasible. However, despite these advancements the realization of the IoT potential requires overcoming significant business and technical hurdles. This includes several system aspects, including dynamicity, scalability, heterogeneity, and E2E security and privacy [46–48], as they are described below.

IoT are dynamic, ever-evolving and often unpredictable environments. This relates to both IoT infrastructures as a whole (e.g. rapid development of new smart objects and IoT applications introducing new requirements to existing infrastructures and networks) and individual IoT applications (e.g. new users and types of objects connecting to said applications). This necessitates dynamically adaptive behaviour at runtime, at the IoT infrastructure, the IoT applications, and locally at the smart objects integrated by them. Intrinsic requirements (e.g. scale, latency) dictate the need for, at least, semi-autonomic adaptation at all layers.

The fast-growing number of interconnected users, smart objects and applications requires high scalability of the IoT infrastructure and network layers. At the network, the vastly increased demands require highly efficient programmable connectivity, service provisioning and chaining in ways that guarantee the much-needed end-to-end (E2E) optimizations, addressing dynamic IoT application requirements. Scalability at the IoT infrastructure level requires seamless discovery and bootstrapping of smart objects, as well as highly efficient orchestration, event processing and analytics and IoT platform integration.

Despite advancements in standardization, there is still limited semantic interoperability within IoT applications and platforms. Semantic interoperability requires three key abilities: (a) to recognize and balance the heterogeneous capabilities and constraints of smart objects, (b) to interpret data generated by such objects correctly, and (c) to establish meaningful connections between heterogeneous IoT platforms.

Smart objects, IoT applications, and their enabling platforms are often vulnerable to security attacks and changing operating and context conditions that can compromise their security [49]. They also generate, make use of, and interrelate massive personal data in ways that can potentially breach legal and privacy requirements [49]. Preserving security and privacy properties remains a particularly challenging problem, due to the difficulty of: (a) analysing vulnerabilities in the complex E2E compositions of heterogeneous smart objects, (b) selecting appropriate controls (e.g., different schemes for ID and key management, different encryption mechanisms, etc.), for smart objects with heterogeneous resources/constraints, and (c) preserving E2E security and privacy under dynamic changes in IoT applications and security incidents, in the context of the ever-evolving IoT threat landscape [50].

The above challenges give rise to significant complexities and relate to the implementation and deployment stack of IoT applications to address them. The overall aim is: demands without considering the data volume. Taking into consideration this ratio, green IT technologies have important environmental and economic benefits. Circular Economy (CE) advocates a continuous development cycle that reforms the currently prominent 'take-make-dispose' linear economic mode by preserving and enhancing the natural capital. SEMIoTICS will also provide the intelligence analytics capabilities and Information Communication Technologies (ICT) that are required for turning IoT data into a worthy asset for CE-centric businesses (e.g. [51]).

7.6.3 Vision

The main goal of the SEMIoTICS project is to develop a pattern-driven framework, built upon existing IoT platforms. The proposed framework will enable and guarantee the secure and dependable actuation and semi-automatic behaviour in IoT/IIoT applications. Specifically, the SEMIoTICS vision in delivering smart, secure, scalable, heterogeneous network and data-driven IoT is based on two key features:

- **Pattern-driven approach:** Patterns are re-usable solutions to common problems and building blocks to architectures. In SEMIoTICS, patterns encode proven dependencies between security, privacy, dependability and interoperability (SPDI) properties of individual smart objects and corresponding properties of orchestrations (composition) involving them. The encoding of such dependencies enables: (i) the verification that a smart object orchestration satisfies certain SPDI properties, and (ii) the generation (and adaptation) of orchestrations in ways that are guaranteed to satisfy required SPDI properties. The SEMIoTICS approach to patterns is inspired from similar pattern-based approaches used in service-oriented systems [52, 53], cyber physical systems [54] and networks [55, 56].
- **Multi-layered Embedded Intelligence:** Effective adaptation and autonomic behaviour at field (edge) and infrastructure (backend) layers depends critically on intelligent analysis and learning the circumstances where adaptation actions did not work as expected. Intelligent analysis is needed locally for semi-autonomous, prompt reaction, but taking into account IoT smart objects limited resources (thus requiring specialized lightweight algorithms) [55, 57]. It should also be possible to fuse local intelligence to enable and enhance analysis and intelligent behaviour at higher levels (e.g. using results of local analysis of "thing events" to globally predict and anticipate failure rates) [58].

7.6.4 Objectives

The SEMIoTICS project will target IoT applications with heterogeneous smart objects, various IoT platforms and different SPDI requirements. Seven main objectives are identified by the SEMIoTICS project including:

- The development of patterns for orchestration of smart objects and IoT platform enablers with guaranteed SPDI properties
- The development of semantic interoperability mechanisms for smart objects, networks, and IoT platforms, like semantic information broker that resolve the semantics of correlated ontologies and common APIs that enable cross-platform programming and interaction
- The development of dynamically and self-adaptable monitoring mechanisms, supporting integrated and predictive monitoring of smart objects in a scalable manner

- The development of core mechanisms for multi-layered embedded intelligence, IoT application adaptation, learning and evolution, and E2E security, privacy, accountability and user control
- The development of IoT-aware programmable networking capabilities based on adaptation and Software-Defined Networking (SDN)/Network Function Virtualization (NFV) orchestration
- The development of a reference prototype open architecture demonstrated and evaluated in both IIoT (renewable energy) and IoT (healthcare), as well as in a horizontal use case bridging the two landscapes (smart sensing), and delivery of the respective open API
- The adaptation of EU technology offerings internationally

These objectives are accomplished, considering the intrinsic requirements of three main use case scenarios for an industrial wind park, an e-health system, and a smart sensing setting.

7.6.5 Technical Approach

Figure 7.12 shows our initial vision of the logical architecture of SEMIoTICS framework and how it relates to smart objects, IoT applications, and existing IoT platforms, and how does it map onto a generic deployment infrastructure consisting of private and public clouds, networks, and field devices. Within the figure, blue boxes show components of the framework that are to be developed by SEMIoTICS; white boxes indicate components of IoT applications managed by the framework. The key role of the SEMIoTICS framework in the IIoT/IoT implementation stack is to support the secure, dependable and privacy-preserving connectivity and interoperability of IoT applications and smart objects used by them, and the management, monitoring and adaptation of these applications, objects and their connectivity.

7.6.5.1 Enhanced IoT aware software defined networks

The sheer number of smart objects that are expected to connect to the Internet by 2020 (more than 50bn smart objects) will increase network traffic dramatically and introduce more diversity of network traffic (from elephant flows to mice flows). This makes the development of networking techniques that are significantly more scalable and agile than today's networks an absolute necessity. Networks will need to dynamically reconfigure their resources and maintain network connectivity. Also, applications running on top of smart connected devices will need to be resource and network-aware, in order to

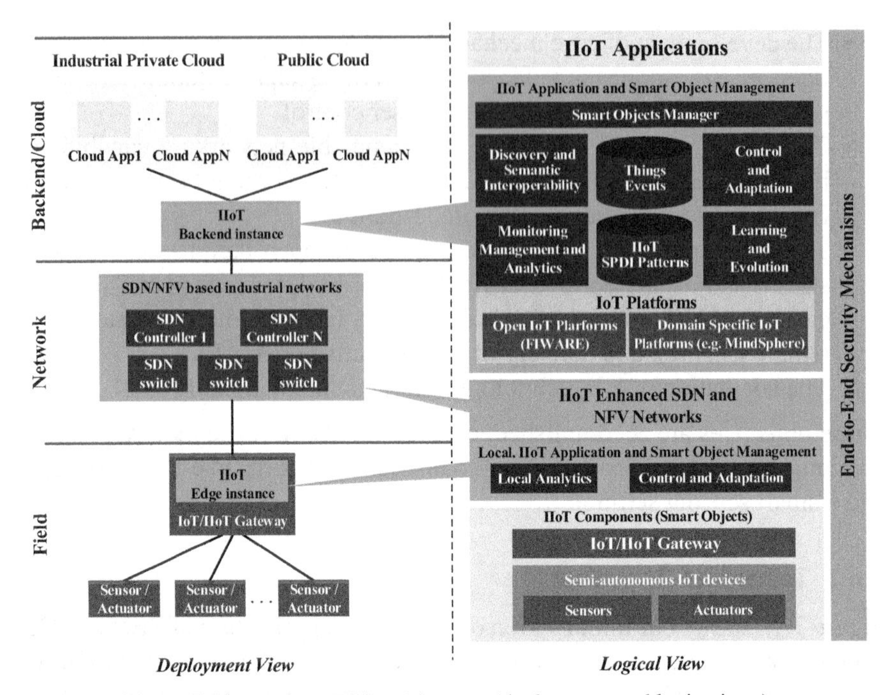

Figure 7.12 SEMIoTICS architecture (deployment and logic views).

take full advantage of underlying network programmability. In summary, **IoT requires more agile networks**.

SDN can provide a solution to this problem. It allows **network programmability**, which can be used to decouple network control from the forwarding network (aka data) plane and to make the latter directly programmable by the former. Integrating IoT and SDN will **increase network efficiency** as it will make it possible for a network to respond to changes or events detected at the IoT application layer through network reconfiguration. If a spontaneous concentration of people in a specific place is detected by an IoT application, for example, the application can send a request to the SDN controller to reconfigure the network and provide more bandwidth to the area before network congestion occurs. As another example, consider an IoT application where sensor readings are transmitted periodically. In such cases, network resources on the path connecting the sensors to the backend IoT application can be reserved during the reporting cycles to enable efficient flows and released outside them. SEMIoTICS aims to develop a middleware layer between the IoT applications and the SDN-controlled field network, abstracting the underlying protocol implementations and SDN APIs. This will allow **IoT applications** to **trigger** the **network reconfiguration** through **pattern-driven adaptations**. In this view, SDN becomes another component in the IoT implementation stack which, like other components, can be dynamically configured through SPDI patterns [56].

7.6.5.2 Localized analytics for Semi-Autonomous IIoT operation

An IDC FutureScape report [59] for IoT reported that by 2018, 40 percent of IoT data will be stored, processed, analysed and acted where they are created before they are transferred to the network. There are two main reasons for this: *big data volume* and *fast reaction*.

First of all, IoTs/IIoTs are generating an unprecedented volume and variety of data depending on the vertical use case. Not all these data need to be sent always to the cloud for storage and processing. Indeed, the volume of the data makes it in many cases extremely difficult to process them globally in an efficient manner and hinders learning the relations that are hidden in the data. For this reason, we need to enrich the generated and collected data with semantic information at the source and intermediate stations, process them locally with machine learning algorithms to extract the most important features of the data and only then transfer the learned local features to the cloud for further, global, processing and feature analysis. Hence, new approaches, techniques, and corresponding designs need to be developed to store, analyse, and derive insight from these data sets. This has already been identified as a challenge by the industry, e.g. Forrester [60] highlighting the need of IoT applications for distributed analytics since centralized analytics cannot cope for many IoT usage scenarios, and Gartner [61] emphasizing the importance of IoT edge architecture and IT/OT integration for achieving such distributed and layered data analysis.

The second reason driving the need for localized analytics is fast reaction. By the time the data makes its way to the cloud for analysis and some analysis results have been obtained and transferred back to the field layer, so much time has passed that the opportunity to act effectively on the obtained analysis results at the field layer (e.g. smart actuation) is usually long gone. Again, this is a crucial requirement for the industry – Forbes and Moor Insights & Strategy (MI&S) [62] expects that machine learning-enabled reaction to changes in the current environmental/system context to be essential for IoT solutions. By 2020 MI&S believes that the machine learning at edge combined with central machine learning in cloud arrangements will exist in a large number of solutions and will account for a great deal of the innovation in IoT world – giving a substantial market advantage to the providers of such solutions. By doing a fast analysis on the local data (whose volume is much reduced compared to the entire data produced by the IIoT/IoT system and thus should be analysable with substantially fewer resources), an IIoT/IoT system can react quickly to context changes and adapt to them, in ways that optimise the use of both its own resources and the environment's, and eventually improving the overall user experience. SEMIoTICS will develop localized analytics at the edge for semi-autonomous operation with smart actuation and use the results of the localized analytics to help improve the subsequent, global analysis that will be performed on the cloud for learning across the whole system and extraction of global patterns – itself a task whose results can be used by local analytics mechanisms to improve their performance and be able to proactively react to situations that had not been observed at that local point in the past but had occurred at other parts of the system.

7.6.6 Security Architecture Concept

As aforementioned, the SEMIoTICS vision is articulated around the development of a framework for smart object and IIoT/IoT application management based on trusted patterns, monitoring and adaptation mechanisms, enhanced IoT centric networks and multi-layered embedded intelligence. These core elements of our approach are described below.

7.6.6.1 Pattern-based trustworthy IIoT/IoT

The key element enabling the SEMIoTICS approach is the use of architectural SPDI patterns. These **patterns define generic ways of composing** (i.e., establishing the connectivity between) and **configuring** the heterogeneous **smart objects and software components** that may exist at all layers of the IoT applications implementation stack, including: sensors and actuators; smart devices; software components at the network, cloud, IoT enabling platforms and/or other middleware layer; as well as software components at the IoT application layer. To do so, **patterns specify** abstract and **generic smart object interaction and orchestration protocols**, enhanced (if necessary) by transformations to ensure the semantic compatibility of data. Furthermore (and more importantly), the smart object interaction and orchestration protocols encoded by the patterns must have proven ability (i.e., an ability proven through formal verification or demonstrated through testing and/or operational evidence) to achieve not only a semantically viable interoperability between the smart objects that they compose but also specific

SPDI properties, which may be required of compositions. The **compositions** defined by patterns are both **vertical** and **horizontal**, i.e., they can involve smart objects at the same (horizontal) or different layers (vertical) layer of the IoT implementation stack. As an example of a pattern that guarantees "data integrity" – i.e., absence of unauthorized modifications of data – consider the **integrity preserving cascade composition pattern** discussed in [63, 64]. According to this pattern in a sequential composition of processes $P_1, \ldots, P_n$ where the input data of P_i are meant to be the output data of P_{i-1}, and the communication between P_{i-1} to P_i (i=2,..., n) is based on an orchestrator O which facilitates data transfers from P_{i-1} to P_i, **overall data integrity is preserved** if data integrity is preserved within each P_i, within O and across all communications from P_{iS} to O and vice versa. The integrity cascade composition pattern applies both to horizontal compositions (e.g., in software services workflows as in [63, 64]) and vertical composition (e.g., in transfer of data in invocation of operations of IoT enabling middleware).

Another (more complex) example of a pattern fitting the SEMIoTICS vision is the **synchronously controlled distribution line (SCDL) pattern** discussed in [54]. SCDL guarantees that a distributed asynchronous sensor system installed upon a physical pipeline (e.g., a pipeline of an electricity distribution network) will operate in virtual synchrony and provide **a guaranteed density of readings** (i.e., a bounded minimum number of readings per distant and per time unit). The pattern suggests a composition consisting of: (i) sensors connected to a controller through a middleware component that realizes a bounded reliable message delivery protocol; (ii) a controller with the capability to authenticate sensors, store readings received from them in fixed length intervals, and **substitute missing or corrupted sensor readings** with synthetic readings computed through the linear interpolation of readings from their closest adjacent sensors and the end of reading intervals. The application of the **SCDL pattern** is proven to **guarantee** the **consumption of readings** at the end of the reading interval where they fit, make them available in a synchronous manner, filter out illegitimate readings and produce **readings of the required density** for the pipeline. In SCDL pattern, these properties are **guaranteed** even **in the presence of missing or corrupted raw data**, as long as there is a minimal number of legitimate sensor readings. Examples of additional patterns have been given in [52] and [56]. These include patterns for confidentiality in service orchestrations and patterns for availability in Software Defined Networks, respectively.

Inspired by these earlier works, SEMIoTICS patterns will develop patterns specifying:

- **Composition structures** for integrating smart objects and components of IoT enabling platforms (e.g., platform enablers) in a manner that guarantees SPDI properties.
- The **E2E SPDI properties** that the compositions expressed by the pattern preserve.
- The **component level SPDI properties** that the types of smart objects and/or components orchestrated by the pattern, must satisfy in order to preserve the end-to-end SPD properties.
- **Additional conditions** that need to be satisfied for guaranteeing end-to-end SPDI properties. These may, for example, include configuration conditions that need to be satisfied by the IoT platforms and the networks providing the connectivity between them, for guarantying the end-to-end availability properties of IoT application (composition).
- **Monitoring checks** that must be monitored at runtime in order to verify that any assumptions about the individual smart objects and components that are orchestrated by a pattern or other operational conditions, which are critical for the preservation of the end-to-end SPDI properties of the pattern, hold at runtime.
- **Adaptation actions** that may be undertaken to adapt IoT applications, which realise the composition structure of the pattern, at runtime. Such actions may, for example, include the replacement of individual smart objects within a composition; the adaptation of the process realizing the composition; the modification of the configuration of the network services used to connect the smart objects of the composition and/or the deployment platforms upon which these objects run. Adaptation actions are specified along with guard conditions determining when they can be executed (guards are monitored, and adaptation is triggered when they are satisfied).

SEMIoTICS will also develop a generic engine supporting the execution of patterns at runtime to realize the overall process of monitoring, forming, adapting and managing smart object orchestrations in IoT applications.

7.6.6.2 Monitoring and adaptation

The SEMIoTICS framework will support **evolving runtime** management and **adaptation** of IoT applications and smart objects [55–58]. Adaptation will be triggered by monitoring the guard conditions of the patterns used by the IoT application of interest, and applying the actions defined in the patterns when such conditions are satisfied. The SEMIoTICS framework will also **monitor** and analyse the **effectiveness of patterns** and the **adaptation**

actions undertaken in reference to the contextual and operational conditions in which they were undertaken. This will be to **identify** deficiencies or **failures** in applying the patterns, to **diagnose** the reasons which may have caused **deficiencies** or **failures** and avoid the application of the same pattern(s) under the same conditions in subsequent phases. The use of a specific type of network connectivity or a specific type of sensor object amongst alternative options may, for example, prove to be a non-optimal option for network performance or sensor signal reliability under particular conditions. Similarly, certain data transformations may prove excessively time consuming for achieving the required scalability in an IoT application. Monitoring will also be necessary to ensure that any component level SPDI properties assumed by the pattern are upheld whilst the pattern is active (i.e., in use) in an IoT application.

Beyond the basic monitoring of the contextual circumstances surrounding the operation of different smart objects and IoT applications, the SEMIoTICS framework will incorporate **learning** and **evolution mechanisms** supporting the analysis of any adaptation and configuration actions undertaken for an IoT application. This will be necessary in order to identify whether the application of patterns is effective over time (e.g., it does indeed prevent the occurrence of breaches of SPDI properties) and what might be the reasons for not being effective when this is the case.

7.6.7 Use Cases

SEMIoTICS will target three IoT application scenarios: two verticals in the areas of energy and health care and one horizontal in the areas of intelligent sensing. These scenarios have been selected since they involve: (a) different and heterogeneous types of smart objects (i.e., sensors, smart devices, actuators) and software components; (b) different vertical and horizontal IoT platforms; and (c) different types of SPDI requirements. Due to these dimensions of variability, our scenarios provide comprehensive coverage of technical issues, which should be accounted for in developing the SEMIoTICS approach and infrastructure, and to this end provide an effective way for driving the R&D work programme of SEMIoTICS and evaluating and demonstrating its outcomes.

7.6.7.1 Renewable energy – Wind energy

Current state of the art of Wind Turbine Controller in a Wind Park control network is typically an embedded or highly integrated operating system, which follows rigorously development and pre-qualification prior to deployment in the real world. As a result of this slow process, new features, adding new sensors, actuators and related advancements require several months or even years to be fully matured and operational in the field.

- **Taking local action on sensing and analysing structured data to find the inclination of a steel tower** – When the nacelle is turned during a cable untwisting event (Sensing), the gravity acceleration (Ag) component measured by an accelerometer in longitude direction (Ay) will vary as a function of the inclination (Inc) of the steel tower. O&M personnel in remote control center wants to know the inclination of all the steel towers on a number of specific wind farms, as these details will have to be shared with the customer to monitor the deformation and fatigue of the steel. To find the inclination of a steel tower, a full cable-untwist procedure has to be activated. This happens, depending on wind conditions, 3–4 times a month. It is also possible to manually instruct the wind turbine to perform the unwind procedure. At the time of the unwindingprocedure a hi-frequency set of data is recorded. A relatively large amount of data is required to calculate the inclination. This datasheet needs to be sent back to the remote control center to model and calculate the inclination. In SEMIoTICS, localized edge analytics will be applied which will result in semi-autonomous IIoT behavior as only the container containing the algorithm and result of the inclination calculation is transferred to between the wind turbine and the remote control centre. The unnecessary data traffic between each turbine and remote control centre is greatly reduced. Without the localized analytics functionality, all the hi-frequency acceleration and nacelle position data should have transferred to remote control centre resulting in suboptimal operation.
- **Smart Actuation by sensing unstructured video/audio data** – Within the turbine, there are many events which can be captured by IIoT sensors such as Grease leakage detection during normal operation or unintended noise detection when the turbine rotor is changing the direction in the line of wind to maximize energy production. The sensing of this unstructured data and acting locally to prevent any damage to the parts of the turbine in the long run will be of key importance. Localized analytics, as proposed in SEMIoTICS, which will lead in smart actuation to protect the critical infrastructure of renewable energy resources.

SEMIoTICS implements:

- Industrial Things semantic discovery, Bootstrapping of IIoT devices and Gateway
- Inventory of the things at the SDN controller
- REST-based Intent interface for network-agnostic cloud applications
- Security at every layer
- Local data analytics at the Sensors, Actuators and Gateway

7.6.7.2 Healthcare

This healthcare use case is an attempt to come up with usable, acceptable and sustainable IoT solution for assisted mobility through falls prevention leading to active and healthy ageing. Falls in older adults are a significant cause of morbidity and mortality and are an important class of preventable injuries. This use case specifically focuses on advanced fall prevention and management solution aimed at both senior citizens and adults with Mild Cognitive Impairment or mild Alzheimer's disease and their (informal) caregivers. The objective of this scenario is to extend the existing IoT platform like AREAS with Assisted Mobility Module (AMM) which is a dedicated module for the management of, and integration of information from, a network of IT services and hardware devices constituting an advanced fall prevention and management solution aimed at both senior citizens and adults with Mild Cognitive Impairment or mild Alzheimer's disease and their (informal) caregivers. Given the figures introduced at the beginning, the social dimension of the solution is reflected in the improved quality of life for people that are susceptible to falls, given that AMM will prolong the time they can work and live independently. The envisaged evolution of the AMM will see the inclusion of additional robotic elements, in particular, the system will include a:

- Robotic Assistant (RA) connected to a network of embedded **IoT devices and services for monitoring** (and maintaining a diary of) a **patient's activities, health status and treatment**, as well as for supporting cognitive skills training, notifying/reminding the patient of upcoming treatments (e.g. medication schedules) and visits.
- Personal assistant robots may help the patients with their **daily activities like walking trail and other routine**.

SEMIoTICS will contribute in the:

- Integration of distributed IoT devices with higher degree of autonomy (i.e. robotic devices)
- Exploitation of computational resources both in the cloud and at the edge
- Security and privacy of patient data, safety of a patient

7.6.7.3 Generic IoT & smart sensing

Today's IoT embedded devices are often described as smart devices. "Smart" usually shall be associated to some Things that show some form of intelligence, bright behaviour during their operations. Unfortunately, current meaning and their reality is that they are locally programmable and always connected to some cloud infrastructure (e.g. typically through a wireless connection such as Wi-Fi or Bluetooth Low Energy) to send raw data. Therefore, these devices transmit sensed data to the cloud without any analytic being performed locally and without showing remarkable forms of computational intelligence. An IoT thing is intelligent is it has capabilities **to learn from and act upon the data (at least without supervision)** it is sensing. Sometimes, they also receive back from the cloud some form of actuation (control) instructions, which are determined by a centralized server-based analysis of sensed and other data. A typical example, on domotic applications, is the one where several sensing nodes stream some relevant raw data at given interval (e.g. temperature, humidity, pressure) to a cloud service. An example is the Microsoft Azure or IBM Bluemix cloud platforms and related ecosystem. In this scenario, the intelligent data processing always resides remotely, and the communication channel is (implicitly) assumed to be always present and open.

The use case provides:

- Evolution of platform technologies enabling local analytics computing (i.e. edge computing)
- Enhanced IoT system scalability and increased robustness
- Open market enhanced middleware portfolio for intelligent embedded devices and innovative businesses opportunities

SEMIoTICS's research efforts focus in the:

- Support for tight integration at device level of sensing and computational elements in close tight cooperation on dedicated embedded HW (i.e. edge computing)

- Increased system scalability and computational partitioning to enhance system responsiveness and stability by exploiting self-adapting online learning mechanisms
- Enhanced architectural models redefining system from bottom to top for handling the continuous and discrete sensing.

7.6.8 Summary

SEMIoTICS aims to develop an open IIoT/IoT framework, interoperating with existing IIoT/IoT platforms (e.g. FIWARE, MindSphere) and programmable networking, through their exposed APIs. The SEMIoTICS framework will also integrate IIoT and IoT sensing and actuating technologies. A core element of the SEMIoTICS approach is the development and use of patterns for orchestration of smart objects and IoT platform enablers in IoT applications with guaranteed SPDI properties. Patterns constitute an architectural concept well founded in software systems engineering. SEMIoTICS advocates the patterns approach to systems engineering, but uses a novel pattern type (i.e., SPDI patterns) to guarantee semantic interoperability, security, privacy and dependability in large scale IIoT/IoT applications integrating smart objects. Said patterns will be supported by mechanisms featuring integrated and predictive monitoring of smart objects of all layers of the IoT implementation stack in a scalable manner, as well as core mechanisms for multi-layered embedded intelligence, IoT application adaptation, learning and evolution, and end-to-end security, privacy, accountability, and user control. This approach will enable and guarantee secure and dependable actuation and semi-autonomic behaviour in IoT/IIoT applications, supporting cross-layer intelligent dynamic adaptation, including heterogeneous smart objects, networks and clouds.

7.7 SerIoT

The Internet of Things or Internet of Everything envisages billions of physical things or objects (sensors and actuators) connected to the Internet via heterogeneous access networks. IoT is emerging as the breakthrough technology introducing the next wave of innovations, including revolutionary applications, significantly improving and optimizing our daily life.

The IoT is capable to create a complex Network of Networks system through IP protocol and Mobile Network connectivity, allowing "things" to be read, controlled and managed at any time and at any place. This brings such technical issues as the lack of a shared infrastructure, lack of common standards, problems with the flexibility, scalability, adaptability, maintenance, and updating the IoT devices, etc.

Especially important are security concerns, resulting from all of the listed technological aspects [77, 78]. In case of lack of the IoT related security standards and commonly accepted technological solutions, every vendor creates their own solutions. Moreover, the solutions currently used in IT systems are mostly unsuitable for direct application in IoT, e.g. authentication based on central server that works well for small scale systems but does not provide sufficient mechanisms for future large scale IoT ecosystems. On the other hand, attacks on the IoT platforms will have significant economic, energetic and physical security consequences, beyond the traditional Internet lack of security.

7.7.1 SerIoT Vision and Objectives

SerIoT aims to conduct research for the delivery of a secure, open, scalable and trusted IoT architecture. The solution will be implemented and tested as a complete, generic solution to create and manage large scale IoT environment operating across IoT platforms and paying attention on security problems.

A decentralized approach, based on peer to peer, overlay communication is proposed [69]. SerIoT will optimize the security of IoT platforms in a cross-layered manner. The concept of Software Defined Networks (SDN) is used and SDN controllers are organized in hierarchical structure [74, 75]. The objectives of SerIoT include to provide the prototype implementation of a self-cognitive [66–68], SDN based core network, easily configurable to adapt to any IoT platform, including advanced analytics modules, self-cognitive honeypots and secure routers. The solution will be supported by appropriate technologies such as Decision Support System (DSS) supplementing controller's functionality. The DSS will be able to detect the potential threats and abnormalities. The system will be supplemented with comprehensive and intuitive visual analytics and mitigation strategies that will be used according to the detected threats. It will be validated in the final phase of the project through representative use cases scenarios, involving heterogeneous EU wide SerIoT network system.

7.7.2 SerIoT Architecture Concept

The SerIoT architecture [65] is based on a software-managed network implementing SDN technology and is divided into the following layers and modules (See Figure 7.13.).

The **IoT Data Acquisition layer** is comprised of the low-level IoT-enabled components that create the infrastructure backbone, including honey-pots, dedicated engines and storage capabilities and the SDN secure routers. The SDN routers will use OpenFlow communication and will be based on Open Switch implementation being significantly extended to cooperate with related SerIoT modules and security mechanism.

The backbone network will be divided into domains (subnets). Every subnet constitutes an autonomic SDN network, controlled by the SDN controller and extended according to SerIoT needs. Controllers will be organized into hierarchical structure [76]. The first level controller is responsible for the routing within the subnet using gathered data. It will be also able to route packets to neighbouring subnets (via the appropriate border node). In the case of destinations outside their own subnet and neighbouring subnets, routing requests will be sent to a second (or third, fourth, etc.) level controllers. The controllers will continuously gather information to feed the analytics module.

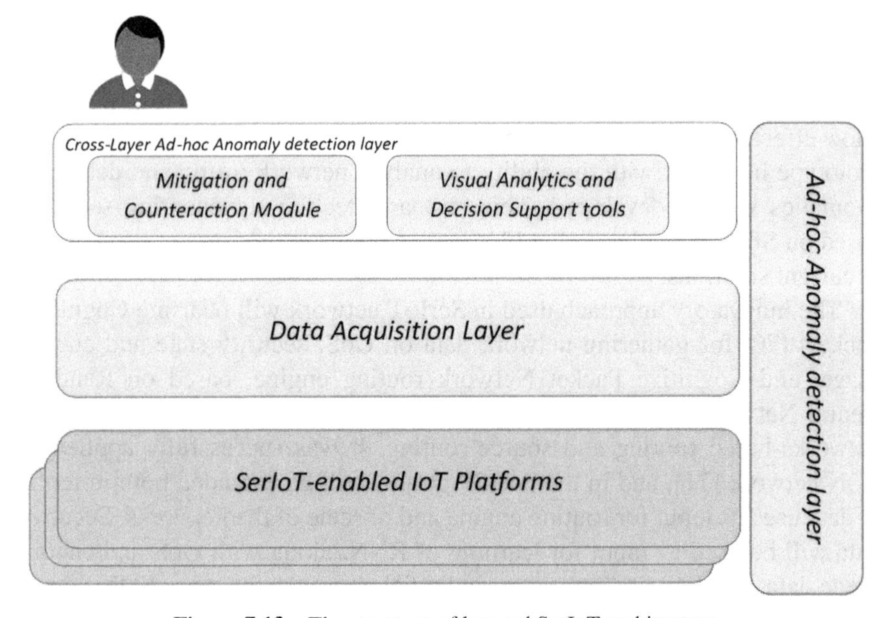

Figure 7.13 The structure of layered SerIoT architecture.

These components will be connected to visual analytics module and support decision making system.

The **Ad-hoc Anomaly detection layer** will provide a number of security mechanisms, executed across IoT devices, honeypots and SDN routers. Anomaly detection techniques based on local traffic characteristics (as dynamic changes in queue lengths) will be regularly probed by smart "cognitive packets" sent by the SDN controller and feeding the controller routing decisions. The controller will have the ability to detect suspicious and risky paths, and re-schedule the routing paths over secure, preferable connections according to secure aware routing, but also energy and Qality of Service (QoS) aware routing [71–73].

The **Visual Analytics and Decision Support tools** will deal with the interactive decision support applications that will be delivered to the end-users, able to effectively detect potential abnormalities at different levels of the network. The end-user tool will be developed together with a novel visual analytics framework, dealing with the effective management and visualization of data.

The **Mitigation and Counteraction Module** will be responsible for implementing decisions taken by the Decision Support tools. The module will use dedicated software and network components as SDN routers, honeypots and IoT devices.

The SerIoT platform will ensure the separation of enterprise and private data. The system will provide monitoring mechanisms and anomaly detection techniques, using a cross-layer data collection infrastructure that will allow effective information transmission and data aggregation for analysis. A prototype honeypot with the ability to analyse network traffic and detecting anomalies will be developed. This new architecture for ensuring security, based on SDN technology, should bring a significant progress in comparison to current solutions.

The innovatory approach used in SerIoT network will be using Cognitive Packets [70] for gathering network data on QoS, security state and energy usage, and Cognitive Packet Network routing engine, based on Random Neural Networks (RNN) [79, 80]. The concept is a combination of neural-networks-based routing and source routing. It was successfully applied in SDN network [71], and in the SerIoT project will be extended both in terms of data used as input for routing engine and of scale of the networks. Security data will be used as input for learning of RNN, along with QoS and energy usage data, to allow finding secure and efficient routes for every SDN flow.

7.7.3 Use Cases

The solutions of the SerIoT project will be evaluated in individual laboratory test-beds and also in an integrated EU wide test-bed which will interconnect significant use cases developed by SerIoT industry partners.

SerIoT aims to design and to deploy four innovative use cases arising from three significant for the global economy domains where the use of IoT is rapidly increasing: (i) Smart Cities domain will be covered by two ambitious use cases where Surveillance and Intelligent Transportation IoT networks will be evaluated, (ii) Flexible Manufacturing domain with the detection of physical attacks on wireless sensor networks, and finally (iii) a novel Food Chain Scenario will be exploited demonstrating mobility security issues.

Each of the use cases considers one or several scenarios. A scenario is intended to describe and specify the system behaviour according to a specific situation, or in other words to describe the situation in which a specific system should work and how the system works and interacts with the different users:

- Use Case 1 (Surveillance) scenarios:
- Facilities monitoring
- Embedded intelligence in buses
- Use Case 2 (Intelligent Transport Systems ITS in Smart Cities) scenarios:
- Automated driving
- Public transport maintenance
- Public transport security
- Road side ITS stations
- Use Case 3 (Flexible Manufacturing Systems) scenarios:
- Wireless robots in warehouse
- Critical infrastructure protection
- Use Case 4 (Food Chain) scenario:
- Fresh food deadline control

7.7.4 Industrial and Commercial Involvement

SerIoT has strong support regarding industrial know-how and implementation. Among the Consortium partners there are eight industrial or small/medium size enterprises (SME) with diverse and complementary technological and research expertise, covering the full spectrum of research and innovation activities anticipated in the project [65]. Six of these partners are large industrial societies able to support the multi-disciplinary topics introduced in SerIoT, i.e. IoT telecom/network infrastructure & Industry 4.0 Use Cases by DT/T-Sys, IoT anomaly detection by ATOS, IoT applications & platform by DT/T-Sys., design-driven & cross-layer analytics by ATOS. Moreover, SMEs involved in the consortium are among the leading and innovative companies in their sectors. Hence, a large amount of innovation foreseen in the project will be also carried by SMEs. What all SerIoT SME partners share in common is their proven ability to apply research results into successful and well established commercial products (e.g. HOP Core, Wear & Extended innovative solution by HOPU). Having in mind their strong commitment in delivering new services in their customers, industrial & SME partners have identified complementary private investments to support the SerIoT business perspectives.

Moreover, specific dissemination actions will be carried out, through already established communication channels, networks and working groups in order to ensure that the new & open solutions of the project will be conveyed to major stakeholders in Europe and Worldwide.

7.7.5 Summary

In this paper we outline the EU H2020 SerIoT project that addresses IoT security challenges. As a scientific project, SerIoT will provide a new approach to understand the threats to IoT based infrastructures and deliver methods to solve the security problems in the IoT. Pioneering research and development based on holistic approaches will be conducted. A generic IoT framework based on an adaptation of the concept of Software Defined Networks with Cognitive Packets will be developed as well as the new methods for intrusion detection with the use of a cross-layer approach. Visual analytics tools for analysing threats in IoT ecosystem will be used.

7.8 SOFIE – Secure Open Federation for Internet Everywhere

The **main goal** of the SOFIE [83] project is to *enable diversified applications from various application areas to utilise heterogeneous IoT platforms and autonomous things across technological, organisational and administrative borders in an open and secure manner, making reuse of existing infrastructure and data easy*. SOFIE is guided by the needs of three pilot use cases with diverse business requirements: food supply-chain, mixed reality mobile gaming, and energy markets. Furthermore, we will explore the synergies among these areas, building a foundation for cross-application-area use of existing IoT platforms and data.

SOFIE will design, implement and pilot a systematic, open and secure way to establish new business platforms that utilise existing IoT platforms and distributed ledgers. With *"openness"*, we mean flexible and administratively open business platforms, as well as technically decentralised federation to enable the interoperability of different IoT platforms, ledgers, and autonomous devices. To realise this vision, SOFIE brings together large system vendors and integrators (ENGINEERING and Ericsson), high tech SMEs delivering highly innovative products and solutions (GuardTime, Synelixis) and prestigious universities (Aalto University and the Athens University of Economics and Business). The results of the project will be guided by these three use cases and will be tested in an equal number of real-life trials. For this purpose, the consortium includes ASM TERNI S.p.A., a public multi-utility company and Emotion who will trial SOFIE developments in the energy sector, OPTIMUM, a leading SME in the area of supply chain IT systems, which (together with SYNELIXIS) will trial SOFIE in the realisation of a farm-to-fork scenario and Rovio Entertainment Corporation, which will lead the SOFIE trial in a mixed reality mobile gaming context.

7.8.1 Objectives

- The SOFIE consortium has broken down the high-level goal into the following specific and tangible objectives:
- Define a secure, open, decentralised and scalable IoT federation architecture for sensing, actuation, and smart behaviour. In order to stay open and interoperable, emerging standard interfaces should be used between the components and towards the outside world.
- Make IoT data and actuation accessible across applications and platforms in a secure and controlled way. SOFIE must provide the means to reuse data, within the limits set by its owner, across applications.
- Develop a solution to provide integrity, confidentiality and auditability of IoT data, events and actions. SOFIE shall define and implement ledger-independent transactions that can be simultaneously entered into various closed and open blockchains and other persistent ledgers.
- Develop an IoT federation framework to facilitate creation of IoT business platforms. The framework can be used to create business platforms, including those for the three pilot use cases.

- Deploy and evaluate the SOFIE federation framework in three field trials.
- Evaluate the commercial viability of the SOFIE federation approach based on the three field trials and research on business models.
- Establish the SOFIE IoT federation approach as a major enabler for the IoT industry through dissemination, standardization, education, workshops and pilots.

7.8.2 Technical Approach

SOFIE combines several IoT platforms and distributed ledgers into a federated IoT platform supporting the reuse of existing IoT infrastructure and data by various applications and businesses. Figure 7.14 illustrates the overall architectural approach.

SOFIE achieves decentralization of business platforms through the use of DLTs. Since the properties of various DLTs, such as scalability, throughput, resilience, and openness, are significantly different, SOFIE relies on using multiple different DLTs in parallel. To allow transactions to be recorded into multiple blockchains or other ledgers, SOFIE will design and implement the inter-ledger transaction layer. We will build upon existing leading-edge work, including the W3C-associated Inter-ledger Protocol (ILP), applying the results to the IoT domain, and developing them further. The transactions will be implemented as multi-stage smart contracts whose resolution depends on

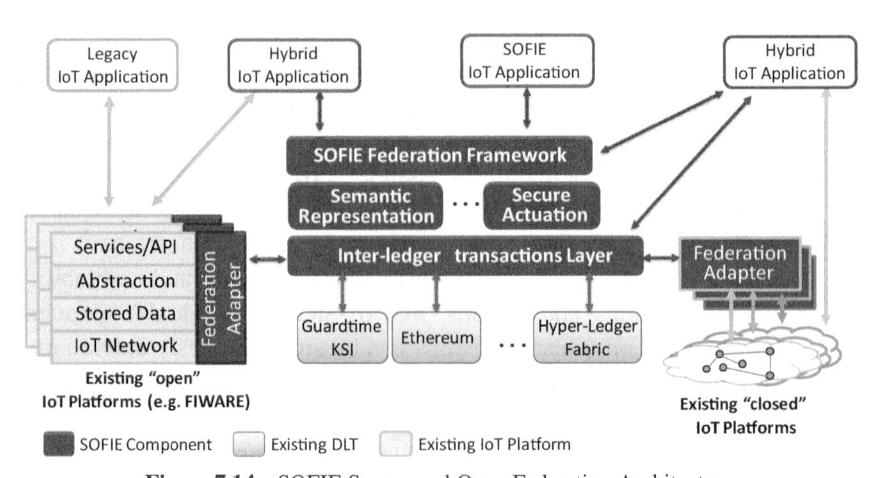

Figure 7.14 SOFIE Secure and Open Federation Architecture.

the transactions being correctly recorded in all the participating ledgers, but without requiring that all the ledgers support smart contracts.

The inter-ledger transaction layer will be used for three main purposes:

- **Describe the ("things") data in the existing IoT platforms,** enabling financially tied IoT actuation between organisations and storing security-related data.
- **Enable secure and traceable IoT actuation.** The idea is to negotiate and use smart contracts that may span multiple ledgers to record intention or desire to actuate, to trigger actuation, to permanently record both actuation instances and the related sensor values, and to trigger any financial transactions, thereby supporting smart behaviour.
- **Enable interoperability between diverse existing IoT platforms.** This is achieved by augmenting the existing IoT platforms with a federation adapter.

These together allow applications to: discover what data and things are available in the IoT platforms; acquire the necessary permissions for access (e.g. by promising to pay or placing a pledge); access the data and/or request actuation in a secure, recorded, and compensated manner; and verify whether the requested actuation took place or not. Beneath the inter-ledger transaction layer are distributed ledgers. These include commonly used blockchains such as Ethereum, and private commercial blockchains such as KSI Blockchain developed by SOFIE partner Guardtime [81].

The SOFIE federation approach is designed to be technology-agnostic, allowing systems with different APIs and data formats to interoperate to the extent allowed by the applicable security policies. Some of the existing IoT platforms already support interoperability across different protocols and standards. Examples of this include FIWARE through its IoT adapters [84], such as the already existing LWM2M and oneM2M adapters, and W3C WoT, where the IoT servient concept supports both proprietary APIs and various protocol adapters. While most of the data will reside within existing IoT systems, a key aspect of SOFIE is the so-called **smart contract**, available in some blockchains, such as Ethereum. From the SOFIE point of view, a smart contract is simply a computer program and its associated computational state that "lives" in a blockchain.

7.8.3 Security Architecture

The SOFIE security architecture provides end-to-end security (confidentiality and integrity), identification, authentication and authorization, and supports users' privacy and control over their data. Most existing solutions already provide decent end-to-end security within the system and system-specific authentication. Therefore, SOFIE concentrates on innovating in the areas of data sovereignty, privacy and federated key management, authentication, and authorization.

IoT data can often be personal and therefore governed by a new EU's GDPR legislation. Ensuring compliance with the GDPR is a major design requirement for the SOFIE security architecture. SOFIE plans to use MyData [85] together with Sovrin Foundation identity blockchain [86] to allow individuals to better control how their personal data is used.

In order to support data sovereignty and privacy, SOFIE adopts a three-level approach to the storage of data. First, there is a private data store managed entirely by the stakeholder. A private blockchain (such as Guardtime's KSI Blockchain) forms the second level data that is shared between collaborating stakeholders (for examples producer, reseller, and supermarket in the food chain use case). Finally, some data (such as hashes of transaction trees from the lower level) will be stored in a public blockchain, such as Ethereum or Bitcoin. Such an approach allows fine grained control of the data, from total openness (e.g. to bring transparency to certain public services) to very tight access control (e.g. to protect trade secrets or the privacy of people). In either case, integrity and non-repudiation of the data is guaranteed.

7.8.4 Use Cases

The SOFIE approach will be tested in three different use cases described below. The **food chain pilot** aspires to demonstrate the field-to-fork scenario towards security in food production and consumption. SOFIE applications and realization of a community-supported heterogeneous end-to-end agricultural food chain will be demonstrated and evaluated. The use case will combine multiple types of ground, micro-climate, soil, leaf and other information stations, existing IoT platforms, mobility, location-based services (LBS), food tracking information, smart micro-contracts, and decentralized autonomous organizations implemented with smart contracts. The consumer may trace the entire history of the product based on the QR or RFID tag on the package, even in the shop before buying the product. Consumers can reliably verify not only the farmer from whom the product originates, but also the entire production and supply chain history associated with each food item, starting from the source of the seeds, the quality of the soil and the air in the producer's premises, the amount of water that has been consumed, the fertilization process, the method and time of growing, the weather conditions, the transportation mode and distance, the storage conditions etc. This gives consumers the ability to make decisions about their food based on health and ethical concerns, including environmental sustainability, fair labour practices, the use of fertilizers and pesticides, and other similar issues.

In the **Mixed Reality Mobile Gaming Pilot, virtual and real worlds will be combined.** Mobile gaming is a rapidly growing market, popular games, such as *Pokémon Go*, are already taking advantage of augmented reality and SOFIE aims to take such interaction further. SOFIE will integrate a mobile game with the real world using a federated IoT platform aiming to: a) enable the gamers to interact with the real world via sensors and actuators, b) take advantage of existing and emerging IoT infrastructure (e.g. building automation), c) enable payments in virtual and real currencies between the gamers, games, and other parties, and d) create new business opportunities for various parties, including gaming companies, as well as the owners of buildings and public spaces (e.g. malls) and various businesses (e.g. shops and cafés). The gamers will be both moving in the physical world and interacting with it through the games. Existing IoT infrastructure, for instance movement sensors and control of lighting and passage, will be included in the game world through the federated SOFIE platform. Owners of spaces and businesses will be able to bring their existing or new IoT infrastructure into the gaming world, while the blockchain-based marketplace will allow for all kinds of business models, including In-Game Assets (IGA) trading.

The **energy pilot** aims at optimized Demand Response and at supporting electricity marketplaces and micropayments. The energy pilot consists of two parts: first, a **real-field** pilot will demonstrate the capability of creating smart micro-contracts and micro-payments in a fully distributed energy marketplace, located in Terni, Italy. The pilot will cover the end-to-end scenario form electricity production, distribution, storage and consumption. During the scenario electricity produced by renewable sources (PVs) will be fed into the low voltage (LV) electricity network. Most of this electricity will be normally consumed by energy customers (i.e. houses, offices, etc). However, the surplus of the generated power would generate reverse power flows through the LV distribution network substation. The electricity distribution network is designed to handle only unidirectional electricity flows, thus reverse flows may generate significant problems. To avoid this abnormal operation, electrical vehicles (EVs) will be offered significant promotional benefits to match their EV charging needs with the network time and space balancing requirements. The EV chargers will be communicating with the EV drivers, with the car battery management system, the local energy generation and consumption, and the smart meters to predict if the requested charging service/network grid stabilization will be available in due time. Second, a **laboratory and interoperability pilot** based on real-data from smart energy meters deployed in the greater area of Tallinn, Estonia. The trial will be based on the Estfeed open software platform [87] for energy consumption monitoring and management from the customer (consumers/prosumers) side, which is capable of interacting with the power network and to provide data feeds for efficient use of energy.

To assess **cross-SOFIE interoperability**, SOFIE pilots will be federated as shown in Figure 7.15. In the cross-pilot, the emphasis will be on the demonstration of the exploitation of data stored/cached in different locations to be accessed across different platforms, as well as the development of

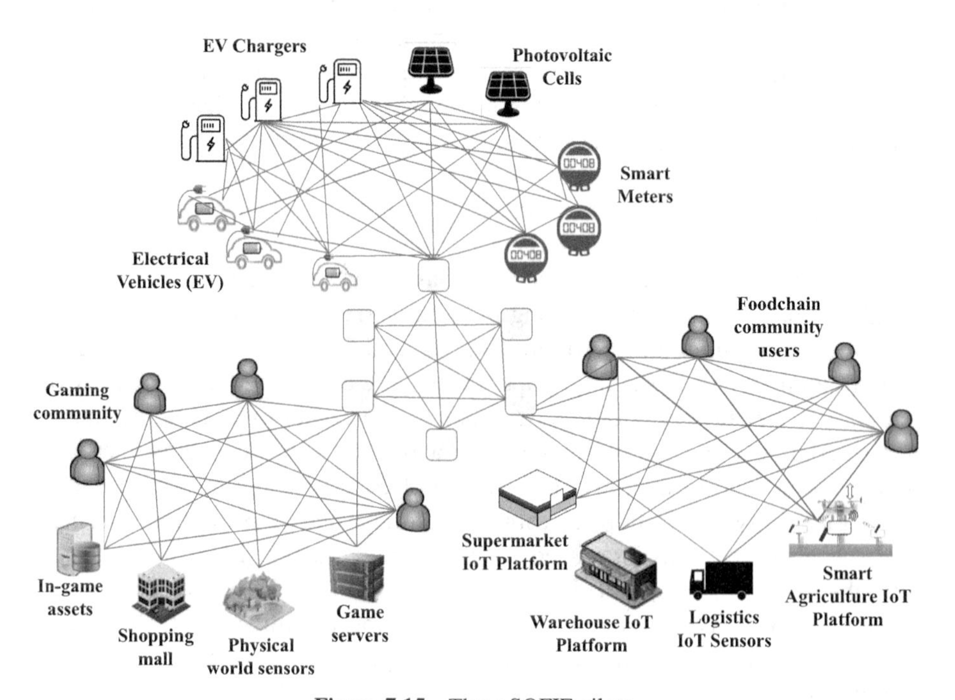

Figure 7.15 Three SOFIE pilots.

applications exploiting different underlying infrastructures. The SOFIE interfaces abstraction will allow virtual entities in one platform to be exploited by applications from a different platform, while data semantics and analytics will facilitate the data exploitation. Initial consideration of scenarios to be tested include: Energy and gaming pilots exploiting data protection/privacy (e.g., for building access), energy pilots (EV) exploiting smart agriculture data with respect to environmental conditions and payments and contracts across pilots (e.g., getting food discounts from gaming achievements).

7.8.5 Conclusions

The SOFIE federation approach will help make the existing siloed IoT platforms interoperable, enabling cross-platform applications and reuse of data in a secure and scalable manner. SOFIE will offer data sovereignty in GDPR-compliant way, giving users more control of their data. Through the usage of distributed ledgers, SOFIE will promote open business platforms, allowing creation of new kinds of decentralised open marketplaces, which no single entity – public or private – can technically control and thus exercise sole pricing power over them. This in turn will lower the barrier of entry for small businesses and individuals. The SOFIE federation framework will be released as open-source and SOFIE partners have the capacity to deliver and boost the penetration of SOFIE offerings in the market and relevant standardization bodies.

List of Notations and Abbreviations

Notations	Abbreviations
AAA	Authentication, Authorisation and Accounting
ABE	Attribute-Based Encryption
CP-ABE	Ciphertext-Policy Attribute-Based Encryption
DHT	Distributed Hash Table
IoT	Internet of Things
JSON-LD	JavaScript Object Notation for Linked Data
KPI	Key Performance Indicator
QoS	Quality of Service
QoI	Quality of Information
OMB	Overlay Management Backbone
RDF	Resource Description Framework
RDQL	RDF Data Query Language

Notations	Abbreviations
TEEs	Trusted Execution Environments
API	Application Programming Interface
bD	by-Design
CE	Circular Economy
E2E	End-to-End
GDPR	General Data Protection Regulation
EU	European Union
ICT	Information Communication Technologies
IoT	Internet of Things
IIoT	Industrial IoT
ML	Machine Learning
NFV	Network Function Virtualization
SDN	Software-Defined Networking
SPDI	Security, Privacy, Dependability and Interoperability

References

[1] International Telecommunication Union (ITU), report on Climate Change, Oct. 2008.
[2] G. Koutitas, P. Demestichas, 'A review of energy efficiency in telecommunication networks', Proc. In Telecomm. Forum (TELFOR), pp. 1–4, Serbia, Nov. 2009.
[3] Gartner Report, Financial Times, 2007.
[4] I. Cerutti, L. Valcarenghi, P. Castoldi, 'Designing power-efficient WDM ring networks', ICST Int. Conf. on Networks for Grid Applic., Athens, 2009.
[5] W. Vereecken, et. al., 'Energy Efficiency in thin client solutions', ICST Int. Conf. on Networks for Grid Applic., Athens, 2009
[6] J. Haas, T. Pierce, E. Schutter, 'Datacenter design guide', White Paper, The Green grid, 2009.
[7] Intel, 'Turning challenges into opportunities in the data center', White Paper, online at: www.intel.com/Intel.pdf
[8] Adel S. Elmaghraby, Michael M. Losavio, "Cyber security challenges in Smart Cities: Safety, security and privacy", Journal of Advanced Research Volume 5, Issue 4, Pages 491–497, July 2014.
[9] CHARIOT Grant Agreement number 780075, Annex 1, Part A.
[10] CHARIOT Grant Agreement number 780075, Annex 1, Part B.
[11] VESSEDIA Project website, https://vessedia.eu/ (last access May 2018).

[12] CHARIOT Project website, http://www.chariotproject.eu/ (last access May 2018).

[13] How To Make 2017 The Year Of IoT Security, William H. Saito, Forbes, 2017.

[14] With the Internet of Things, we're building a world-size robot. How are we going to control it?, Bruce Schneier, New York Magazine, January 2017.

[15] The Internet of Things becomes the Internet that thinks with Watson IoT, https://www.ibm.com/internet-of-things, (last access May 2018).

[16] IEC: IoT 2020: Smart and secure IoT platform. IEC white paper (2016)

[17] NESSI: Cyber physical systems: Opportunities and challenges for software, services, cloud and data. NESSI white paper (2015).

[18] NESSI: SOFTWARE CONTINUUM: Recommendations for ICT Work Programme 2018+. NESSI report (2016).

[19] Humble, J., Farley, D.: Continuous Delivery: Reliable Software Releases through Build, Test, and Deployment Automation. Addison-Wesley Professional (2010).

[20] Taivalsaari, A., Mikkonen, T.: A roadmap to the programmable world: software challenges in the iot era. IEEE Software 34(1) (2017) 72–80.

[21] Morin, B., Fleurey, F., Husa, K.E., Barais, O.: A generative middleware for heterogeneous and distributed services. In: 19th International ACM SIGSOFT Symposium on Component- Based Software Engineering (CBSE), IEEE (2016) 107–116.

[22] I. Stoica, R. Morris, D. Karger, M. F. Kaashoek, and H. Balakrishnan, "Chord: A scalable peer-to-peer lookup service for internet applications," in Proceedings of the 2001 Conference on Applications, Technologies, Architectures, and Protocols for Computer Communications, (New York, NY, USA), pp. 149–160, ACM, 2001.

[23] P. Maymounkov and D. Mazieres, "Kademlia: A peer-to-peer information system based on the xor metric," in Proceedings of the First International Workshop on Peer-to-Peer Systems, (London, UK), pp. 53–65, Springer-Verlag, 2002.

[24] L. Cheng, et al., "Self-organising management overlays for future internet services," in Proceedings of the 3rd IEEE International Workshop on Modelling Autonomic Communications Environments, (Berlin, Germany), pp. 74–89, Springer-Verlag, 2008.

[25] G. Klyne and J. J. Carroll, "Resource Description Framework (RDF): Concepts and Abstract Syntax," 2004. http://www.w3.org/TR/rdf-concepts/

[26] M. Cai, M. Frank, B. Yan, and R. MacGregor, "A subscribable peer-to-peer RDF repository for distributed metadata management," Web Semantics: Science, Services and Agents on the World Wide Web, vol. 2, no. 2, pp. 109–130, 2004.
[27] A. Seaborne, "RDQL – A Query Language for RDF," 2004. http://www.w3.org/Submission/RDQL/
[28] E. Prud'hommeaux and A. Seaborne, "SPARQL Query Language for RDF," 2008. http://www.w3.org/TR/rdf-sparql-query/
[29] Jan Henrik Ziegeldorf, Oscar García Morchon, Klaus Wehrle. "Privacy in the Internet of Things: threats and challenges," Security and Communication Networks 7(12): 2728–2742, 2014.
[30] S. Nakamoto, "Bitcoin: A P2P Electronic Cash System," 2009.
[31] Hernandez-Ramos, J.L.; Pawlowski, M.P.; Jara, A.J.; Skarmeta, A.F.; Ladid, "L. Toward a Lightweight Authentication and Authorization Framework for Smart Objects," IEEE J. Select.Areas Commun., 33, 690–702, 2015.
[32] José L. Hernández-Ramos, Antonio J. Jara, Leandro Marín, and Antonio F. Skarmeta Gómez, "DCapBAC: embedding authorization logic into smart things through ECC optimizations," International Journal of Computer Mathematics, 93(2): 345–366, 2014.
[33] N. G. Weiskopf and C. Weng, "Methods and dimensions of electronic health record data quality assessment: enabling reuse for clinical research," Journal of the American Medical Informatics Association, vol. 20, no. 1, pp. 144–151, 2013.
[34] P. N. Mendes, H. Muhleisen, and C. Bizer, "Sieve: Linked data quality assessment and fusion," in Proceedings of the 2012 Joint EDBT/ICDT Workshops, ser. EDBT-ICDT '12. New York, NY, USA: ACM, 2012, pp. 116–123.
[35] F. Lecue, S. Tallevi-Diotallevi, J. Hayes, R. Tucker, V. Bicer, M. Sbodio, and P. Tommasi, "Smart traffic analytics in the semantic web with star-city: Scenarios, system and lessons learned in dublin city," Web Semantics: Science, Services and Agents on the World Wide Web, vol. 27, pp. 26–33, 2014.
[36] N. Bissmeyer, S. Mauthofer, K. M. Bayarou, and F. Kargl, "Assessment of node trustworthiness in VANETs using data plausibility checks with particle filters," in Vehicular Networking Conference (VNC), 2012 IEEE, Nov 2012, pp. 78–85.
[37] N. Bissmeyer, J. Njeukam, J. Petit, and K. M. Bayarou, "Central misbehavior evaluation for VANETs based on mobility data plausibility," in

Proceedings of the Ninth ACM International Workshop on Vehicular Inter-networking, Systems, and Applications, ser. VANET '12. New York, NY, USA: ACM, 2012, pp. 73–82.

[38] R. Zafarani and H. Liu, "Evaluation without ground truth in social media research," Communications of the ACM, vol. 58, no. 6, pp. 54–60, 2015.

[39] R. Toenjes, D. Kuemper, and M. Fischer, "Knowledge-based spatial reasoning for IoT-enabled smart city applications," in 2015 IEEE International Conference on Data Science and Data Intensive Systems. IEEE, 2015, pp. 736–737.

[40] Antonio F. Skarmeta, et al. "IoTCrawler: Browsing the Internet of Things" in IEEE 2018 Global Internet of Things Summit (GIoTS).

[41] Bröring, A., S. Schmid, C.-K. Schindhelm, A. Khelil, S. Kaebisch, D. Kra-mer, D. Le Phuoc, J. Mitic, D. Anicic, E. Teniente (2017): Enabling IoT Ecosystems through Platform Interoperability. IEEE Software, 34(1), pp. 54–61.

[42] OpenFog Consortium "OpenFog Reference Architecture for Fog Computing", February 2017.

[43] International Standardization Organization, "ISO 27001: Information Se-curity Management System Requirements", Geneva, Switzerland, 2013.

[44] B. Nagpal, N. Singh, N. Chauhan and R. Murari, "A survey and taxonomy of various packet classification algorithms," 2015 International Conference on Advances in Computer Engineering and Applications, Ghaziabad, 2015, pp. 8–13. doi: 10.1109/ICACEA.2015.7164675.

[45] Z. Ma, A. Hudic, A. Shaaban and S. Plosz, "Security Viewpoint in a Reference Architecture Model for Cyber-Physical Production Systems," 2017 IEEE European Symposium on Security and Privacy Workshops (EuroS&PW), Paris, 2017, pp. 153–159. doi: 10.1109/EuroSPW.2017.65

[46] A. Botta, W. De Donato, V. Persico, A. Pescapé, "Integration of Cloud computing and Internet of Things: A survey," Futur. Gener. Comput. Syst., vol. 56, pp. 684–700, 2016.

[47] M. A. Razzaque, M. Milojevic-Jevric, A. Palade, S. Cla, "Middleware for internet of things: A survey," IEEE Internet Things J., vol. 3, no. 1, pp. 70–95, 2016.

[48] I. Lee, K. Lee, "The Internet of Things (IoT): Applications, investments, and challenges for enterprises," Bus. Horiz., vol. 58, no. 4, pp. 431–440, 2015.

[49] Kert M. et al., "State of the Art of Secure ICT Landscape," NIS Platform WG 3, V2, April 2015

[50] ENISA, "Threat Landscape Report." 2016, online at: https://www.enisa.europa.eu/publications/enisa-threat-landscape-report-2016

[51] G. Hatzivasilis, et al., "The industrial Internet of Thins as an enabler for a Circular Economy Hy-LP: a novel IIoT protocol, evaluated on a wind park's SDN/NFV-enabled 5G industrial network." Computer Communications, Elsevier, vol. 119, pp. 127–137, April 2018.

[52] L., Pino, "Pattern Based Design and Verification of Secure Service Compositions." IEEE Transactions on Services Computing (2017).

[53] L. Pino, G, Spanoudakis, A. Fuchs, S. Gurgens, "Discovering Secure Service Compositions," 4th International Conference on Cloud Computing and Services Sciences (CLOSER 2014), Barcelona, Spain, April 2014.

[54] A. Maña, E. Damiani, S. Gürguens, G. Spanoudakis, "Extensions to Pattern Formats for Cyber Physical Systems," Proceedings of the 31st Conference on Pattern Languages of Programs (PLoP'14. Monticello, IL, USA. Sept. 2014.

[55] K. Fysarakis, et al. "RtVMF: A Secure Real-Time Vehicle Management Framework," in IEEE Pervasive Computing, vol. 15, no. 1, pp. 22–30, Jan.-Mar. 2016. doi: 10.1109/MPRV.2016.15.

[56] N. Petroulakis, G. Spanoudakis, I. Askoxylakis, "Patterns for the design of secure and dependable software defined networks," Computer Networks 109 (2016): 39–49.

[57] G. Hatzivasilis, I. Papaefstathiou, C. Manifavas, "Real-time management of railway CPS," 5th EUROMICRO/IEEE Workshop on Embedded and Cyber-Physical Systems (ECYPS), IEEE, Bar Montenegro, June 2017.

[58] G. Hatzivasilis, I. Papaefstathiou, D. Plexousakis, C. Manifavas, N. Papadakis, "AmbISPDM: managing embedded systems in ambient environment and disaster mitigation planning," Applied Intelligence, Springer, pp. 1–21, 2017.

[59] IDC FutureScape, "Worldwide Internet of Things 2017 Predictions," Nov 2016.

[60] Forrester, "IoT Applications Require Distributed Analytics, Centralized Analytics Won't Work For Many IoT Use Cases," March 29, 2017, online at: https://www.forrester.com/report/IoT+Applications+Require+Distributed+Analytics/-/E-RES133723

[61] Gartner, "7 Technologies Underpin the Hype Cycle for the Internet of Things, 2016, The challenges of creating, implementing and preparing for the IoT," Nov 2016, online at: http://www.gartner.com/smarterwithgartner/7-technologies-underpin-thehype-cycle-for-the-internet-of-things-2016/
[62] Forbes and Moor Insights & Strategy, "The Internet of Things and Machine Learning," 2016, online at: https://www.forbes.com/sites/moorinsights/2016/03/16/the-internet-of-things-and-machine-learning/#a83d2d13fb16
[63] L, Pino, G. Spanoudakis, "Finding Secure Compositions of Software Services: Towards A Pattern Based Approach," 5th IFIP Int. Conference on New Technologies, Mobility and Security, Istanbul, Turkey, May 2012.
[64] L. Pino, G. Spanoudakis, A. Fuchs, S. Gurgens, "Discovering Secure Service Compositions," 4th International Conference on Cloud Computing and Services Sciences (CLOSER 2014), Barcelona, Spain, April 2014.
[65] J. Domanska, E. Gelenbe, T. Czachorski, A. Drosou, D. Tzovaras Research and Innovation Action for the Security of the Internet of Things: The SerIoT Project, to appear in Proceedings of the ISCIS2018 Security Work- shop, Springer CCIS, 2018.
[66] E. Gelenbe, Zhiguang Xu, and Esin Seref, Cognitive packet networks, Tools with Artificial Intelligence 1999. Proceedings. 11th IEEE International Conference on, pp. 47–54, 1999.
[67] E. Gelenbe, Cognitive Packet Network, US Patent 6,804,201, 2004.
[68] E. Gelenbe, Steps toward self-aware networks, Commun. ACM 52(7): 66–75, 2009.
[69] O. Brun, L. Wang, and E. Gelenbe, Big Data for Autonomic Intercontinental Overlays, IEEE Journal on Selected Areas in Communications 34(3): 575–583, 2016.
[70] L. Wang and E. Gelenbe, Real-Time Traffic over the Cognitive Packet Network, Computer Networks Conference 2016: 3–21
[71] F. Francois and E. Gelenbe, Towards a cognitive routing engine for software defined networks, ICC 2016: 1–6, IEEE Xplore, 2016.
[72] E. Gelenbe and T. Mahmoodi, Energy-aware routing in the cognitive packet network, ENERGY, 7–12, 2011.
[73] E. Gelenbe and C. Morfopoulou, A Framework for Energy-Aware Routing in Packet Networks, Computer Journal 54(6): 850–859, 2011.

[74] M. Jammal, T. Singh, A. Shami, R. Asal, and Y. Li, Software defined networking: State of the art and research challenges, Computer Networks, vol. 72, pp. 74–98, Oct. 2014.
[75] W. Stallings, Foundations of modern networking: SDN, NFV, QoE, IoT, and Cloud, Pearson Education, 2015.
[76] Y. Liu, A. Hecker, R. Guerzoni, Z. Despotovic, and S. Beker, On optimal hierarchical SDN, Proc. IEEE Int. Conf. on Communications (ICC), pp. 5374–5379, June 2015.
[77] N. Zhang, S. Demetriou, X. Mi, W. Diao, K. Yuan, P. Zong, F. Qian, X. Wang, K. Chen, Y. Tian, C. A. Gunter, K. Zhang, P. Tague, Y. Lin, Understanding IoT Security Through the Data Crystal Ball: Where We Are Now and Where We Are Going to Be, CoRR, arXiv:1703.09809, 2017.
[78] Symantec Internet Security Threat Report, vol. 23, Symantec Corporation, Tech. Rep., March 2018.
[79] E. Gelenbe, Reseaux neuronaux aleatoires stables, Comptes Rendus de l'Academie des Sciences. Serie 2, 310 (3): 177–180, 1990.
[80] E. Gelenbe, Learning in the recurrent random neural network, Neural Computation, 5 (1): 154–164, 1993
[81] Ahto Buldas, Andres Kroonmaa, Risto Laanoja, "Keyless Signatures' Infrastructure: How to Build Global Distributed Hash-Trees", In: Hanne Riis Nielson, Dieter Gollmann., Secure IT Systems – 18th Nordic Conference, NordSec 2013, Proceedings. LNCS 8208, Springer, 2013.
[82] https://ec.europa.eu/digital-single-market/en/internet-of-things
[83] https://www.sofie-iot.eu
[84] https://www.fiware.org/
[85] https://mydata.org
[86] https://www.sovrin.org
[87] http://estfeed.ee/en/analyses/
[88] BRAIN-IoT Project website,http://www.brain-iot.eu/ (last access June 2018).

8

CREATE *Your* IoT

Luis Miguel Girao[1], So Kanno[2], Maria Castellanos[3] and Patricia Villanueva[4]

[1]Artshare – Investigacao, Tecnologia e Arte, Portugal
[2]Artist, Japan
[3]uh513, Spain
[4]LaBoral Centrode Arte y Creación Industrial, Spain

Abstract

This chapter describes the background and origin of the artistic series CREATE Your IoT. The series is a very special one as it is being created and hosted at the heart of the Large-Scale Pilots Programme of the European Union (LSPs). The series puts artworks at the core and makes them the motive for dialogue between all actual and potential stakeholders in use-cases of the LSPs aiming at pointing out ways of how other innovative actions can be implemented on top of the developments made available by LSPs. The text looks at how art and technology are intrinsically related, how art practices historically expanded their field of action to make the world and life a canvas and how more recently the influence of artistic ideas in the creation of new products, services and processes is irrevocable. More specific examples of this connection between technology and the arts in the field of ICT and IoT are presented. Finally, an updated report on ongoing artistic actions in the context of the CREATE IoT coordination and support action are presented.

8.1 Introduction

Technologies and the arts have always been closely related. Indeed, this relationship is invoked with every mention of the word technology, which has its origins in the Ancient Greek tékhnē, meaning art. In this project, we will explore the contemporary relationship between technology and the arts, reflecting on how they can influence each other and the conditions under which their synergies can flourish. New technologies have shaped artistic practices since the dawn of history. Demand for tools to accomplish specific tasks has compelled technology to develop in new directions.

Potentially, the first tool one can conceive *Homo erectus* to have created, after winning the fight between the weight of their brain and gravity, was the invention of the stick – to more easily pick fruit from trees. The stick as an extension of the arm. The paint brush as an extension of the stick – an artistically driven technological innovation – is naturally conceivable as well. More recently, Andy Clark and David Chalmers conceived the iPhone as an extension of the mind [1]. Understanding Steve Jobs as the most contemporary artist of the past century then becomes key to pursuing the transformations of the timeless intertwining between technologies and the arts.

> *"The ability to produce art was an indication that humans had begun to think in more abstract terms. It's a thought process that enabled us to come up with the science and technology that enabled our species to become so successful."*
>
> *BBC article by Pallab Gosh, Oct. 2014* [2]

This quote comes from BBC Science correspondent Pallab Gosh, who was reporting on recent discoveries in a rural area on the Indonesian island of Sulawesi, where cave art from 40,000 years ago was found. The discoveries are the first of their kind outside the European continent, thus putting into question the positioning of Europe, and Western culture for that matter, as pioneering human development [3].

"The emergence of art marks the beginning of a surge in the development of human intelligence. The people who produce art are able to reflect their thoughts in the form of pictures and symbols", reports Gosh. Indeed, the ability to transform "abstract knowledge" into "knowledge of perception" is a unique characteristic of human intelligence. This ability fulfils the human need for making sense of what happens/happened by creating narratives. There is also a need to freeze moments in time: the need for creating images one can grasp and hold on to, the need for making sense of life, the need for giving meaning to life – meaningfulness.

The attribution of meaning to technologies is a relevant aspect for understanding the intertwining of technologies and the arts. For example, the invention of photography in the 19th century is possibly the event that had the most impact in the course of the history of art. On the one hand, it liberated painters from the duty of portraying personalities and started a movement of abstraction in painting that gave origin to great diversity of styles in the 20th century. On the other hand, it created a new tool for expression that is nowadays one of the most established forms of artistic expression. In sum, new meanings were attributed to the technique of painting and to a new technology – photography – which lead to new forms of images with its associated novel techniques. For instance, Pointillism can be interpreted as the first step towards a digital format of images, similar to what we nowadays call pixels.

8.2 CREATE *Your* IoT

Artistic practices are thinking processes as well as they generate reproducible knowledge. The peculiarity of the knowledge generated by the arts using technology is that by reverse engineering the final products of those creations, one can fully understand its functionality. One of the characteristics of the practice of art is that artists act first and rationalize later. The actual context of the relationship between ICT and art allows for an unprecedented integration of subjectivity in the context of technological research.

8.2.1 The Practice of Art as a Thinking Process

Urgency is a condition *sine qua non* in the attribution of the artistic quality to a practice. For a practice to be considered as artistic, it has to originate in that primordial urgency. Karl Phillip Moritz (1756–1793), in his writings *Artistic Imitation of the Beautiful*, defined this urgency or *artistic impetus* as drive and not as idea, concept or a representation [4]. This reverses the Leibniz–Wolffian hierarchy of human faculties by valuing the artistic, by considering the irrational and subconscious as the true source of human agency. Philosophers such as Schopenhauer, Nietzsche and others support that culture initiated by Moritz in which the "dark and undefined" balances with and is as relevant as the "clear and distinct" [5].

According to Landgraf, Moritz sees *urgency* as being about the productivity of nature that serves as media. It links the artist and the artwork as well as driving the creative process. Artistic creativity allows for the mediation between an undefined non-representational stimulus, i.e. *realisation*, and the artistic objectification or communication of the stimulus, i.e. *manifestation*.

Artistic is that which is created by an artistic practice, and that distinguishes it from any other knowledge generation practice, such as scientific practice. The core of artistic practice is the *urgency* for creation, composed of two poles: *realisation* and *manifestation*. They are the indivisible components of *artistic urgency*. *Realisation* is the need to make things happen while *manifestation* is the need to create beings.

Realisation is the core of practice itself. It is the action of making. It is movement, the energy of exteriorization. It is embodying in an outward form.

Manifestation is the core of creation. It is openness to revelation. It is recognising a being, the energy of interiorization. It is embodying in an inward form.

The Portuguese poet, Helder, expresses an extraordinary image of *Manifestation*:

> *I now dive and ascend as a glass.*
> *I bring up that image of internal water.*
> *Poem pen dissolved in the primordial direction of the poem.*
> *Or the poem going up the pen,*
> *passing through its own impulse,*
> *poem returning.*

Extract from *Sumula* (sum and substance), by Herberto Helder translated by Luis Miguel Girao (not published).

Herberto Helder (1930–2015), in the excerpt above, describes the bipolar coexistence of *Realisation* and *Manifestation* with a special focus on the latter. He describes the inwards embodiment of the pen by a poem. The "primordial direction of the poem" is towards the pen and the poet himself. "The poem going up the pen, passing through its own impulse" represents the impulsive nature of the need of making of the poet, *Realisation*. By "passing through its own impulse", the poem embodies the pen and reveals itself to the poem which, in turn, writes it on the paper. *Manifestation* nurtures *Realisation*, which in turn nurtures *Manifestation*, in a non-starting and non-ending cycle of *urgency*.

Helder understands the artwork, the poem, as a being. The poem has its own life and manifests itself through the poet, the pen and the paper: "poem returning". The return of the poem is the process of *Manifestation* that, however, is dependent on the poet's need for objectification: *Realisation*, his need to make things happen, in order to materialise as a form.

In a particular way, Helder expresses how "Nietzsche saw thinking itself: as a dance of concepts and the pen", as pointed out by Roy Ascott (1934–) in

Telematic Embrace [6], when describing telematic networks as a "planetary field for the dance of data". Telematics, as envisioned by Ascott, allows for the disappearance of "senders" and "receivers", so that they all become "users", creative participants. He established the concept of "distributed authorship" in digital networks following up on the ideas of Barthes' "dispersed authorship" in his *Le Plaisir du Text* [7] and Derrida's free play of sense.

Seconding the notion of thinking of culture as started by Moritz and followed by Nietzsche is Agostinho da Silva (1906–1994). In one episode of the TV series C*onversas Vadias*, broadcast by the Portuguese national broadcaster, RTP, between 8th March and 31st May 1990, the Portuguese philosopher stated:

> *We could carry on our shoulders a machine that thinks, or rather a machine that detects ideas that roam around the world.*
>
> Agostinho da Silva, 1990.[1]

This statement by da Silva is not an affirmation, but rather a proclamation of doubt. It was made after the interviewer, the writer Armando Baptista-Bastos (1933–2017), asked Professor da Silva why he normally advised his students not to think. His answer was the above quoted proclamation of doubt. According to da Silva, "we still don't know" whether we produce thoughts or whether thoughts come to our minds. In case of doubt, his choice was not to think.

Da Silva, in a communicative way aimed at addressing the masses, pointed out, as Morris did, that 'detecting ideas' is also valid for the generation of knowledge. He was trying to bring to the general public a discussion that has been going on for centuries about *noumena* and *phenomena.*

One of the high points of the discussion about *noumena* and *phenomena* is the critique by Arthur Schopenhauer (1788–1860) of Immanuel Kant's (1724–1804) use of the word *noumena*:

> *But it was just this distinction between abstract knowledge and knowledge of perception, entirely overlooked by Kant, which the ancient philosophers denoted by noumena and phenomena. (See Sextus Empiricus, Outlines of Pyrrhonism, Book I, Chapter 13, 'What is thought (noumena) is opposed to what appears or is*

[1] https://arquivos.rtp.pt/conteudos/conversa-com-baptista-bastos/ from minute 22 onwards. (last accessed on 13/08/2017) Translated by Luis Miguel Girao.

> *perceived (phenomena).') This contrast and utter disproportion greatly occupied these philosophers in the philosophemes of the Eleatics, in Plato's doctrine of the Ideas, in the dialectic of the Megarics, and later the scholastics in the dispute between nominalism and realism, whose seed, so late in developing, was already contained in the opposite mental tendencies of Plato and Aristotle. But Kant who, in an unwarrantable manner, entirely neglected the thing for the expression of which those words phenomena and noumena had already been taken, now takes possession of the words, as if they were still unclaimed, in order to denote by them his things-in-themselves and his phenomena"* [8].

Schopenhauer again makes a clear distinction between "what is thought" and "what appears or is perceived". This distinction has been fundamental to the discussion on the *practice of art as a thinking process* and its potential contributions for knowledge-generation systems[2]. It is relevant to understand and recognise the opposed concepts of *noumena* and *phenomena* in order to understand the uniqueness of the practice of art in making both concepts coexist simultaneously, as expressed above in the definition of *artistic urgency*.

Robert Pepperell, author of *Post Human Condition* [9], has resumed the discussion on *noumena* and *phenomena* by proposing the concept of *phenoumenon*. The notion of *phenoumenon* as the basic assumption for understanding the practice of art as a thinking process that "includes (all) our thoughts about reality which are part of a continuous *phenoumenon*" [10].

The great contribution of the practice of art for the generation of knowledge is transforming *noumena* into *phenomena* through *Realisation* and *Manifestation* by being both simultaneously: a *phenoumenon*. In other words, art is simultaneously embodying inwards and outwards. This means transforming "abstract knowledge" into "knowledge of perception" by producing technology-based artworks whose reasoning can be reversed. That is, making things happen and creating beings.

[2]Knowledge-generation systems refer to organisms that bring together contributions from different fields of study to research a specific subject. The overall vision is that the integration of the practice of art in these interdisciplinary, multidisciplinary or transdisciplinary groups of researchers is crucial for the development of research. www.starts.eu (last accessed on 14/08/2017).

8.2.2 Art is Life (Integration)

"Art is Life, Life is Art"

Wolf Vostell (1932–1998)

Vostell was a German painter and sculptor and is considered a pioneer of *happening* and Fluxus. Fluxus was an international and interdisciplinary group of artists that, in the late 1960s, produced performance "events", *happenings*, including concrete poetry, visual art, urban planning, architecture, design, literature and publishing. Fluxus has sometimes been described as intermedia, a category into which composers such as Niblock fall. The ideas and practices of John Cage influenced Fluxus, especially his understanding of the work as a site of interaction between artist and audience [11].

For Vostell, a human's physical *action*, the handling of things, was already considered art. What happens, the *happening*, is already art only if one wants it to be and one affirms it. Artworks no longer need an envelope or frame. Art steps out of its frame, normally the gallery, and melts immediately into the stream of life. The dilution of the boundaries between daily life and the places determined for art was one of the main objectives of the Fluxus movement. This was partially achieved, especially at the beginning of the movement. Yet, arguably, the institutionalisation of art has been unavoidable, and it naturally took advantage of this sort of movement to expand its area of action. Nonetheless, nowadays we can experience very interesting forms of art practice, such as street theatre, viral theatre, pop-rock bands and performances, and even some fashion industry-related events, happenings or products such as flashmobs, which somehow integrate art in daily life contexts.

One of the most common Fluxus happenings is the hammering of a piano as symbol of the destruction of the institutionalisation of the arts. This thesis builds upon a further development of this act, which is the destruction of desktop or laptop computers, as a symbol of the institutionalization of digital technologies. It is also the expression of the idea that ubiquity, in the form of the Internet of Things, amongst others, could allow the spread of artistic ideas, works or concepts embedded in new technologies themselves. Such a movement would allow for a worldwide dissemination of artistic ideas impregnated in society and economy via technological innovation. This idea is further developed in artistic interventions in the European Commission *STARTS Initiative* and the *Internet of Things Large-Scale Pilots.*

Those interventions are an expansion of some activities already happening in the field of arts and ubiquity. In that context, it has become common practice to organise workshops in which ideas to be congregated in public participation are developed.

Contextualising digital practices within architectural spaces and exploring the opportunities of experiencing and perceiving domestic environments with the use of media and computing technologies have been used as methods for the design of reflexive and intimate interiors that provide informational, communicational, affective, emotional and supportive properties according to embedded sensorial interfaces and processing systems. To properly investigate these concepts, a fundamental criterion is magnified and dissected: dwelling, as an important ingredient in this relationship entails the magical power to merge physical environment with the psyche of inhabitants. For this reason, a number of views providing necessary conditions to include matters of affectivity, ubiquity and layering complexity of interior space have been highlighted [12].

Integrative art is the integration of artistic practice into daily life. The way it is envisaged in this chapter is through the technology described above. Nowadays, integrative art is not a common practice amongst artists in the sense that is envisaged here. However, some relevant examples are emerging.

An example of artistic critical approach in IoT is the work of artist James Brindle who is trying to build his own self-driving car and published all the code developed in pursuit of the DIY self-driving car[3]. Brindle says:

> *"Self-driving cars bring together a bunch of really interesting technologies – such as machine vision and intelligence – with crucial social issues such as the atomization and changing nature of labour, the shift of power to corporate elites and Silicon Valley, and the quasi-religious faith in computation as the only framework for the production of truth-and hence, ethics and social justice.(. . .) The attempt to build my own car is a process of understanding how the dominant narratives of these technologies are produced, and could be changed."*

[3]https://github.com/stml/austeer

8.2.3 ICT and Art

Golan Levin, one of the most prominent individuals of the emerging field of ICT and art, very clearly demonstrated how artistic projects presented ICT solutions well before they became known:

Figure 8.1 Myron Krueger's Video Place (1974), and the Sony EyeToy (2003).

Figure 8.2 Michael Naimark & MIT ArchMac's Aspen Movie Map (1978–1980), and Google Street View (2007–).

Figure 8.3 Jeffrey Shaw's Legible City (1988) and E-fitzone exercise equipment (2008).

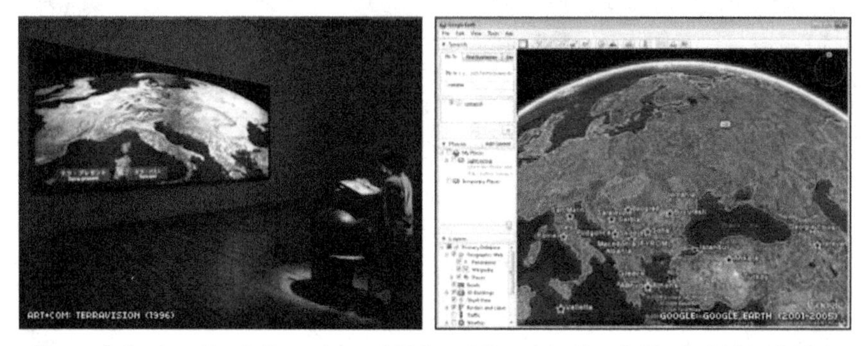

Figure 8.4 Art+Com's Terravision (1996) and Google's Google Earth (2001, 2005–).

The artist and technologist states that he wrote his article *New Media Artworks: Prequels to Everyday Life*,[4] as consequence of the following:

> *"I struggled to justify the value of new-media arts research to an audience of Silicon Valley business people; while simultaneously, some new-media artist friends of mine discovered that their work had been 'appropriated' by a large corporation."*

This example reveals one of the most important gaps in the generation of new businesses models in global markets: the one existing between creativity and business. The US, moreover, is home of crucial players in the field, such as the most relevant academic publisher in the field, *Leonardo*, and SEAD, the network for Sciences, Engineering, Arts and Design. However, the European context nurtures the development of institutions such as Ars Electronica that distinguished Golan Levin with its Prix; the same institution also distinguished Linus Torvalds, considering the collective process that

[4] http://www.flong.com/blog/2009/new-media-artworks-prequels-to-everyday-life/

led to Linux as an artistic expression. It is this same European context that recognized the emergence of ICT & ART so as to allow for the worldwide establishment of STARTS, the European Commission initiative in Science, Technology and the Arts, as a recognized field of Research and Technological Development.

A considerable number of organisations, institutions and programmes promoting activities linking ICT and art proliferate in the European Union. Some of these institutions are worldwide leaders such as Ars Electronica (AT), ZKM (DE) or IRCAM (FR), to name but a few. However, it is from small organisations and individuals that the most innovative projects or actions originate. As an example, the Finnish artistic/researcher Laura Beloff, who has been operating as an individual focused on the development of wearable technologies, has recently been appointed the Head of Section–Interaction Design and Computer Game Development at IT University in Copenhagen, Denmark. The Finnish Bioart Society that she founded and directs was one of the participants of a workshop on bioart, promoted by the FBI in California, USA. Laura is one of many artists that are becoming institutionally prominent not in the field of the arts but in the field of ICT.

Small organisations seem in fact to be the strategic focus of promoters, as funding in the field is mostly directed more to groups of people than to individuals. Medialab-Prado (ES), Kitchen Budapest (HU), F.A.C.T. (GB), Pervasive Media (UK), iDAT (UK), iMAL (BE) and CIANT (CZ) seem to be good examples of organizations promoting more relevant activities. More and more ideas of collaboration, co-creation, shared knowledge and participation are present in their initiatives. The concept of lab, from OpenLab, to FabLab and Living Lab, has been instrumental in the diffusion of techniques of digital fabrication and physical prototyping, allowing everybody to go, learn and create. From pieces of 3D printing to lines of code for multimedia installations, activities linking ICT and art seem to follow a model of establishing an artistic context for creative participation, be it in the form of interactive installations or workshops to learn how to make or to create. It seems that we are moving from models of engaging the arts to illustrate and communicate science, such as the one implemented at CERN, to the ideas of living labs, such as the iMinds' own iLab.o or Barcelona Laboratori Cultural, promoted by Josep Perelló, who was previously responsible for the Science Area on behalf of the University of Barcelona at Arts Santa Mònica centre in Barcelona (SP).

There are already a considerable number of small and medium businesses developing around these activities, such as Libelium (ES). The concept of providing creative learning platforms as a new business model is actually expanding as a strategy. What started as an artistic project is becoming the standard for rapid prototyping in physical computing: Arduino (IT). This development platform created and has been maintaining a large community of developers around itself, based on the establishment of an easy-to-use programming language, a playful set of online tutorials and an active online form. The same model has been applied in Processing or openFrameworks. The community started with a majority of artists and expanded to become of a majority of technologists. Almost every electronics store in big European cities sells Arduino and related products. The expansion of this model is becoming visible in big companies such as Farnell and its community platform Element 14 or the DIGI. Also, in this last case, the new XBee project gallery is a result of the collaboration of Rob Faludi (US) with the electronics corporation. At the University of Cambridge's Computer Laboratory (GB), what is becoming the next platform for development was created: The Raspberry Pi.

In education, the most interesting model seems to be related to the concepts of ubiquity and the internet of things. i-DAT24 of the University of Plymouth (UK) has been promoting exemplary initiatives such as the *Confluence* Project: a group of students of schools located at North Devon's UNESCO Biosphere Reserve, in collaboration with artists and technologists, developed and implemented remote wireless networks from which they created online data visualisations.

In research, the most relevant worldwide network of researchers in the field of Art, Science, Technology and Consciousness Studies is the Planetary Collegium. The network has nodes in Lucerne, Trento and Shanghai. The main hub is at the University of Plymouth.

A considerable number of conferences on the crossings of ICT and arts happen all over the world, the most relevant being, for example, ISEA, Ars Electronica, Siggraph, HCI International and Transmediale.

At the level of social innovation, the growing intersection between the application of ICT and art in the field of disability is notable. The Artabilitation (DK) group has been joining a relevant number of researchers in this area, including the exemplary case of Rolf Gehlhaar. Gehlhaar developed a number of digital interfaces for musical expression, some of them recently integrated into the British Paraorchestra. The orchestra opened the Queen's Christmas Speech of 2012 and played at the Paralympics closing ceremony in London, in 2012.

The European Commission has been supporting a number of projects engaging the arts as described in the call for tender. However, the most relevant recent activities come from DG CONNECT which promoted the ICT ART CONNECT workshops and related events, which have been dedicated to better understanding how to integrate the arts with ICT. The COST Arts and Technologies Event took place in Zagreb, from the 25th to the 28th of November 2013.

The COST Arts & Technologies (CAT) workshop assumes that there are large potential gains in integrating arts on the one hand with technologies on the other, to a larger extent than has been done so far. Combining artistic creativity with technological expertise should in itself have a great potential to lead to new products, services and social innovations. The workshop aimed at enabling innovative integration of arts and multi-, inter-, and transmedia technologies and their actual and potential integration with industries and society as a way of enhancing competitiveness and creativeness of European innovation in arts and technologies.[5]

The CAT workshop gave rise to a relevant collective white paper entitled *Organisms for Change and Transformation.*[6]

DG CONNECT of the European Commission has been promoting key initiatives in the context of the Digital Single Market (DSM), under the umbrella of the STARTS Initiative.

However, bearing in mind long-term targets such as 2050, it will be in the context of the now developing Framework Programme Horizon Europe of the European Commission that further development of STARTS will have to develop. In order to find conditions for the nurture of these future activities, areas of opportunity need to be found within this context. The present understanding seems that regional development will be instrumental. The reason behind this assumption is that the regions of Europe strategically dedicated to this area of innovative development will be determined to a large extent by this programme. The context of the Cultural and Creative Industries (CCIs) seems to be the ideal host for the ideas forthcoming from the potential research results from the future of STARTS. Nonetheless, it seems that the focus of this emerging field should probably not lie in the utilization of ICT for digital content, cultural industries and creativity. The utilization of the arts as a means to communicate aspects of science on its own also does not seem to be innovative enough for the purposes of the emerging field in question: this practice has emerged and spread worldwide, as these activities have been

---

[5] http://www.cost.eu/events/cat

[6] http://www.cost.eu/download/47808

happening worldwide since the last century and are already quite established as described.

The engagement of the arts with ICT are also instrumental in allowing the active participation of a large number of European citizens to create and live their own lives in a better way. Protocols such as Open Data and Open Source allow for digitally mediated forms of social innovation both at the level of opinion-making participation as well as at the level of self-employment. From this perspective, the creation and establishment of new business models and entrepreneurship becomes an active form of social innovation. This implies nurturing not only the visionary and exploratory characteristics of artistic practices, but also furthering their wider capability of research and development.

These ideas are clearly aligned with actual objectives such as Inclusive Societies by which "The European cities have to be at the heart of policies aiming to create growth, jobs and a sustainable future" and "the increasing socio-economic importance of digital inclusion, research and large-scale innovation actions will promote inclusive ICT solutions and the effective acquisition of digital skills leading to the empowerment of citizens and a competitive workforce." In her report on H2020, MEP Maria da Graça Carvalho proposes "education and science, arts and humanities as fundamental drivers of social and economic progress and well-being".

Social innovation generates new goods, services, processes and models that meet societal needs and create new social relationships. It is important to understand how social innovation and creativity may lead to change in existing structures and policies and how they can be encouraged and scaled up. Grass-roots online and distributed platforms networking citizens and allowing them to collaborate and co-create solutions based on an extended awareness of the social, cultural, political and environmental context can be a powerful tool to support the objectives of Europe 2020.

Moreover, aspects of participation are also at the core of the programme: "...address social-network dynamics and crowd-sourcing and smart-sourcing for co-production of solutions addressing social problems, based on open data sets. They will help to manage complex decision-making, in particular the handling and analysis of huge quantities of data for collaborative policy modelling, simulation of decision-making, visualisation techniques, process modelling and participatory systems (...) as well as to analyse changing relationships between citizens and the public sector. Increased levels of complexity, the implications of questions posed by technology, advanced computation, life sciences and bio-engineering impinge upon areas of knowledge traditionally related with human studies, such as philosophy, theology, and legal, political and economic thought should be addressed. It is important to combine art, science and entrepreneurship; new forms of urban expression; knowledge, art and entrepreneurialism related to the integration of multiculturalism and integration of migratory flows; multilingualism."

The same applies for creativity and innovation: "Exploring processes which provide a favourable background to creativity and innovation. Providing a better understanding of the social, cultural, economic and political context for innovation shall be a priority. In particular, the role of youth perception of the opportunities for innovation in the current economic environment of high unemployment in many EU regions shall be carefully understood in relation to education and to the risk of brain-drain."

Finally, cultural heritage and European identity are also important: "The aim is to contribute to an understanding of Europe's intellectual basis [...] European collections, including digital ones, [...] should be made accessible through new and innovative technologies and integrated information services to researchers and citizens to enable a look to the future through the archive of the past and to contribute to the European participative intelligence."

In sum, an understanding of the crossroads of science, technology and the arts on all these levels is crucial to fostering post-crisis processes of recovery in the European Union.

At the heart of the Fourth Industrial Revolution lies the outstanding feature of the automation of mechanisation. Artificially intelligent computerised machines liberate humans from mechanistic tasks in chain with machines. The actual concept of a hybrid system integrates humans into industrial chains where subjectivity is a need. In this general context, disciplinary specialisation in research is giving way to transversal and holistic approaches, with the assurance that intelligent machines are in place to perform highly precise and effective tasks. Humans are no longer needed to perform complex operations but instead, they become indispensable to trigger and correlate highly complex operations, where knowledge convened by subjective processes is crucial for the achievement of results.

Research and development practices are no longer methodological processes of confirming expectations or hypothesis, but now become flexible processes of discovery due to the availability of easier means of experimentation and repetition.

Sciences of cognition set a very good example of that expressed above. They often cross knowledge from different disciplines, they depend on high-tech imaging and measuring equipment and their results are mostly dependent on subjective reports. Therefore, they demand an articulation of many different disciplines and their experiments are led by enquiry on subjective aspects of perception. Already in this field of research, the integration of artistic practices is an emerging factor.

Art practices are transdisciplinary by nature, independently of the channels of expression used. Throughout history artists have specialised in developing technologies and implementing techniques in the design of artistic experiences. Therefore, in the present context of experimental integration of subjectivity, artists are emerging as relevant contributors in research and development of technology. Technologies being a consequence of scientific developments, artistic practices become interesting experimental methods for the generation of new knowledge. Strategies are needed for the integration of these new ways of thinking amongst different scientific communities, leading to true social and economic innovation.

Historically narrowing our perspective over more recent events, one can say that the existing collaborations among artists and scientists are a consequence of the work of Frank Malina. He was at the origin of NASA's Jet Propulsion Laboratories, of which he was the first director. In 1968, in Paris, as a way of pursuing his interest in kinetic art, he founded the Leonardo Journal, which is still the leading publication on the crossings of arts, sciences and technologies. His son, Roger Malina, continuing his father's work, is one of the most prominent agents in the field, by running the Leonardo project as well as triggering other actions such as SEAD. SEAD is looking into congregating best practices of collaborations in science, engineering, arts and design. However, all actions in the field tend to make the old-fashioned model prevail, where every single actor of the collaborative system conserves and develops his or her own speciality, which of course enhances political aspects of real and productive collaboration. Their results are mostly limited to theoretical papers, and in cases that could result in practical applications, issues of generating economic value such as intellectual property generation and protection are generally dismissed. Not to mention how far these practices stay from aspects of creation of new products, services, and of the business aspects of generating new jobs and self-employment. Nurturing the expansion of fields of action of each discipline and solving conflicts resulting from their overlapping of functions are the instruments to achieve transdisciplinary. However, the main question still remains: How to integrate arts and sciences in truly productive ways, both in the direction of the generation of new philosophical knowledge, as well as in the direction of the creation of new technology-based businesses, in order to make the European Union the world leader of emerging markets and creating future ones?

The adoption of artistic practices such as research methodologies is instrumental for the integration of subjectivity in the production of scientific, reproducible knowledge, allowing for a holistic approach, not only to the emergence of future sciences, but also of future technologies, leading to close-to-market results, especially concerning the intersection of the arts and information and communication technologies.

Therefore, the main benefits of STARTS are in its features of nurturing socio-economic innovation by mediation of digital information and communication technologies. By promoting the intersection of scientific and artistic practices, inevitably leading to new methodologies and processes of generating new knowledge, STARTS aims to transform the way research and development communities face their own research targets: at one stage, to make them more open to novelty, exploration mechanisms, creativity and imagination; at another, to make them focus on concrete research outputs, in the form of close-to-market prototyping.

The main targets of STARTS should be to disperse the idea that blue-sky, thinking-based research can generate added value, not just because of its inherent novelty, but because this novelty, by being tracked at intermediate states of development, will lead to new scientific and technological developments.

Innovation at the social level where scientists and artists interact will lead to both new knowledge and new technologies, in accordance with actual demands of society and markets. In other words, STARTS aims to benefit European research communities by merging scientific and artistic research and innovation (R&I) practices into producing new philosophical knowledge and new technologies, as well as making R&I practitioners aware that having both of the above combined might allow for the creation of new markets, based on new business models, new products and new services.

At the time writing, STARTS has three dedicated projects running and a relevant intervention in the Large Scale Pilots Initiative. *Vertigo* is promoting the integration of artists in knowledge generation systems by attributing grants for artistic residencies in research projects. *Wearsustain* is promoting the creation of new artistic driven prototypes. The *STARTS Prize* attributes yearly two distinctions for technological innovation through the arts. CREATE-IoT, the coordination project of the Large-Scale Pilot projects (LSPs) of the EU, also integrates the arts. CREATE-IoT has a crucial role in STARTS by promoting the notion of co-creation based on artistic practices within the LSPs and by introduction of the Experience Readiness Level indicator.

8.2.4 Next Things_Next Starts

Between December 2017 and April 2018, in Gijon, Spain, the exhibition Next Things_Next Starts showed for the first time the results of the research and production residency programme called Next Things organised by LABoral Centro de Arte y Creación Industrial in conjunction with Telefónica R+D over a five-year period with the mission to forge new connections and collaborations between art, science, technology and society. Following an open call issued to artists and other creatives, the most innovative ideas and projects related with the Internet of Things were chosen. The award consisted of organizing and funding a six-month residency – two months at LABoral, in Gijon and four at Telefónica I+D, in Barcelona – to materialise their ideas and projects.

The exhibition was centred on the critical role played by creativity and social involvement in processes of innovation. Along with scientific and technical know-how and learning, art is a catalyst that helps transform knowledge into objects or processes. The showcased projects presented the paths which are opened up when combining creative thinking with the possibilities of open technology.

The five critical and innovative projects chosen for the programme rethink and open a debate on contemporary situations stemming from technological advances. Through the creative use of new technologies, these projects propose prototypes for new solutions and working spaces.

The exhibition was a first for STARTS in learning from previous experiences on artistic residencies in research contexts. The artists awarded with the prize were: Laura Malinverni and Lilia Villafuerte; Lot Amorós, Cristina Navarro and Alexandre Oliver; Sam Kronick; María Castellanos and Alberto Valverde; Román Torre and Ángeles Angulo. Here we present only three of those works as they represent a useful spectrum as examples for the integration of artists in the context of the IoT European Large-Scale Pilots.

Environmental Dress represents the tendency of artists to engage with global matters such as climate change. Furthermore, the project demonstrates the capability of artists to technically implement elaborate systems as well as reiterating the importance of open source code and hardware.

Figure 8.5 *Environmental Dress*, by María Castellanos and Alberto Valverde.

"We are surrounded by polluting agents and other factors that have a direct impact on our everyday lives, our mood and, ultimately, on our behaviour. Variations in noise, temperature, atmospheric pressure, ultraviolet radiation or amounts of carbon monoxide are some of the challenges we have to face on a daily basis. At the end of the day, they are agents that influence our temper and our behaviour with others.

Environment Dress is a piece of smart clothing that uses a number of sensors to measure the aggressiveness of our surrounding environs, detecting environmental variables and alerting us to them. Our body's natural sensors are unable to measure and anticipate factors such as an increase in ultraviolet radiation, dust or noise, and others.

The interface geo-locates environmental analyses and allows users to register their mood through a smartphone app. In consequence, we can establish the relationship between both variables and determine whether an increase in ultraviolet radiation can make the person who wears the dress feel better or whether an increase in noise level can make him or her feel more uncomfortable in a certain place. Finally, all these data are shown on an emotional map, pinpointing the most pleasant and unpleasant areas in a city." (Source: catalogue of NEXT THINGS_NEXT STARTS Exhibition, LABoral, Gijon, Spain).

Flone is representative of the need of artists to make technology accessible. In a subject matter such drones, with profound implications on security and privacy, the exposure and dissemination of how drones' function is crucial, from an artistic point of view.

Figure 8.6 *Flone, the flying phone by Lot Amorós, Cristina Navarro, Alexandre Oliver.*

"Flone, The Flying Phone, is a platform to make smartphones fly, involving an innovative drone which combines digital manufacturing, personal empowerment and the use of a smartphone to remotely control the device.

Flone is a self-built, low-cost biodegradable drone, conceived as an open source digital design. Some of its design elements (shape, size, material, lack of screws) make it accessible and adaptable for many people to conquer air space.

The use of open software and documentation and the simplicity of making it democratise the knowledge needed to manufacture a drone and claim air space as a common domain. Flone aims at opening up the range of applications of air social robotics. This multimedia drone, a mobile multipurpose machine, moves through the public air space thanks to various smartphone sensors (camera, microphone, GPS, accelerometers, gyroscopes) and actuators (LED flash and speaker) together with wireless connections (Bluetooth, Wi-Fi and 4G).

The members of this project have imparted workshops in countless schools, art centres and universities in several countries. Dozens of individuals have replicated this project worldwide and made a flone for themselves." (Source: catalogue of NEXT THINGS_NEXT STARTS Exhibition, LABoral, Gijon, Spain).

Thero reiterates the urgency of giving control to end users of the decision of being connected. Having the option to consciously connect or disconnect to the internet is nowadays extremely important mainly because a great number of people are not aware that they are constantly connected through their devices.

Figure 8.7 *THERO by Román Torre and Ángeles Angulo.*

"As a concept, THERO wishes to raise our digital privacy to the status of a precious and sacred object. Accordingly, the object has been given a highly aesthetic treatment, with the geometry and clean lines of an idol or talisman endowed with a value beyond its material qualities: the value of freedom and the right to digital anonymity.

THERO is presented as a heavy sculpture which contains a device that blocks and encrypts our digital communications by allowing the user to directly manipulate the object. By manually rotating its structure, THERO is capable of managing our digital contact with the outside space.

The piece basically consists of a router to which we can wirelessly connect all our digital devices. It can be handled physically to offer various levels of privacy: blocking pages we do not want to visualise or which demand excessive attention from us, encrypting our communication by using the TOR network, completely blocking access to the network, cutting all communication with the outside in order to only browse locally.

The piece opens up a space for reflection on our actions and their subsequent traces and significance in the net. THERO tries to lower the abstract barrier of the digital tool by means of a number of physical actions that make us more aware of our use of the Internet.

The presence of THERO in our homes would give corporeity to the need for privacy in our digital interactions. In essence, THERO gives us the power to decide when we want and when we do not want to be visible." (Source: catalogue of NEXT THINGS_NEXT STARTS Exhibition, LABoral, Gijon, Spain).

The most important conclusion from the Next Things programme is the management of intellectual property in this type of context of collaboration of tech companies and artists: keep it all open source.

8.2.5 Artists and the IoT European Large-Scale Projects

In the CREATE-IoT project a methodology for integrating ICT and the arts – or better put: to include artistic practices in the ICT development cycle – was designed to be fully adaptable. Its implementation in the LSPs will result from specific combinations of its methods according to the specificities, not only of each one of the LSPs they will be tailored for, but of each of the particular LSPs' use cases.

The methodology is designed to be applied in the specific areas of innovation of the IoT LSPs initiative: food and farming, healthy aging, public mass events, self-driven vehicles and smart cities. The basic principles are implemented in the ICT framework through a sequence of actions that will be selected from the range of artistic related activities and their correlation with the ICT cycle.

The actions of integration of artistic practices in the LSPs are being done mostly around their use cases. The reason for this option is to demonstrate that artistic practices are useful in connecting humans and technology, towards a human-centred approach to technology as an enabler of better lives in general.

The methodology is developed around the development of artworks to trigger dialogue with the LSPs and raise some questions that might help improve their final solutions. The underlying idea is that the services provided by the LSPs can trigger socio-economic innovation if made available to SMEs and individuals.

The first step was the integration of an artist in residency in the CREATE-IoT project. So Kanno, a Japanese artist proposed creating the *The ideal showroom of IoT.*

The ideal showroom of IoT is a two-part composition, a participatory installation. It shows the possibility of sensing, recognising and determining the world through the perspective of objects. A living room full of IoT devices is set out to let visitors experience this shift: sensors and cameras are interweaved into a well-known environment. The second part is providing a new point of view to perceive a post-IoT age perspective onto things and technology.

Figure 8.8 *The ideal showroom of IoT.*

The installation is set up in two parts:

1. The first is a living room with many small computers, cameras and sensors installed. Most of them are not obvious and are hidden. These systems try to capture information of the visitors. A robot in the room will welcome the visitors. It will introduce and explain the context of the work as well as trying to have a conversation with visitors.

2. In second room, there's a laboratory set up, with a desk and VR headset. Visitors will experience the living room now from different perspectives. When putting on the VR headset, the visitor will have the view from the hidden cameras or robot.

Experiencing the same situation again through an object-related perspective should give the visitors a new perspective on IoT and personal robots.

In this residency between CREATE-IoT and So Kanno, a new artistic work is being developed, challenging the fundamental issues of interest in the Internet of Things. CREATE-IoT provides access to the artist for key people, companies, concepts and technologies associated with Trust in the Internet of Things. Key elements will be made available to the artist regarding the development of a trusted environment for the development of IoT and comprehensive technical and non-technical solutions regarding privacy, security and trust issues.

The development of the new artwork involves various levels of research and development. Existing IoT products are explored and researched and selected regarding the functions they include for the installation. Technology used for the project are IoT devices with hidden cameras, smart speaker systems, personal robots and VR technology. In the development of the art work, the consumer products will be manipulated and adjusted for the artistic purpose.

The developed system will integrate the video stream of hidden IoT security cameras. The IoT devices and robot will be accessed and controlled through a VR headset experience.

The second step is the development of model of artistically mediated co-creation process around an artwork.

Towards the creation of exemplary case studies, the LSPs IoF2020, ACTIVAGE, SYNCHRONICITY are being developed in order to realise artistic-led co-creation hackathons as a support to some of the use cases of these pilots. The aim of these hackathons is to artistically enhance the context of those use cases and stimulate creativity of all participants.

The concrete target is to better understand the role of artists in pushing for innovative approaches either in the technology in question or its applications. Impact on uptake, adoption and acceptance will also be observed, as well as the potential of new businesses built on top of the technologies made available by the LSPs under study.

At the moment, the use cases that are being considered for action are part of IoF2020:

Figure 8.9 *Added-value weeding data.*

This use-case collects location-specific camera data to provide insights on the number of vegetables growing on the field, the plants' growth status and best harvesting moment, weed prevalence, nutrient shortages and drought stress. From an artistic point of view, it is interesting to understand how agriculture is becoming less anthropocenic.

Figure 8.10 *City farming for leafy vegetables.*

IoT technology in city farming enables the production of high-quality vegetables in a predictable and reliable manner, unaffected by plant diseases, free from pesticides and independent of seasonal influences. From an artistic point of view, it is interesting to imagine better lives that could allow free-time to have contact with the vegetables we eat.

The focus of this use case is mainly on the growth of poultry with respect to animal welfare. This starts with an adequate environment in which the birds

Figure 8.11 *Poultry chain management.*

feel comfortable, as well as good-quality feed and water. These are extremely important aspects from an artistic point of view. Some years ago, the artistic community started to be concerned with this type of challenges, especially after the film *Baraka.*

An example of the possible impact of these actions in the IoF2020 would be to see in its open calls a focus on more human-centred technology based on the technologies made available by the project. That is one of the underlying principles of the choices use cases to work with.

8.2.6 CREATE *Your* IoT

The present result of the work undertaken is a series of works entitled CREATE *Your* IoT. Drawing inspirations from the title of the coordination and support action to the LSPs, CREATE-IoT, the series aims at expanding it by pointing out ways of how other innovative actions can be implemented on top of the developments made available by LSPs. It emphasizes the co-creative aspect of the all LSPs but in an alternative sense than that of citizen participation as promoted by the U4IoT CSA. In the series, artworks are the core and are motive for dialogue between all actual and potential stakeholders in use-cases.

The CREATE *Your* IoT Series is at the moment composed of two artworks under development. *The Connected Hennery* and *The Migrant Home.* The artworks are being designed to allow the integration of multiple technologies made available by the LSPs. For example, *The Migrant Home* could host technologies from all LSPs MONICA, AUTOPILOT, IoF2020, ACTIVAGE, SYNCHRONICITY.

The Connected Hennery is a reflection about the use case of the Poultry chain management of IoF2020. Inspired by the motto of that use case to respect animal welfare, the artwork starts by giving the chicken the control of the location of their home. It follows recent tendencies of permaculture, within which mobile henneries are substitutes for tractors in the cleaning of agricultural land. In permaculture, chicken inhabiting a defined piece of land clean it and fertilize it. Farmers, by simply moving the hennery around their land, make it ready for cultivation. The digital system of *The Connected Hennery* analyses the position and movement of chicken inside the hennery and predicts in which direction they would like to progress next, freeing the farmer from that work task. Furthermore, other sensors implemented in the hennery allow easier monitoring for the farmer in order to simplify and more effectively manage her/his intervention in the maintenance of the hennery.

The CREATE *Your* IoT Series is looking at decentralized models of production of chicken and at its potential as added value for the associated use cases of the LSPs. Food suppliers are looking at how consumers are more and more interested in biological and organic products and how can they adapt to keep their leadership of the supply markets. This leads these suppliers to create their own production experiments in order to better understand how to create new products the fit customers' demands. It is for this sort of context that works such *The Connected Hennery* are being developed in order to promote the LSPs towards end users.

The Migrant Home is still in the early phases of concept development. At this stage it is looking at how an IoT mobile house can be transformed into a home for migrants for short-term jobs/enterprises connected with urban and rural niche developments, for instance recovery of rural and urban cultural heritage.

Preliminary experience of the development of the CREATE *Your* IoT Series reveals that Open Standards and Architectures in the LSPs are crucial to make the technologies developed accessible and allow for the development of new business models.

8.3 Conclusion

The actual context of the relationship between ICT and art allows for an unprecedented integration of subjectivity in the context of technological research. The integration of subjective approaches is fundamental in making human-centred technological innovation. Human-centred technological applications fill in the gap between what is possible from a pure technological point of view and what people can encounter as useful in their daily lives to make those lives better.

The LSPs represent a unique opportunity for the spread of creative approaches to technological solutions and those approaches can help to potentiate the results of the LSPs in terms of new applications and associated business models. It is this reciprocal relationship that will allow on one side for an expansion of the field of action of the LSPs and on the other for the LSPs of potential fields for innovation to be informed.

The instruments of those actions are the co-creation hackathons. They will develop around the artworks of the *CREATE Your IoT* Series to trigger new solutions based on the technologies made available by the LSPs.

References

[1] R. Menary, The Extended Mind, online at: https://mitpress.mit.edu/books/extended-mind

[2] P. Ghosh, Cave paintings change ideas about the origin of art, online at: http://www.bbc.com/news/science-environment-29415716. The quote is a transcription of the video of the article and not from its text.

[3] M. Aubert, Brumm, A., Ramli, M., Sutikna, T., Saptomo, E. W., Hakim, B., Morwood, M. J., van den Bergh, G. D., Kinsley, L., Dosseto, A., 2014. Pleistocene cave art from Sulawesi, Indonesia. Nature 514, 223–227.

[4] J. M. Bernstein, (Ed.), 2003. Classic and romantic German aesthetics, Cambridge texts in the history of philosophy. Cambridge University Press, Cambridge, UK, New York.

[5] E. Landgraf, 2011. Improvisation as art: conceptual challenges, historical perspectives, New directions in German studies. Continuum, New York, NY.

[6] R. Ascott, E. A. Shanken, 2007. Telematic embrace: visionary theories of art, technology, and consciousness, 1. paperback print. ed. Univ. of California Press, Berkeley.

[7] R. Barthes, 1973. Le plaisir du texte, Points Essais. Éd. du Seuil, Paris.

[8] A. Schopenhauer, 1969. The world as will and representation. Dover Publications, New York.

[9] R. Pepperell, 1995. The post-human condition. Intellect, Oxford, England.

[10] A. P. Almeida, 2007. O universo dos sons nas artes plásticas, Teses. Edições Colibri, Lisboa.

Index